Polyphenolic Compounds in Wine and Beer

Polyphenolic Compounds in Wine and Beer

Editor

Mirella Nardini

MDPI • Basel • Beijing • Wuhan • Barcelona • Belgrade • Manchester • Tokyo • Cluj • Tianjin

Editor
Mirella Nardini
CREA Research Centre for
Food and Nutrition
Council for Agricultural
Research and Economics
(CREA)
Italy

Editorial Office
MDPI
St. Alban-Anlage 66
4052 Basel, Switzerland

This is a reprint of articles from the Special Issue published online in the open access journal *Molecules* (ISSN 1420-3049) (available at: www.mdpi.com/journal/molecules/special_issues/polyphenolic_wine_beer).

For citation purposes, cite each article independently as indicated on the article page online and as indicated below:

LastName, A.A.; LastName, B.B.; LastName, C.C. Article Title. *Journal Name* **Year**, *Volume Number*, Page Range.

ISBN 978-3-0365-6128-8 (Hbk)
ISBN 978-3-0365-6127-1 (PDF)

Contents

About the Editor

Mirella Nardini

Dr. Nardini is a biologist with a PhD in Biochemistry. She is employed as technologist in the Council for Agricultural Research and Economics (CREA). The results of her studies have been published in many international scientific journals. Her field of interest is the nutritional quality of food and the role of food in the modulation and prevention of oxidative stress and related pathologies, particularly cardiovascular diseases. She has gained extensive experience in antioxidant and polyphenol compounds, through many studies concerning the identification, characterization, biological activity, bioavailability, and metabolism of phenolic compounds.

Article

Beer Phenolic Composition of Simple Phenols, Prenylated Flavonoids and Alkylresorcinols

Anna Boronat [1], Natalia Soldevila-Domenech [1,2,3], Jose Rodríguez-Morató [1,2,4], Miriam Martínez-Huélamo [5], Rosa M. Lamuela-Raventós [4,5] and Rafael de la Torre [1,2,4,*]

[1] Integrative Pharmacology and Systems Neuroscience Research Group, Neurosciences Research Program, IMIM-Institut Hospital del Mar d'Investigacions Mèdiques, Dr. Aiguader 88, 08003 Barcelona, Spain; aboronat@imim.es (A.B.); nsoldevila@imim.es (N.S.-D.); jrodriguez1@imim.es (J.R.-M.)

[2] Department of Experimental and Health Sciences, Universitat Pompeu Fabra (CEXS-UPF), Dr. Aiguader 80, 08003 Barcelona, Spain

[3] Medtep Inc., 08025 Barcelona, Spain

[4] Spanish Biomedical Research Centre in Physiopathology of Obesity and Nutrition (CIBERObn), Instituto de Salud Carlos III (ISCIII), 28029 Madrid, Spain; lamuela@ub.edu

[5] Department of Nutrition, Food Science and Gastronomy, School of Pharmacy and Food Sciences and XaRTA, Institute of Nutrition and Food Safety (INSA-UB), University of Barcelona, 08921 Santa Coloma de Gramanet, Spain; mmartinezh@ub.edu

[*] Correspondence: rtorre@imim.es; Tel.: +34-933-160-484

Academic Editor: Mirella Nardini
Received: 8 May 2020; Accepted: 29 May 2020; Published: 2 June 2020

Abstract: Beer is a fermented beverage with beneficial phenolic compounds and is widely consumed worldwide. The current study aimed to describe the content of three families of phenolic compounds with relevant biological activities: prenylated flavonoids (from hops), simple phenolic alcohols (from fermentation) and alkylresorcinols (from cereals) in a large sample of beers ($n = 45$). The prenylated flavonoids analyzed were xanthohumol, isoxanthohumol, 6- and 8-prenylnaringenin. The total prenylated flavonoids present in beer ranged from 0.0 to 9.5 mg/L. The simple phenolic alcohols analyzed were tyrosol and hydroxytyrosol, ranging from 0.2 to 44.4 and 0.0 to 0.1 mg/L, respectively. Our study describes, for the first time, the presence of low amounts of alkylresorcinols in beer, in concentrations ranging from 0.02 to 11.0 μg/L. The results in non-alcoholic beer and the differences observed in the phenolic composition among different beer types and styles highlight the importance of the starting materials and the brewing process (especially fermentation) on the final phenolic composition of beer. In conclusion, beer represents a source of phenolic compounds in the diet that could act synergistically, triggering beneficial health effects in the context of its moderate consumption.

Keywords: beer; antioxidants; prenylated flavonoids; tyrosol; hydroxytyrosol; alkylresorcinols

---

1. Introduction

Beer is a fermented alcoholic beverage containing unique kinds of phenolic compounds. Its basic ingredients are water, barley or wheat malt, hops and yeasts. Based on the type of fermentation, beer can be divided into two broad types: ale and lager. Beer has become the most prevalent form of alcohol consumption in Europe, accounting for 40% of the total alcohol consumed [1]. In general, the evidence suggests a J-shaped curve relationship between alcohol consumption and cardiovascular disease (CVD) morbidity and mortality, indicating that moderate drinkers are at lower risk than abstainers and heavy drinkers [2]. Other more specific studies observed that cardiovascular protection was only observed with moderate consumption of fermented alcoholic beverages containing phenolic compounds such as wine or beer. Nonetheless, the protective effect was not observed following moderate consumption of spirits [3,4]. In the specific case of beer, low-to-moderate consumption (up to one drink/day in women and two drinks/day in men) reduces the risk of CVD and represents no harm in relation to major chronic conditions [3,4]. Evidence suggests that beer's beneficial health effects result from an additive effect between beer's alcohol content and beer's phenolic compounds [5]. Beer's phenolic compounds derive from hops (about 30%), from barley malt (about 70%) and from the chemical transformations that these compounds undergo during the brewing process [6]. Changes in the type and proportion of each ingredient have an impact on the phenolic content which, in turn, influences the quality parameters of the resulting beer (e.g., flavor, flavor stability, color and clarity) and gives rise to different styles [7]. The total phenolic content of beer is slightly higher than in white wine and lower than in red wine [8], but it may vary according to the raw material used and brewing process parameters [7]. At the same time, its alcohol content is lower compared to other popular alcoholic drinks. Therefore, its low alcohol content together with its phenolic composition suggest beer to be a potential trigger of positive health effects while minimizing the detrimental effects associated with alcohol consumption.

An extensive variety of phenolic compounds had been described in beer including simple phenols, phenolic acids, catechins, proanthocyanins, prenylated flavonoids α- and iso-α-acids, among others [9]. To identify them, several studies have used a wide range of techniques, such as high-performance liquid chromatography (HPLC) coupled with electrochemical [10–14] or diode [15] array detectors, and a minority have used couplings with high resolution mass spectrometry [16]. Nevertheless, there are some gaps in the literature regarding the quantitative characterization of these compounds present in beer [7].

In terms of beer's phenolic compounds and its potential biological activity, phenolic acids, prenylated flavonoids, α- and iso-α-acids have been the most studied. These phenolic compounds had been associated with relevant biological activities such as antioxidant, anti-inflammatory, antidiabetic and estrogenic activities [17]. However, beer can also be a source of compounds with potential toxic and pro-carcinogenic properties at higher concentrations such as carbonyl compounds and furan derivates [18].

Due to a worldwide increase in beer consumption, there is a need to characterize beer's antioxidant profile to unveil the potential health effects attributed to moderate beer consumption. A better understanding of the phenolic composition of different types of beers is key to (i) identify the antioxidants which could be potentially responsible for the health effects attributed to moderate beer consumption and (ii) to evaluate the impact of raw material choices and brewing technology in the resulting chemical composition of beer. The aim of the present study was to explore the potential of beer as a source of antioxidant compounds in the diet, characterizing the differences between ale, lager, and non-alcoholic beers. In order to achieve this objective, we screened 45 commercially available beers for their prenylated flavonoid content, specifically those from hops (xanthohumol (XN), isoxanthohumol (IX), 6-prenylnaringenin (6PN) and 8-prenylnaringenin (8PN)), alkylresorcinols (ARs) from cereals (AR-C17:0, AR-C19:0, AR-C21:0, AR-C23:0, and AR-C25:0) and the simple phenols from tyrosine fermentation (tyrosol (TYR) and hydroxytyrosol (HT)).

2. Results

2.1. Beers Characterization

A total of 45 different beers were analyzed in the current work. The individual characteristics of analyzed beers are available in Supplementary Table S1. Beers analyzed included 18 ales, 22 lagers and five non-alcoholic beers. Within each type of beer, a further sub-classification was made in terms of their style. Information regarding alcoholic content and international bitterness units (IBU) were obtained from the manufacturer. The mean (SD) alcoholic content was 5.10 (2.15) v/v % and the mean (SD) of the IBU was 26.41 (13.11).

2.2. Prenylated Flavonoids

The present study analyzed the concentrations of the prenylated chalcone IX, and the prenylated flavanones XN, 8PN and 6PN. The amount of total prenylated flavonoids in the analyzed beers ranged from 0.0 to 9.47 mg/L with mean (SD) values of 0.62 (1.51) mg/L. The specific prenylated flavonoid present in largest concentrations was IX, with a mean (SD) of 0.34 (0.41) mg/L, followed by XN 0.17 (0.87) mg/L, then 6PN 0.08 (0.03) mg/L and, finally, 8PN 0.03 (0.10) mg/L. Between ale and lager beers, no statistical differences were observed in either individual or total prenylated flavonoid concentration (Table 1). However, non-alcoholic beers presented lower concentrations of IX, reaching borderline significance compared to both ale and lager ($p = 0.06$ for both) (Table 1).

Table 1. Beer phenolic composition of prenylated flavonoids, simple phenols tyrosol (TYR) and hydroxytyrosol (HT) and alkylresorcinols (ARs) according to beer type (ale vs. lagers vs. free).

Compound	All Beers	Type			p [a]	p [b]	p [c]
		Ale	Lager	Non-Alcoholic			
IX (mg/L)	0.34 (0.41)	0.42 (0.54)	0.33 (0.31)	0.08 (0.08)	0.76	0.06	0.06
XN (mg/L)	0.17 (0.87)	0.39 (1.38)	0.03 (0.04)	0.01 (0.02)	0.98	0.55	0.55
8PN (mg/L)	0.03 (0.10)	0.06 (0.15)	0.02 (0.03)	0.00 (0.00)	0.78	0.55	0.49
6PN (mg/L)	0.08 (0.34)	0.17 (0.53)	0.02 (0.03)	0.01 (0.01)	0.84	0.50	0.57
Total PN (mg/L)	0.62 (1.51)	1.04 (2.33)	0.40 (0.37)	0.11 (0.98)	0.80	0.17	0.17
TYR (mg/L)	11.45 (10.55)	13.53 (12.94)	11.58 (8.75)	3.38 (2.60)	0.87	**0.01**	**<0.01**
HT (mg/L)	0.03 (0.03)	0.04 (0.03)	0.02 (0.02)	0.01 (0.01)	**0.02**	**0.01**	**0.05**
Total simple phenols (mg/L)	11.5 (10.5)	13.6 (12.9)	11.6 (8.8)	3.4 (2.6)	0.90	**0.01**	**0.01**
AR-C17:0 (μg/L)	0.00 (0.01)	0.01 (0.02)	0.00 (0.00)	0.00 (0.00)	0.08	0.32	0.70
AR-C19:0 (μg/L)	0.07 (0.15)	0.13 (0.20)	0.03 (0.07)	0.00 (0.00)	0.44	0.44	0.44
AR-C21:0 (μg/L)	0.19 (0.40)	0.39 (0.55)	0.07 (0.18)	0.02 (0.03)	0.23	0.50	0.91
AR-C23:0 (μg/L)	0.17 (0.36)	0.30 (0.49)	0.09 (0.21)	0.03 (0.02)	0.27	0.45	0.68
AR-C25:0 (μg/L)	0.58 (1.19)	1.04 (1.69)	0.31 (0.59)	0.15 (0.11)	0.34	0.58	0.83
Total ARs (μg/L)	1.01 (2.03)	1.87 (2.84)	0.50 (1.04)	0.2 (0.15)	0.25	0.58	0.93

Results shown as mean (SD); $p = p$-value from Kruskal-Wallis test, comparing (a) ales vs. lagers; (b) non-alcoholic vs. ales; (c) non-alcoholic vs. lagers. Bold values denote statistical significance at the $p < 0.05$ level. Standard deviation (SD). Tyrosol (TYR). Hydroxytyrosol (HT). Alkylresorcinols (ARs). Sum of AR-C17:0, AR-C19:0, AR-C21:0, AR-C23:0, AR-C25:0 (total AR). Xanthohumol (XN). Isoxanthohumol (IX). 8-prenylnaringenin (8PN). 6-prenylnaringenin (6PN). Sum of XN, IX, 8PN and 6PN (total PN).

2.3. Simple Phenols

The phenols TYR and HT were determined in all samples. The presence of TYR in beer ranged from 0.2–44.4, while HT concentration ranged from 0.0 to 0.13 mg/L. No significant differences were found in TYR levels between ale and lager (Table 1). In the case of HT, ale presented significantly greater concentrations than lager ($p < 0.05$) (0.04 (0.03) mg/L for ale and 0.02 (0.02) mg/L for lager). Regarding the phenolic content of non-alcoholic beer, TYR and HT levels were significantly lower than in ale and lager beers ($p < 0.05$ for both) (Table 1).

2.4. Alkylresorcinols

The present study described the presence of ARs in beer in concentrations varying from 0.02 to 11.04 μg/L. We measured a total of five ARs differing on the length of the alkyl chain, from the AR-C17:0 to AR-C25:0. The most abundant AR in all the analyzed beers was AR-C25:0 with a mean (SD) of 0.58 (1.19) μg/L (Table 1). No significant differences were found between ale, lager and non-alcoholic beers.

2.5. Correlation Study

Figure 1 represents the correlation matrix between all analyzed compounds, alcoholic content and beer bitterness of all beers. Total prenylated flavonoids and total simple phenols exhibited a moderate correlation with beer's alcoholic content ($p < 0.001$) with correlation coefficients of 0.53

and 0.62, respectively. Beer bitterness (IBUs) presented a modest correlation with total prenylated flavonoids with a coefficient of 0.36. Weaker or non-significant correlations were observed among the three families of phenolic compounds analyzed. Significant correlations were obtained between compounds belonging to the same family.

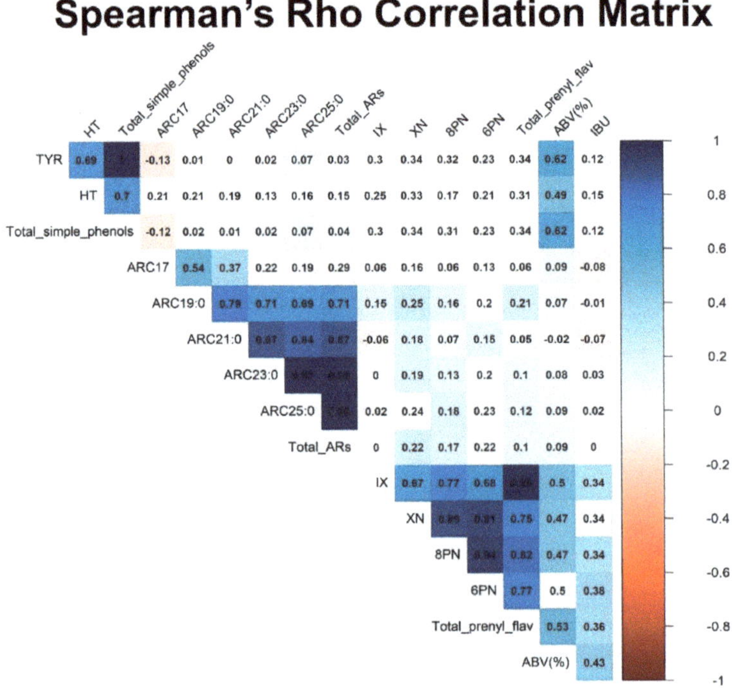

Figure 1. Spearman's Rho correlation matrix of the studied phenolic compounds in the overall beer sample (n = 45). Blank squares represent correlations of $p > 0.05$. Tyrosol (TYR). Hydroxytyrosol (HT). Sum of TYR and HT (total simple phenols). Alkylresorcinols (ARs). Sum of AR-C17:0, AR-C19:0, AR-C21:0, AR-C23:0, AR-C25:0 (total AR). Xanthohumol (XN). Isoxanthohumol (IX). 8-prenylnaringenin (8PN). 6-prenylnaringenin (6PN). Sum of XN, IX, 8PN and 6PN (total prenylated flavonoids). Percentage of alcohol by volume (%ABV). International bitterness units (IBU).

2.6. Beer Styles

Figure 2 outlines the different families of phenolic compound concentrations across the beer's styles. The beer styles with the highest concentrations of prenylated flavonoids were stout and Indian Pale Ale (IPA) with a mean (SD) of total prenylated flavonoids of 2.19 (3.10) and 1.98 (3.68) mg/L, respectively. In terms of total phenols, Belgian strong and blonde ale presented the highest concentrations: 29.2 (14.3) and 24.3 and 28.4 mg/L, respectively. Finally, stout, with a 7.84 (4.52) µg/L, was the beer style with the highest total ARs content. No statistical analysis was performed due to the low number of samples within certain beer styles.

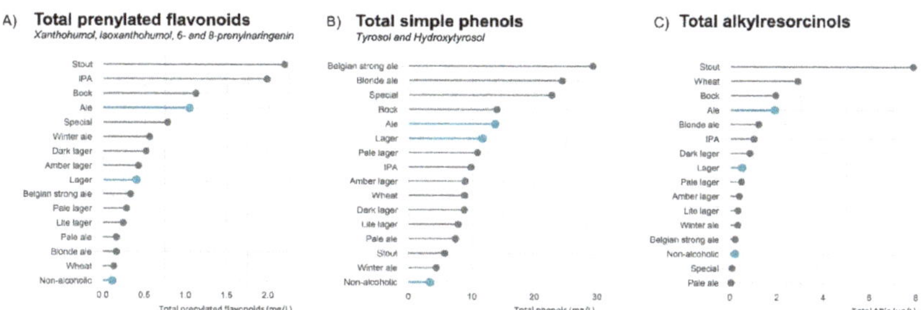

Figure 2. Beer's phenolic content (**A**) total prenylated flavonoids, (**B**) total simple phenols and (**C**) total alkylresorcinols distributed by beer type (blue) in ale ($n = 18$); lager ($n = 22$) and non-alcoholic ($n = 5$) and in beer styles (grey) in amber lager ($n = 2$); Belgian strong ale ($n = 3$); blonde ale ($n = 2$); bock ($n = 1$); dark lager ($n = 1$); IPA ($n = 6$); lite lager ($n = 1$); pale ale ($n = 1$); pale lager ($n = 15$); special lager ($n = 2$); stout ($n = 2$); wheat ($n = 2$) and winter ale ($n = 1$). Results are expressed as mean concentration in descending order.

3. Discussion

The present work studied a broad group of beers to describe their content of three families of phenolic compounds that have been associated with a wide range of potential biological activities and protective health effects. Specifically, this study characterizes beer's antioxidant composition, showing that it is an important dietary source of prenylated flavonoids and the simple phenols TYR and HT. Moreover, our study reports, for the first time, the presence of low amounts of ARs in beer.

A distinctive ingredient of beer is the hop flower (*Humulus lupulus L*), which is added during the brewing process for its preserving properties and for its organoleptic characteristics. Beer is considered a unique source of these prenylated flavonoids in the diet. Urinary IX is used as a unique and accurate biomarker of beer consumption [19], which is in agreement with our analysis, pointing out IX as the main prenylated flavonoid present in beer. The type of fermentation, classifying beer into ale or lager was not associated with the prenylated flavonoid concentrations. Prenylated flavonoids are of interest due to their display of antibacterial, anti-inflammatory, antioxidant and other biological effects [20]. In particular, the compound XN is being closely studied for its potential chemopreventive properties. In the case of IX, and especially 8PN, these compounds are characterized by their strong phytoestrogen activities [21,22].

Malt phenolic compounds represent the main source of bioactive substances found in beer [23]. The most abundant are phenolic acids, phenolic alcohol, phenolic amines, phenolic amino acids and finally α-acids and iso-α-acids [24]. In the present study, we have focused on the analysis of the phenolic alcohols TYR and HT. The presence of the simple phenol TYR in relatively high concentrations in beer has been previously reported [25–27]. We confirmed the presence of TYR in beer and, additionally, we described, for the first time, the presence of HT in certain beers. TYR and HT are not present in beer as raw components, they are formed during the fermentation process. TYR is produced as a product of tyrosine metabolism generated by yeast in the Ehrlich pathway. A minor part of the TYR formed can be later hydroxylated to give rise to HT [28]. Based on the concentrations observed, beer is a relevant source of TYR in the diet. TYR average content in beer is comparable to white wine. Nevertheless, certain beers exhibited TYR levels at the same range as red wine, considered a good source of TYR, whose concentrations have been reported to be between 20.5 and 44.5 mg/L [29]. The contribution of beer as a direct source of HT is negligible. Moreover, the presence of TYR in beer is relevant, since it has been demonstrated that dietary TYR is converted in humans into HT [30,31]. Both TYR supplementation and its biotransformation into HT are capable of triggering relevant beneficial effects on the cardiovascular system [30]. HT is considered one of the strongest dietary antioxidants, with anti-inflammatory, antiproliferative, antiplatelet and proapoptotic activities [32]. Therefore, beer would represent an indirect source of HT via TYR hydroxylation. Consequently, beer should also be considered a relevant source of TYR and HT, together with the traditional dietary sources of extra virgin olive oil and wine.

ARs are a group of phenolic lipids that contain a resorcinol (a benzene ring with two hydroxyl groups in positions 1 and 3 and an odd-numbered alkyl chain in the range of 15–25 carbons at position 5). They are present in the outer part of certain grains and in the products produced from them [33,34]. They have been described in barley, wheat, rye, oats, rice and other cereal grains, and the relative abundance of the different homologues varies depending on the type of cereal. AR-C25:0, the most abundant AR found in beer, is (accordingly) the dominant AR homologue in barley [35]. ARs are being studied for their potential biological activities. They have shown antioxidant activity [36], protecting against LDL oxidation [37], and also improving glucose and cholesterol metabolism [38]. Nevertheless, it is important to contextualize the sources of ARs in the diet and to understand that, although we describe the presence of ARs in beer in trace amounts, their contribution to total AR dietary intake would be almost negligible. The intake of ARs in countries with a high consumption ranges between 12 and 18 mg/day [30]. Based on our results, a glass of an average beer of 330 mL (equivalent to one standard drink) could represent an intake of 0.3 µg of total ARs and, therefore, a minor contribution to the overall amount of ARs ingested.

Beer's alcoholic content was positively correlated with prenylated flavonoid, TYR and HT concentration. It has been described that, during fermentation, the presence of phenols with antioxidant activity within the wort protect yeast viability against the stress generated by high levels of ethanol [39]. Therefore, a high concentration of prenylated flavonoids with their inherent antioxidant activity would contribute to yeast stability, enhancing the fermentation process and increasing alcohol content. TYR and HT are byproducts of this fermentation. Another fact confirming the importance of fermentation in the phenolic profile of beer is that the variety of yeast strain used for beer brewing is capable of triggering differences in the antioxidant activity and total phenolic composition of the beer produced [40,41]. In the case of ARs, their amount was not correlated with the alcoholic content of beer, nor with any of the analyzed groups of phenolic compounds. Given that ARs are biomarkers of whole grain intake [34], and that there are known differences in AR composition depending on the cereal type [33], the presence of ARs in beer most likely derives from the cereals used for the elaboration of the beer and is independent of other beer compounds.

Non-alcoholic beer's popularity has risen due to a concern about alcohol abuse and its health consequences. The production of beer with a limited alcohol content can be achieved by two approaches: limiting the fermentation process, and hence the alcohol production, or by using physical methods to remove the alcohol at the end of brewing [42]. On one hand, the concentration of the prenylated flavonoid IX in non-alcoholic beer was borderline significantly lower than ale and lager beers. On the other hand, and in agreement with a previous study [26], non-alcoholic beers presented lower TYR and lower HT content. TYR and HT are produced as byproducts of fermentation and its limitation during the dealcoholizing process is likely to have a negative effect on their accumulation levels. In non-alcoholic beer production, physical methods including thermal processes or inverse osmosis, are often used. These processes could trigger the degradation or the loss of IX, TYR and HT, explaining the lower concentrations observed in non-alcoholic beers. Our results suggest that the non-alcoholic brewing process has a detrimental impact on the content of the simple phenols and IX. In the case of ARs, no statistical differences were observed in the values present in non-alcoholic beer, suggesting the stability of these compounds during dealcoholizing. Finally, in the context of non-alcoholic beer consumption, it is worth to mention the role that has been attributed to alcohol in promoting the bioavailability of phenolic compounds. This has been recently demonstrated with a reduced bioavailability of TYR following non-alcoholic beer consumption [31]. Therefore, non-alcoholic beers, which have a low concentration of phenolic compounds from the outset and an absence of alcohol, would represent a minor source of phenolic compounds.

Eventually, Figure 2 represents an exploratory overview of the concentrations of the analyzed phenolic compounds across different beer styles. In the case of total prenylated flavonoids, stout and IPA styles presented the greatest concentrations in this family. In the case of TYR and HT, Belgian strong and blonde ales exhibited the highest concentrations. Finally, stout beer stands out for its AR content. However, caution must be applied as certain beer styles were under-represented, with a low number of samples being available. Further studies should analyze a larger sample of beers belonging to the mentioned styles to confirm their high concentration of phenolic compounds. Research on the characteristics of the mentioned beer styles was performed to understand the reason behind the high concentrations of selected phenolic compounds. In the case of IPA beers, originally, this beer style was characterized with the greatest proportion of hops, as it is known for its antimicrobial properties that enhance beer stability. Therefore, a high proportion of prenylated flavonoids would be expected. In the case of Stout beers, a distinctive characteristic is the use of roasted barley as a starting material. Based on the high concentration of ARs, this step could facilitate the extraction of ARs from the cereal to the wort during the brewing process. Finally, Belgian strong ale, the beer with the highest concentration in TYR and HT, uses a specific and traditional yeast that could produce higher proportions of TYR [43]. Overall, these observations confirm the importance of the starting materials and the fermentation on the final concentration of phenolic compounds.

Finally, it is important to mention that, despite the interesting beer antioxidant profiles described in the present paper and in the literature, it is important to highlight the importance of a moderate consumption of beer in the context of a healthy dietary pattern, such as the Mediterranean diet [23]. Excessive beer consumption can lead to an excessive body weight, hamper pancreatic function and increase the risk of cancer due to its ethanol content and also due to the low levels of toxic compounds [17].

Our study presents some strengths and limitations. A key strength of the present study is the high number of beers analyzed, including different beer types and styles. The quantitative assessment of three different families of phenolic compounds provides a broad perspective of the phenolic profile of beer. More specifically, we show that TYR, formed during the fermentation process, is a phenolic compound abundantly present in beer. Additionally, prenylated flavonoids that derive from the variety of hops used during the brewing process are present in lower amounts than TYR. Finally, ARs, which most likely come from the malted cereals selected as ingredients for brewing, are only present in trace amounts in beers. However, our analysis was limited by the fact that certain beer styles were under-represented, due to their low availability on the market. Our current research has only quantified three families of phenolic compounds; however, beer is an extremely complex drink with several phenolic compounds whose concentrations have not yet been assessed.

Overall, the exploratory nature of the present research offers some insight into the phenolic composition of beer, highlighting it as an important dietary source of prenylated flavonoids, TYR and, indirectly, HT. Additionally, it extends our knowledge of the levels of phenolic compounds present in different beer styles, different beer types and non-alcoholic beers. This work represents a starting point in understanding beer's antioxidant profile; however, future studies should assess the bioavailability and the potential synergies of the mentioned compounds in the context of moderate beer consumption and its potential health effects.

4. Materials and Methods

4.1. Chemicals and Reagents

TYR, HT, 3-(4-hydroxyphenyl)-1-propanol, XN, IX and 8PN taxifolin and ammonium fluoride were purchased from Sigma-Aldrich (St. Louis, MO, USA). HT-D$_3$, were purchased from Toronto Research Chemicals Inc. (Toronto, ON, Canada). 5-heptadecylresorcinol (AR C17:0), 5-nonadecylresorcinol (AR C19:0), 5-nonadecylresorcinol-D4 (AR C19:0-D$_4$), 5-heneicosylresorcinol (AR C21:0), 5-tricosylresorcinol (AR C23:0), 5-pentacosylresorcinol (AR C25:0) were purchased from ReseaChem GmbH (Burgdorf, Switzerland). Methanol and acetonitrile were obtained from Merck (Darmstadt, Germany). Ultra-pure water was supplied by a Milli-Q® purification system (Darmstadt, Germany).

4.2. Samples and Sample Preparation

A total of 45 different beers were selected for the analysis of phenolic compounds. A 10 mL sample of each beer was stored in a Falcon tube at −20 °C until the analysis. Beer foam was removed from all samples by means of ultrasonication prior to any analysis. All determinations were performed in duplicate.

4.3. Extraction and Analysis of Prenylated Flavonoids: IX, XN, 8PN and 6PN

All the samples were filtered through a 0.45 mm polytetrafluoroethylene filter and 600 ng/mL of taxifolin was added as internal standard. Aliquots of 10 μL were injected into the liquid chromatography coupled to mass spectrometry (LC–MS/MS) system without any other pretreatment. The identification and quantification of IX, XN, 6PN, and 8PN was performed using an Acquity UHPLC® system equipped with a Waters binary pump system (Waters, Milford, MA, USA) coupled to an API 3000™ triple quadrupole mass spectrometer (Sciex, Concord, ON, Canada) with a turbo ion spray source working in a negative mode. Chromatographic separation was performed with a Luna C18® column, 50 mm × 2.0 mm i.d., 5 mm (Phenomenex, Torrance, CA, United States), using 5 mM ammonium bicarbonate buffer adjusted to pH 7.0 as an aqueous mobile phase and acetonitrile and methanol (1:1 proportion) as an organic phase. For the quantification of analytes, the multiple reaction monitoring (MRM) mode was used, monitoring 3 transitions: 353–119 (IX and XN), 339–219 (8PN and 6PN), and 303–285 (taxifolin) [19]. Calibration curves were prepared adding standards to pure water.

4.4. Extraction and Analysis of Simple Phenols TYR and HT in Beer

TYR and HT content were determined by LC–MS/MS following a dilute-and-shoot approach. Samples were diluted 40 times with a mobile phase (65% A: 35% B) and spiked with 10 μL of an internal standard containing 1 μg/mL of 3-(4-hydroxyphenyl)-1-propanol and 1 μg/mL of HT-D$_3$. Mobile phase A contained 0.5 mM of ammonium fluoride in water. Mobile phase B contained 0.5 mM of ammonium fluoride in methanol. All samples were analyzed by LC–MS/MS (Agilent Technologies, Santa Clara, CA, USA). The separation was carried out with an Acquity UPLC® BEH C18 column 1.7μm particle size, 3 mm × 100 mm (Waters, Milford, MA, USA). The following transitions were monitored on the acquisition method in MRM mode: 137–106 (TYR), 151–106 (3-(4-hydroxyphenyl)-1-propanol), 153–123 (HT) and 156–126 (HT-D$_3$). Calibration curves were prepared adding standards of TYR and HT to pure water.

4.5. Extraction and Analysis of ARs in Beer

Extraction of AR from beers was based on a liquid-liquid extraction protocol using ethyl acetate as described for the analysis of ARs in cereals [39]. Briefly, 1 mL of each beer was spiked with 20 μL of internal standard (AR C19:0D$_4$ at 50 ng/mL). ARs were extracted using 4 mL of ethyl acetate for 24 h. Then, the organic layer was evaporated to dryness under nitrogen (40 °C and 15 psi) and reconstituted in 0.25 mL of methanol. Then, it was centrifuged for 5 min at 4 °C (12.000 rpm) to obtain a clear supernatant fraction, which was directly injected into the LC–MS/MS system. Chromatographic separation of five ARs was performed by using an Acquity UPLC® instrument (Waters, Milford, MA, USA) operated using MassLynx 4.1 software. The LC system was equipped with an Acquity UPLC® (BEH C18 1.7 μm 2.1 × 100 mm) column (Waters, Milford, MA, USA). The injection volume was 10 μL, the flow rate was 0.3 mL/min, and the temperature of the column was set at 55 °C. An isocratic method was selected with a solution of 0.5 mM ammonium fluoride in methanol as a mobile phase solvent. The detection was performed with a triple quadrupole mass spectrometer (Xevo® TQS-Micro MS, Waters, Milford, MA, USA) equipped with an orthogonal Z-spray-electrospray ionization source (ESI).

The monitoring and quantification of AR was performed in MRM mode, monitoring the following transitions: 347–305 (AR C17:0), 375–122 (AR 19:0), 403–361 (AR 21:0), 431–389 (AR 23:0), 459–417 (AR 25:0), and 379–337 (AR 19:0-D$_4$). Calibration curves were prepared by adding standards to pure water.

4.6. Statistical Analysis

Statistical analyses and figures were performed using the R software (R Foundation for Statistical Computing, Vienna, Austria), version 3.5.2. The normality of continuous variables was assessed by normal probability plots and non-parametric tests were used if data did not follow a normal distribution. The R packages used were 'corrplot' and 'tidyverse'. The significance level was set at $p < 0.05$.

Supplementary Materials: The following are available online, Table S1: List and characteristics of the analyzed beers.

Author Contributions: R.d.l.T. and A.B. designed the research. A.B., N.S.-D., J.R.-M. and M.M.-H. carried out the laboratory analysis and wrote the manuscript. R.d.l.T. and R.M.L.-R. provided a critical revision of the manuscript. All authors have read and agreed to the published version of the manuscript.

Funding: This research was funded by the Instituto de Salud Carlos III FEDER (PI14/00072) and by grants from DIUE of Generalitat de Catalunya (2017 SGR138). AB is recipient of a PFIS predoctoral fellowship from the Instituto Carlos III FEDER (PFIS-FI16/00106), NSD is recipient of predoctoral fellowship (2019-DI-47) from Agència de Gestió d'Ajuts Universitaris i de Recerca (AGAUR) Generalitat de Catalunya. JMR is recipient of Marie Skłodowska-Curie grant agreement (No 712949) and from ACCIO. CIBER de Fisiopatología de la Obesidad y Nutrición (CIBEROBN) is an initiative of the ISCIII, Madrid, Spain.

Conflicts of Interest: The authors declare no conflict of interest.

References

1. World Health Organization. *Global Status Report on Alcohol and Health 2018*; World Health Organization: Geneva, Switzerland, 2018.

2. Poli, A.; Marangoni, F.; Avogaro, A.; Barba, G.; Bellentani, S.; Bucci, M.; Cambieri, R.; Catapano, A.L.; Costanzo, S.; Cricelli, C.; et al. Moderate alcohol use and health: A consensus document. *Nutr. Metab. Cardiovasc. Dis.* **2013**, *23*, 487–504. [CrossRef] [PubMed]

3. Costanzo, S.; Di Castelnuovo, A.; Donati, M.B.; Iacoviello, L.; de Gaetano, G. Wine, beer or spirit drinking in relation to fatal and non-fatal cardiovascular events: A meta-analysis. *Eur. J. Epidemiol* **2011**, *26*, 833–850. [CrossRef] [PubMed]

4. de Gaetano, G.; Costanzo, S.; Di Castelnuovo, A.; Badimon, L.; Bejko, D.; Alkerwi, A. Effects of moderate beer consumption on health and disease: A consensus document. *Nutr. Metab Cardiovasc Dis* **2016**, *26*, 443–467. [CrossRef] [PubMed]

5. Chiva-Blanch, G.; Arranz, S.; Lamuela-Raventos, R.M.; Estruch, R. Effects of wine, alcohol and polyphenols on cardiovascular disease risk factors: Evidences from human studies. *Alcohol Alcoholism* **2013**, *48*, 270–277. [CrossRef] [PubMed]

6. Jerkovic, V.; Callemien, D.; Collin, S. Determination of stilbenes in hop pellets from different cultivars. *J. Agric Food Chem* **2005**, *53*, 4202–4206. [CrossRef]

7. Wannenmacher, J.; Gastl, M.; Becker, T. Phenolic Substances in Beer: Structural Diversity, Reactive Potential and Relevance for Brewing Process and Beer Quality. *Compr. Rev. Food Sci. Food Saf.* **2018**, *17*, 953–988. [CrossRef]

8. Neveu, V.; Perez-Jiménez, J.; Vos, F.; Crespy, V.; du Chaffaut, L.; Mennen, L. Phenol-Explorer: An online comprehensive database on polyphenol contents in foods. *Database (Oxford)* **2010**, *2010*, bap024. [CrossRef]

9. Gerhäuser, C. Beer constituents as potential cancer chemopreventive agents. *Eur. J. Cancer* **2005**, *41*, 1941–1954. [CrossRef]

10. Nardini, M.; Natella, F.; Scaccini, C.; Ghiselli, A. Phenolic acids from beer are absorbed and extensively metabolized in humans. *J. Nutr. Biochem.* **2006**, *17*, 14–22. [CrossRef]

11. Piazzon, A.; Forte, M.; Nardini, M. Characterization of phenolics content and antioxidant activity of different beer types. *J. Agric. Food Chem.* **2010**, *58*, 10677–10683. [CrossRef]

12. Floridi, S.; Montanari, L.; Marconi, O.; Fantozzi, P. Determination of free phenolic acids in wort and beer by coulometric array detection. *J. Agric. Food Chem.* **2003**, *51*, 1548–1554. [CrossRef] [PubMed]
13. Řehová, L.; Škeříkova, V.; Jandera, P. Optimisation of gradient HPLC analysis of phenolic compounds and flavonoids in beer using a CoulArray detector. *J. Sep. Sci.* **2004**, *27*, 1345–1359. [CrossRef] [PubMed]
14. Jandera, P.; Škeříková, V.; Řehová, L.; Hájek, T.; Baldriánová, L.; Škopová, G. RP-HPLC analysis of phenolic compounds and flavonoids in beverages and plant extracts using a CoulArray detector. *J. Sep. Sci.* **2005**, *28*, 1005–1022. [CrossRef] [PubMed]
15. Nardini, M.; Garaguso, I. Characterization of bioactive compounds and antioxidant activity of fruit beers. *Food Chem.* **2020**, *305*, 125437. [CrossRef]
16. Quifer-Rada, P.; Vallverdú-Queralt, A.; Martínez-Huélamo, M.; Chiva-Blanch, G.; Jáuregui, O.; Estruch, R. A comprehensive characterisation of beer polyphenols by high resolution mass spectrometry (LC-ESI-LTQ-Orbitrap-MS). *Food Chem.* **2015**, *169*, 336–343. [CrossRef]
17. Osorio-Paz, I.; Brunauer, R.; Alavez, S. Beer and its non-alcoholic compounds in health and disease. *Crit. Rev. Food Sci. Nutr.* **2019**, *2019*, 1–14. [CrossRef]
18. Hernandes, K.C.; Souza-Silva, É.A.; Assumpção, C.F.; Zini, C.A.; Welke, J.E. Validation of an analytical method using HS-SPME-GC/MS-SIM to assess the exposure risk to carbonyl compounds and furan derivatives through beer consumption. *Food Addit. Contam. Part. A* **2019**, *36*, 1808–1821. [CrossRef]
19. Quifer-Rada, P.; Martínez-Huélamo, M.; Chiva-Blanch, G.; Jáuregui, O.; Estruch, R.; Lamuela-Raventós, R.M. Urinary Isoxanthohumol Is a Specific and Accurate Biomarker of Beer Consumption. *J. Nutr.* **2014**, *144*, 484–488. [CrossRef]
20. Chen, X.; Mukwaya, E.; Wong, M.-S.; Zhang, Y. A systematic review on biological activities of prenylated flavonoids. *Pharm Biol* **2014**, *52*, 655–660. [CrossRef]
21. Stevens, J.F.; Page, J.E. Xanthohumol and related prenylflavonoids from hops and beer: To your good health! *Phytochemistry* **2004**, *65*, 1317–1330. [CrossRef]
22. Štulíková, K.; Karabín, M.; Nešpor, J.; Dostálek, P. Therapeutic perspectives of 8-prenylnaringenin, a potent phytoestrogen from hops. *Molecules* **2018**, *23*, 660. [CrossRef]
23. Arranz, S.; Chiva-Blanch, G.; Valderas-Martínez, P.; Medina-Remón, A.; Lamuela-Raventós, R.M.; Estruch, R. Wine, beer, alcohol and polyphenols on cardiovascular disease and cancer. *Nutrients* **2012**, *4*, 759–781. [CrossRef] [PubMed]
24. Quesada-Molina, M.; Muñoz-Garach, A.; Tinahones, F.J.; Moreno-Indias, I. A New Perspective on the Health Benefits of Moderate Beer Consumption: Involvement of the Gut Microbiota. *Metabolites* **2019**, *9*, 272. [CrossRef] [PubMed]
25. Li, M.; Yang, Z.; Hao, J.; Shan, L.; Dong, J. Determination of Tyrosol, 2-Phenethyl Alcohol, and Tryptophol in Beer by High-Performance Liquid Chromatography. *J. Am. Soc. Brew. Chem.* **2008**, *66*, 245–249. [CrossRef]
26. Cerrato-Alvarez, M.; Bernalte, E.; Bernalte-García, M.J.; Pinilla-Gil, E. Fast and direct amperometric analysis of polyphenols in beers using tyrosinase-modified screen-printed gold nanoparticles biosensors. *Talanta* **2019**, *193*, 93–99. [CrossRef] [PubMed]
27. Bartolomé, B.; Peña-Neira, A.; Gómez-Cordovés, C. Phenolics and related substances in alcohol-free beers. *Eur. Food Res. Technol.* **2000**, *210*, 419–423. [CrossRef]
28. Hazelwood, L.A.; Daran, J.M.; Van Maris, A.J.A.; Pronk, J.T.; Dickinson, J.R. The Ehrlich pathway for fusel alcohol production: A century of research on Saccharomyces cerevisiae metabolism. *Appl. Environ. Microbiol.* **2008**, *74*, 2259–2266. [CrossRef]
29. Piñeiro, Z.; Cantos-Villar, E.; Palma, M.; Puertas, B. Direct liquid chromatography method for the simultaneous quantification of hydroxytyrosol and tyrosol in red wines. *J. Agric. Food Chem.* **2011**, *59*, 11683–11689. [CrossRef]
30. Boronat, A.; Mateus, J.; Soldevila-Domenech, N.; Guerra, M.; Rodríguez-Morató, J.; Varon, C. Cardiovascular benefits of tyrosol and its endogenous conversion into hydroxytyrosol in humans. A randomized, controlled trial. *Free Radic. Biol. Med.* **2019**. [CrossRef]
31. Soldevila-Domenech, N.; Boronat, A.; Mateus, J.; Diaz-Pellicer, P.; Matilla, I.; Pérez-Otero, M. Generation of the Antioxidant Hydroxytyrosol from Tyrosol Present in Beer and Red Wine in a Randomized Clinical Trial. *Nutrients* **2019**, *11*, 2241. [CrossRef]

32. Robles-Almazan, M.; Pulido-Moran, M.; Moreno-Fernandez, J.; Ramirez-Tortosa, C.; Rodriguez-Garcia, C.; Quiles, J.L. Hydroxytyrosol: Bioavailability, toxicity, and clinical applications. *Food Res. Int.* **2018**, *105*, 654–667. [CrossRef] [PubMed]

33. Landberg, R.; Marklund, M.; Kamal-Eldin, A.; Åman, P. An update on alkylresorcinols–Occurrence, bioavailability, bioactivity and utility as biomarkers. *J. Funct. Foods* **2014**, *7*, 77–89. [CrossRef]

34. Rodríguez-Morató, J.; Jayawardene, S.; Huang, N.K.; Dolnikowski, G.G.; Galluccio, J.; Lichtenstein, A.H. Simplified method for the measurement of plasma alkylresorcinols: Biomarkers of whole-grain intake. *Rapid Commun. Mass Spectrom.* **2020**, *34*, e8805. [CrossRef] [PubMed]

35. Chen, Y.; Ross, A.B.; Åman, P.; Kamal-Eldin, A. Alkylresorcinols as Markers of Whole Grain Wheat and Rye in Cereal Products. *J. Agric. Food Chem.* **2004**, *52*, 8242–8246. [CrossRef] [PubMed]

36. Wang, Z.; Hao, Y.; Wang, Y.; Liu, J.; Yuan, X.; Sun, B. Wheat alkylresorcinols protect human retinal pigment epithelial cells against H_2O_2-induced oxidative damage through Akt-dependent Nrf2/HO-1 signaling. *Food Funct.* **2019**, *10*, 2797–2804. [CrossRef] [PubMed]

37. Parikka, K.; Rowland, I.R.; Welch, R.; Wähälä, K. In vitro antioxidant activity and antigenotoxicity of 5-n-alkylresorcinols. *J. Agric. Food Chem.* **2006**, *54*, 1646–1650. [CrossRef]

38. Oishi, K.; Yamamoto, S.; Itoh, N.; Nakao, R.; Yasumoto, Y.; Tanaka, K. Wheat alkylresorcinols suppress high-fat, high-sucrose diet-induced obesity and glucose intolerance by increasing insulin sensitivity and cholesterol excretion in male mice. *J. Nutr.* **2015**, *145*, 199–206. [CrossRef]

39. Gharwalova, L.; Sigler, K.; Dolezalova, J.; Masak, J.; Rezanka, T.; Kolouchova, I. Resveratrol suppresses ethanol stress in winery and bottom brewery yeast by affecting superoxide dismutase, lipid peroxidation and fatty acid profile. *World J. Microbiol. Biotechnol.* **2017**, *33*, 205. [CrossRef]

40. Capece, A.; Romaniello, R.; Pietrafesa, A.; Siesto, G.; Pietrafesa, R.; Zambuto, M. Use of Saccharomyces cerevisiae var. boulardii in co-fermentations with S. cerevisiae for the production of craft beers with potential healthy value-added. *Int. J. Food Microbiol.* **2018**, *284*, 22–30. [CrossRef]

41. Basso, R.F.; Alcarde, A.R.; Portugal, C.B. Could non-Saccharomyces yeasts contribute on innovative brewing fermentations? *Food Res. Int.* **2016**, *86*, 112–120. [CrossRef]

42. Ross, A.B. Analysis of Alkylresorcinols in Cereal Grains and Products Using Ultrahigh-Pressure Liquid Chromatography with Fluorescence, Ultraviolet, and CoulArray Electrochemical Detection. *J. Agric. Food Chem.* **2012**, *60*, 8954–8962. [CrossRef] [PubMed]

43. Strong, G.; England, K. Guide of Styles 2015 Beer Judge Certification Program. 2015. Available online: https://www.bjcp.org/stylecenter.php (accessed on 25 May 2020).

Sample Availability: Samples of the compounds are not available from the authors.

Article

Caftaric Acid Isolation from Unripe Grape: A "Green" Alternative for Hydroxycinnamic Acids Recovery

Veronica Vendramin [1], Alessia Viel [1] and Simone Vincenzi [1,2,*]

[1] Centre for Research in Viticulture and Enology (CIRVE), University of Padova, Viale XXVIII Aprile 14, 31015 Conegliano (TV), Italy; veronica.vendramin@unipd.it (V.V.); alessia.viel@unipd.it (A.V.)
[2] Department of Agronomy, Food, Natural Resources, Animals and Environment (DAFNAE), University of Padova, Viale dell'Università 16, 35020 Legnaro (PD), Italy
* Correspondence: simone.vincenzi@unipd.it; Tel.: +39-0438453711

Abstract: Phenolic acids represent about one-third of the dietary phenols and are widespread in vegetable and fruits. Several plants belonging to both vegetables and medical herbs have been studied for their hydroxycinnamic acid content. Among them, *Echinacea purpurea* is preferentially used for caffeic acid-derivatives extraction. The wine industry is a source of by-products that are rich in phenolic compounds. This work demonstrates that unripe grape juice (verjuice) presents a simple high-pressure liquid chromatography (HPLC) profile for hydroxycinnamic acids (HCAs), with a great separation of the caffeic-derived acids and a low content of other phenolic compounds when compared to *E. purpurea* and other grape by-products. Here it is shown how this allows the recovery of pure hydroxycinnamic acids by a simple and fast method, fast protein liquid chromatography (FPLC). In addition, verjuice can be easily obtained by pressing grape berries and filtering, thus avoiding any extraction step as required for other vegetable sources. Overall, the proposed protocol could strongly reduce the engagement of solvent in industrial phenolic extraction.

Keywords: hydroxycinnamic acids; caftaric acid; verjuice; FPLC; unripe grape juice

Citation: Vendramin, V.; Viel, A.; Vincenzi, S. Caftaric Acid Isolation from Unripe Grape: A "Green" Alternative for Hydroxycinnamic Acids Recovery. *Molecules* **2021**, *26*, 1148. https://doi.org/10.3390/molecules26041148

Academic Editor: Mirella Nardini

Received: 15 January 2021
Accepted: 11 February 2021
Published: 21 February 2021

Publisher's Note: MDPI stays neutral with regard to jurisdictional claims in published maps and institutional affiliations.

1. Introduction

Phenolic acids constitute about one-third of the dietary phenols and are widespread in vegetable and fruits. Phenolic acids are divided into two subgroups, the hydroxybenzoic (HBAs) and hydroxycinnamic acids (HCAs). HBAs show a C6–C1 structure and include gallic, *p*-hydroxybenzoic, protocatechuic, vanillic, and syringic acids, while HCAs are characterized by an aromatic ring with three-carbon side chain (C6–C3) and are primarily represented by caffeic, ferulic, coumaric, and sinapic acids (Figure 1).

In the past years, HCAs gained attention because of their cosmetic application as anti-tyrosinase, anti-collagenase, and anti-hyaluronidase activity, apart from an interesting photo-protection action [1], and for their possible application as a food additive to prevent oxidation [2]. HCAs are key precursors of several more complex polyphenols, are structural components of the cell wall, are involved in the plant defense system, and act as signaling molecules [3]. In plants, the first HCA produced is the *p*-coumaric acid, which is obtained from phenylalanine or tyrosine. This is then transformed into caffeic acid by hydroxylation. Ferulic and synaptic acid derive from caffeic acid by methoxyl and hydroxyl substitution and, in the case of synaptic acid, from an additional methylation [1]. HCAs are found in several conjugated forms, including amides (conjugated with mono- or polyamines, amino acids, or peptides), esters (mainly esters of hydroxy acids, such as tartaric acid and sugar derivatives), and sugars. Cinnamate esters occur widely in higher plants, while the amides seem to be less present [4]. Caffeic acid (CA)-derivatives is a group of compounds derived from modification of caffeic acid by esterification with organic acids, such as quinic acid (i.e., chlorogenic acid, neochlorogenic acid), glucaric acid (caffeoylglucaric acid) [5], and more frequently with tartaric acid (i.e., chicoric acid [6], caftaric acid, and coutaric acid) [7]. To date, several plants have been studied for their CA-derivatives content, plants belonging to both vegetables and medical herbs, e.g., *Echinacea purpurea* [8], *Cichorium intybus* [9] and *C. endivia* [10], *Lactuca virosa* [11], *Eupatorium perfoliatum* [5], and *Smallanthus sonchifolius* [12]. Nevertheless, other potential sources are known, for example, CA-derivatives were discovered to be quite abundant in fruits and particularly in grape berries [13]. The wine industry is a source of by-products that can be exploited for the recovery of high value compounds. Pomace, lees, and canes have been well explored in the past [14–16] as polyphenols sources, while less attention was dedicated to other products, such as unripe grapes. Cluster thinning is a common practice used to avoid over-cropping in compliance with the production regulations and is deemed a way to accelerate ripening and increase the grape quality, even if this function is still being debated [17]. Several parameters, among others, the grape variety and the weather conditions, which affect the plant production, define the degree of thinning, which could achieve up to 50% of cluster reduction [18]. Two main forms of berries reduction are spread, namely cluster thinning, consisting of the elimination of several complete clusters, and berry thinning, which involves the removal of the tips of the clusters [19]. Additionally, thinning is encouraged from institutional organs in particular cases (called "green harvest"), i.e., for a great imbalance between supply and demand of the international market [20], as in the case of the pandemic disease COVID-19. The clusters (or their parts) are usually left on the ground, and it makes it difficult to have a clear panorama of the effective waste mass. An alternative use of unripe grapes is the production of verjuice, an acidic juice traditionally produced in the Mediterranean area, which is extracted from the mechanical pressing of unripe green grapes. Verjuice has been mainly studied for its physicochemical and sensory properties [21,22], while its bioactive compounds were just recently isolated [23,24]. In grape berries, HCAs are constituted by caffeic, coumaric, and ferulic acids and are present mainly in their ester forms, associated with tartaric acid giving caftaric (CFA), coutaric (CUA), and fertaric (FEA) acids, respectively [25]. HCAs' levels in the juice of different *Vitis vinifera* and *V. labrusca* varieties were recorded as very variable, namely 4339.9–1681.0 and 4154.5–786.7 µg/100 g in the former and in the latter, respectively [26]. The authors identified a strong difference in CFA content depending on *Vitis* species and varieties, while the other HCAs showed different patterns of accumulation amongst varieties, evidencing a general independency of hydroxycinnamic acids metabolisms. Total hydroxycinnamates concentrate mainly in grape berries pulp, and it has been recorded to peak prior to véraison [27]. The successive reduction of HCAs concentration depends on the grapes volume increase and on the engagement of key precursors into the biosynthesis pathways of other phenolic compounds [18,20]. Furthermore, HCAs reduce further during grape juice processing and winemaking, as these compounds are promptly oxidized by endogenous tyrosinase when the grape berries are crushed. Instead, the harvest of green berries and their processing through crushing and maceration could implement HCAs content [24]. Therefore, unripe grape juice (verjuice) represents a rich source of hydroxycinnamic acids and should be considered as raw material for the HCAs extraction.

Several patents aiming to improve the recovery of CA-derivates from plants have been deposited in the past. Among them, the proposed raw material, *E. purpurea* adventitious roots, were recognized as the most suitable for the HCAs production at industrial scale because of their easy management and the high yield [28]. HCAs are extracted mainly in their ester form as chicoric (CCA), caftaric (CFA), and chlorogenic acids (CLA) [29]. Generally, these protocols involve the use of organic solvents, acidification, centrifugation, and the retrieval of final compounds by filtering or, more often, by separation with macroporous adsorption resin, which strongly improved the final HCAs purity, which could pass from 31% up to 72% *w/v* [28,30,31]. Finally, the compounds are further concentrated by crystallization upon acidification of the extract and its cooling or by evaporation at high temperatures (about 90 °C) that could, however, degrade HCAs. This last concentration step permits the achievement of purity values above 90% *w/w* [21,26].

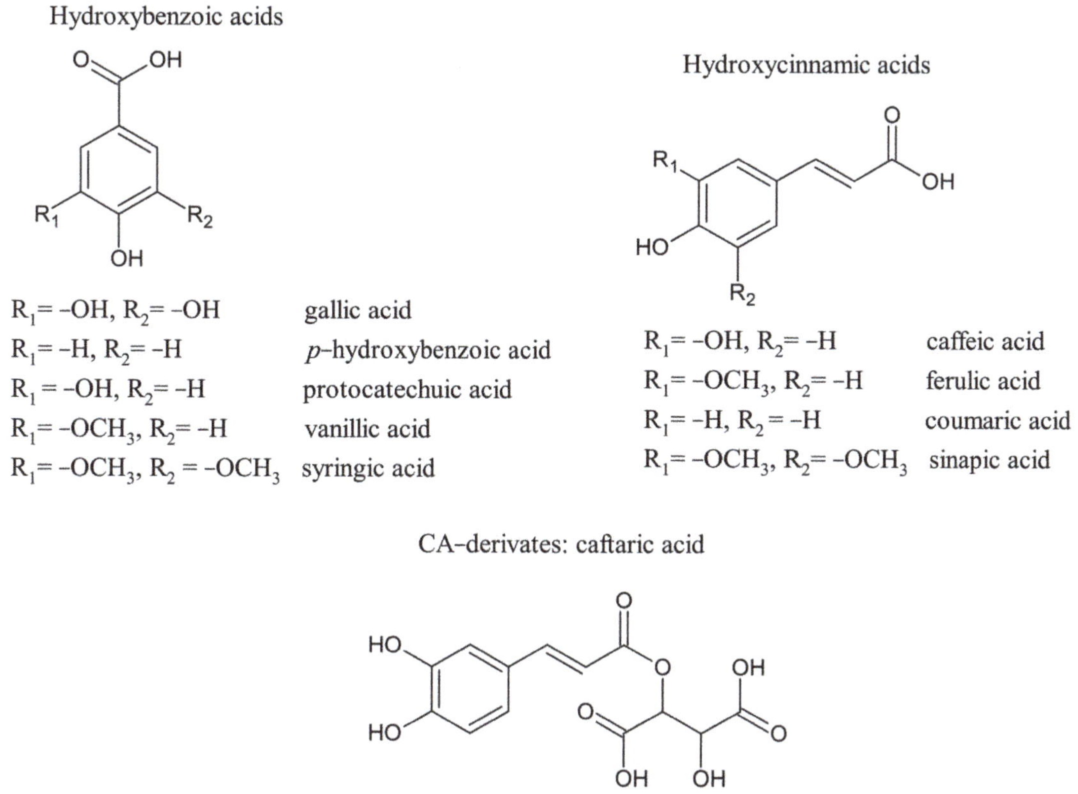

Hydroxybenzoic acids

Hydroxycinnamic acids

R_1= –OH, R_2= –OH   gallic acid
R_1= –H, R_2= –H   *p*–hydroxybenzoic acid
R_1 = –OH, R_2= –H   protocatechuic acid
R_1= –OCH$_3$, R_2= –H   vanillic acid
R_1= –OCH$_3$, R_2 = –OCH$_3$ syringic acid

R_1= –OH, R_2= –H   caffeic acid
R_1= –OCH$_3$, R_2= –H   ferulic acid
R_1= –H, R_2 = –H   coumaric acid
R_1= –OCH$_3$, R_2= –OCH$_3$ sinapic acid

CA–derivates: caftaric acid

Figure 1. Chemical structures of hydroxybenzoic and hydroxycinnamic acids and of caftaric acid.

To obtain pure compounds, due to the coexistence of different HCA in *E. purpurea*, an additional step of high-pressure liquid chromatography (HPLC) separation is necessary. Thus, those procedures led to the production of high volume of pollutants and the requirement of high-pressure liquid chromatography dramatically increase the process costs.

Cluster thinning is commonly applied in different wine production regions, and the unripe berries are today underutilized, so this work proposes a method to valorize unripe grape by their juice as source of HCAs, with a particular attention paid to caftaric acid, which is supposed to have several healthy functions [32]. Verjuice obtained by grape berries manual pressing of five varieties, both international and local, have been compared during four successive weeks between bunch closure and early véraison, revealing that, overall, the highest amount of HCAs is recorded in the premature varieties and at the bunch closure. Additionally, a low-environmental impact chromatographic method that permits the reduction of chemical waste by eliminating the several purification steps has been tuned to separate and recover high purity caftaric acid from verjuice.

2. Results and Discussion

2.1. Caftaric Acid Concentration over Green Grape Berries Maturation

First, for a complete overview of the potentiality of the unripe grape as a caftaric acid source, it was considered important to determine which varieties and which moment of ripeness optimized CFA recovery. Therefore, five varieties, three international and two of the most important Italian ones, have been monitored on CFA production from bunch closure to the early véraison (Table 1).

Table 1. Yield, titratable acidity (TA), sugar content (SC), caftaric (CFA), and coutaric acid (CUA) concentration of verjuice obtained by different grape varieties during green berries maturation. Values represent the mean averages (n = 3) and standard deviation (in brackets).

Variety	BBCH Code	Yield (%)	TA (g/L) [1]	SC (g/L)	CFA (mg/L)	CFA Purity (%) [2]	CUA (mg/L)	CUA Purity (%) [2]
CH	75	57.4 (0.3)	28.2 (0.4)	7.3 (0.6)	412.1 (12.2)	44.9 (3.1)	50.1 (2.8)	11.6 (1.3)
CH	77	57.3 (2.5)	35.1 (1.0)	8.0 (1.0)	230.4 (39.4)	46.8 (2.9)	58.8 (5.6)	11.5 (1.5)
CH	79	63.1 (1.6)	42.6 (3.6)	9. 7 (0.6)	173.1 (16.7)	48.7 (0.6)	14.3 (4.0)	11.1 (1.1)
CH	81	66.1 (1.0)	33.3 (3.6)	53. 7 (16.0)	257.5 (25.7)	47.3 (3.4)	70.5 (4.0)	11.42 (1.3)
GL	73	50.8 (1.1)	32.6 (1.3)	9.0 (0.1)	295.8 (7.6)	61.2 (3.0)	23.5 (9.3)	8.0 (0.2)
GL	75	46.0 (0.7)	41.8 (0.7)	11.7 (0.6)	272.5 (19.6)	67.6 (2.8)	50.2 (10.5)	6.36 (1.4)
GL	77	52.5 (0.9)	40.5 (0.7)	11.7 (0.6)	165.1 (1.4)	62.4 (7.0)	36.6 (7.7)	7.6 (0.5)
GL	79	55.1 (0.3)	39.5 (0.9)	27.0 (2.7)	179.8 (5.8)	63.3 (1.0)	23.0 (2.0)	7.3 (0.2)
ME	75	54.7 (2.8)	29.6(0.6)	8.3 (0.6)	207.0 (40.4)	52.4 (0.9)	54.2 (8.8)	7.2 (0.2)
ME	77	59.7 (0.9)	38.4 (0.1)	10.0 (0.1)	296.1 (34.1)	50.9 (4.1)	18.5 (7.6)	8.2 (0.2)
ME	79	62.3 (1.3)	42.5 (0.7)	11.0 (1.0)	222.8 (20.3)	58.3 (6.6)	28.1 (7.3)	6.2 (0. 8)
ME	81	64.9 (1.6)	34.8 (2.6)	36.3 (5.9)	154.0 (13.1)	48.7 (6.9)	36.5 (9.1)	7.5 (0.9)
PN	77	52.1 (2.3)	27.80 (0.6)	7.0 (0.1)	161.6 (15.9)	59.8 (3.4)	13.3 (2.0)	8.9 (0.6)
PN	79	56.7 (1.4)	38.9 (0.1)	7.7 (0.6)	211.1 (6.3)	58.4 (9.7)	47.2 (5.5)	11.7 (2.1)
PN	81	58.3(2.8)	45.7 (0.2)	21.0 (0.1)	267.4 (33.5)	63.0 (1.4)	21.9 (2.2)	8.2 (0.5)
PN	83	60.7 (1.6)	35.8(1.0)	43.7 (3.1)	192.9 (10.8)	57.8 (4.8)	22.6 (6.1)	8.1 (0.6)
SG	73	55.2 (2.2)	32.4 (1.1)	7.0 (0.1)	119.6 (17.6)	59.7 (4.9)	30.5 (7.4)	16.3 (1.5)
SG	75	57.1 (3.0)	36.5 (0.5)	8.0 (0.1)	118.9 (11.0)	56.3 (6.5)	15.5 (5.2)	13.6 (0.4)
SG	77	56.3 (2.4)	38.3 (0.1)	11.3 (0.6)	182.6 (30.8)	59.8 (1.7)	33.3 (3.1)	12.7 (0.1)
SG	79	60.7 (3.2)	33.7 (1.5)	36.3 (4.0)	212.4 (0.5)	50.8 (6.6)	21.3 (2.8)	12.6 (1.7)

[1] Titratable acidity is expressed as Tartaric acid equivalents. [2] Hydroxycinnamic acid purity is calculated as the ratio between acid peak area and the total of peaks area at 280 nm.

Because caftaric acid concentration is affected by the berry volume increase during maturation, the concentration was adjusted by mass/juice yield and the CFA per kilo of fresh grapes weight (FW) was compared (Figure 2a). Analyses of variance performed on a linear model of the standardized CFA concentration data recognized significant effects of variety ($F_{(4,52)} = 23.2$, $p > 0.01$) and date of sampling ($F_{(3,52)} = 17.8$, $p > 0.01$), while color, as well as time request for maturation, did not statistically affect CFA content, differently from the data reported by Burin and colleagues at the grape technical maturation [26]. However, the interaction between the two main factors was statistically significant, and this suggests that different caftaric acid accumulation is observed depending on the grape variety. Indeed, Figure 2a highlights that not all the varieties were significantly affected by the week of sampling. Moreover, while in Pinot Noir (PN), CFA is strongly reduced between the first and the second week (85 mg/kg in one week), in Merlot (ME), the major reduction was recorded later, between the second and the last week of sampling 30.17 mg/kg (Figure 2a). PN and Glera (GL) revealed the greatest caftaric acid accumulation among red and white varieties, respectively (achieving 412.10 ± 12.28 and 298.86 ± 7.55 mg/L of CFA at the bunch closure). HCAs accumulation is influenced by grape light exposure [33], which is correlated with leaf surface, characters varying among varieties. Concerning CUA, the overall analyses of variance revealed a significant effect of the variety ($F_{(4,52)} = 15.745$, $p > 0.01$) and the sampling date ($F_{(3,52)} = 3.007$, $p = 0.04$) and again the interaction of the two factors was significant. Figure 2b shows that Sangiovese (SG) and Chardonnay (CH) were the major producers of coutaric acid, particularly at the bunch closure, followed by PN. If the total amount of HCAs is considered, it results that the highest amount is accumulated at (or before) the bunch closure, in decreasing order in PN (achieving 241.70 mg/kg), SG (182.58 mg/kg), CH (178.73 mg/kg),

GL (161.96 mg/kg), and ME (102.67 mg/kg). Maturation variables, i.e., acidity, juice yield, and sugar content (SC), were related to HCAs concentrations (standardized per 1 kg FW) and revealed negative significant correlation of CFA with yield ($r = -0.67$, $p > 0.01$, $n = 20$) and SC ($r = -0.48$, $p = 0.045$, $n = 20$), while no significant correlation was found for CUA. Interestingly, the correlation between CFA and CUA results in a positive, not significant, correlation ($r = 0.31$, $p = 0.17$, $n = 20$).

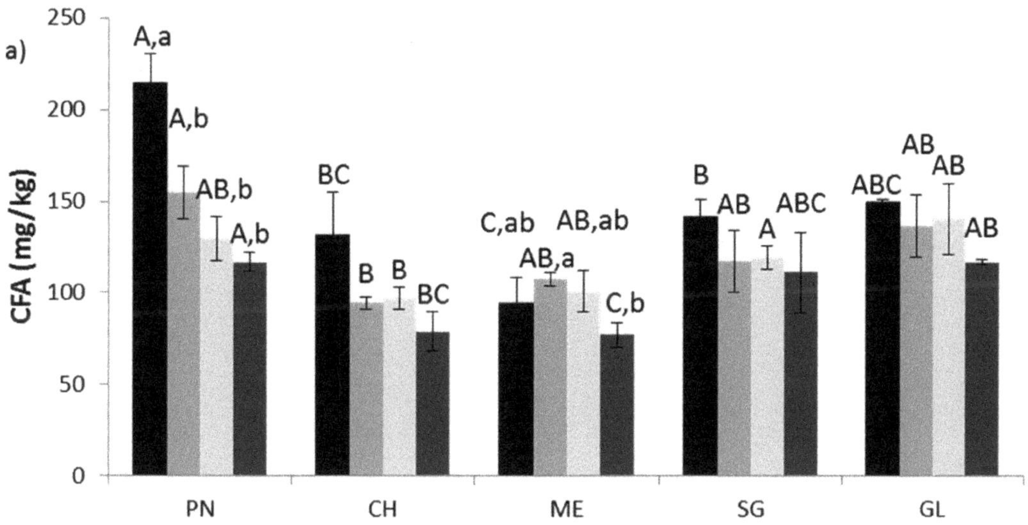

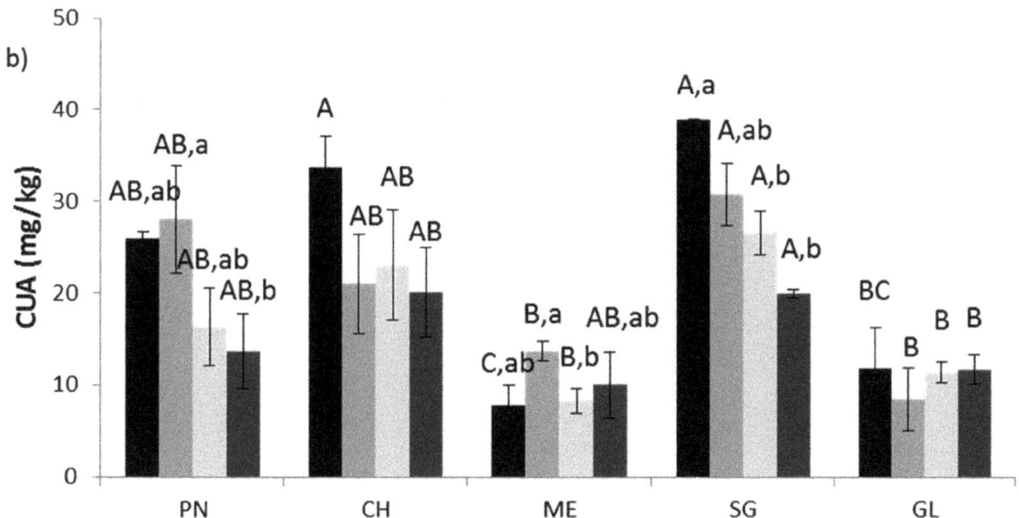

Figure 2. Standardized caftaric acid (**a**) and coutaric acid (**b**) content of verjuice. PN: Pinot noir, ME: Merlot, CH: Chardonnay, SG: Sangiovese, GL: Glera. Black bars: week one of sampling, grey bars: week two of sampling, light grey bars: week three of sampling, dark grey bars: week four of sampling. Different capital letter indicates significant differences among varieties at the same week of sampling; different lowercase letters indicates significant differences among weeks within the same variety.

2.2. Hydroxycinnamic Acid Esters in Verjuice

The natural content in HCAs of unripe grape juice (verjuice) was analyzed using the HPLC method. All the analyzed verjuice revealed a simple peak profile for HCAs, with a great separation of the caffeic-derived acids and a general reduction of other phenolic compounds when compared with *E. purpurea* aerial part (Figure 3a,b). However, because of its availability, a verjuice obtained by pressing Riesling grapes collected in 2020 (BBCH stage 79) was used for purification. Two peaks were well distinguished. The first with a retention time (RT) of 8.54 min represented 74.1% of the total peak area, and a second peak at RT = 9.82 min corresponded to 9.6% of the total area (Figure 3a).

a)

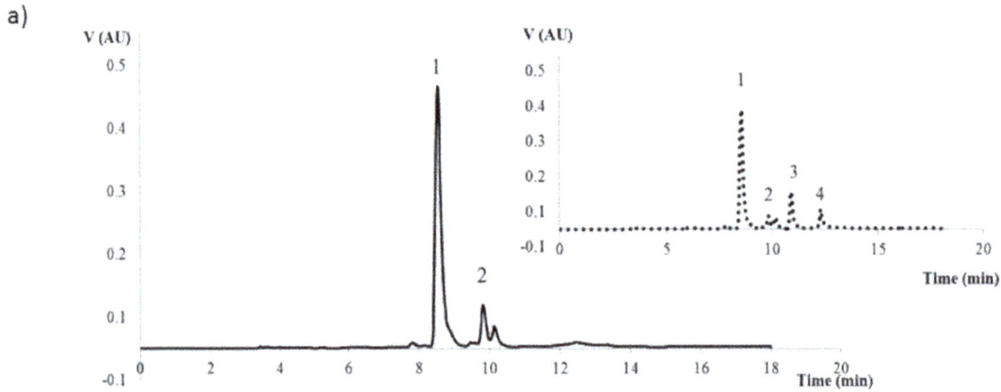

b)

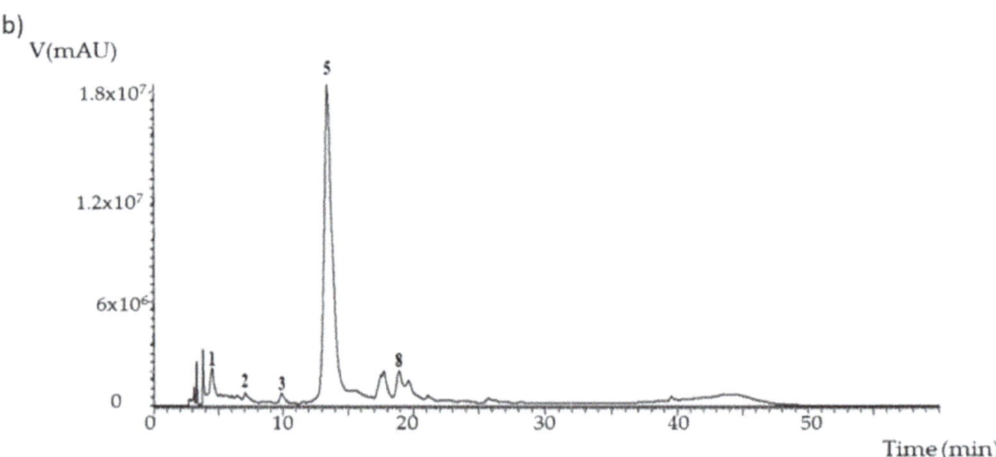

Figure 3. (**a**) HPLC profile of Riesling verjuice at 280 nm, before (continuous line) and after (dashed line) enzymatic treatment obtained in this work. Peak 1: caftaric acid, peak 2: coutaric acid, peak 3: caffeic acid, peak 4: coumaric acid. (**b**) HPLC profile (at 280 nm) of *E. purpurea* aerial part as reported by *Coelho* and colleagues [34]. Peak 1: caftaric acid, peak 2: 5-*O*-caffeoylquinic acid, peak 3: caffeic acid, peak 5: chicoric acid, peak 8: feruloylcaffeoyltartaric acid.

Peaks identification was performed using commercial standards (CFA, RT = 8.54 min, caffeic acid, RT = 11.0 min, and coumaric acid, RT = 12.4 min), while coutaric acid (CUA), which is known to be the second hydroxycinnamic ester in grape for abundance [35], was identified by the comparison of HPLC profiles before and after an enzymatic degradation of ester bounds. The enzymatic reaction induced the appearance of two new peaks, corresponding to caffeic acid and coumaric acid. Additionally, the two original peaks were partially degraded, corresponding to the identified CFA peak and to the peak 2 (Figure 3a), which was consequently assigned to CUA. The analysis did not detect the fertaric acid that was probably present in too low concentration. The comparison between unripe grape juice and *E. purpurea* spectra made clear that while the latter is generally considered the best vegetable matrix for chicoric acid extraction, verjuice should be considered the best raw matrix for caftaric acid as well (Figure 3a,b). HCAs were quantified by the comparison of sample peaks area with a calibrating curve prepared using 25 to 200 mg/L of commercial CFA. Thus, CUA was expressed as CFA equivalents. Riesling grape juice contained 286 mg/L of caftaric acid and 38 mg/L of coutaric acid, namely about three times the maximum HCAs content detected in commercial verjuice [36]. This result could be explained in light of the strong effect of varieties and grape maturation point and because of the easy oxidation of HCAs during commercial verjuice preparation [37]. Considering the average yield of verjuice about 57% *v/m*, the data could be easily transformed into 163.02 and 21.66 mg/kg of fresh grapes, respectively, not dissimilar from the data reported for grape pomace by Kammerer et al. [38]. In the work of Wu and colleagues [29], several conditions were tested in order to evaluate which ones optimize the CA-derivates extraction from *E. purpurea* roots. Authors reported that, growing the adventitious root at 20 °C in an industrial system, 65 g fresh material could be obtained from 1 L of growth medium, corresponding to 10.4 g of dry material. The measured amount of caftaric acid was 4.9 mg/g of dry material corresponding to 784 mg per kilo of fresh roots. Therefore, verjuice could represent a promising source of caftaric acid for its easy preparation that avoids the additional costs of a specific industrial plant.

2.3. Fast Protein Liquid Chromatography Applied to Hydroxycinnamic Acid Esters Separation

The most critical step in HCAs extraction from grapes raw material is represented by the isolation of the phenolic acids from other polyphenols. Chromatography was demonstrated to represent a handle tool for the selective isolation of HCAs ester in grape [39], and several methods have been tuned to obtain polyphenols high resolution peaks from fruits juice [40].

In addition to the traditional HPLC methods, Maier and colleagues [41] developed a method for CA-derivative esters recovery from ripe grape pomace based on the high-speed counter-current chromatography (HSCCC). This chromatography allowed the extraction and to successful separation of caffeic acid, coumaric acid, and ferulic acid esters by the head-to-tail elution mode, where the target compounds were separated from co-extracted polyphenolics and subsequently isolated in a second run. Liquid chromatography required a significantly longer time for separation; thus, CA-derivates separation required up to 390 min for the elution in the second HSCCC run. Additionally, this method involved the preliminary extraction with methanol and ethyl acetate and two mobile phases based on a mixture of hexane/ethyl acetate/methanol/water 3:7:3:7 (*v/v/v/v*) and tertbutyl-methyl ether/acetonitrile/*n*-butanol/water, 2:2:1:5 (*v/v/v/v*), both acidified by 0.5% of trifluoroacetic acid (TFA), which represent high pollutant waste.

The simplicity of phenols profile of verjuice made possible a handy sample manipulation and the use of low-pressure chromatography as separation technology. Filtered juice of unripe berries has been processed without any sample preparation. After some preliminary tests, it has been determined that 50 mL of verjuice was the uploading limit for a column volume of 20 mL. Nevertheless, this limit could be easily overtaken by rearranging the column sizes.

Separation was monitored by means of the UV detector (at 280 nm). After sample loading, the column was washed with deionized water plus 0.5% trifluoroacetic acid (TFA) to remove the unanchored compounds, and then the target molecules were eluted by gradient of water: alcohol that achieves 30% *v/v* of alcohol in 100 min.

Two commonly used solvent have been tested for the fast protein liquid chromatography (FPLC) separation, namely methanol, which is commonly used in HCAs chromatographic analyses [35], and ethanol, which was considered more suitable for further food application. The chromatographic profile revealed that well defined peaks could be obtained by methanol elution (Figure 4a), while ethanol elution evidenced less separation capability (Figure 4b).

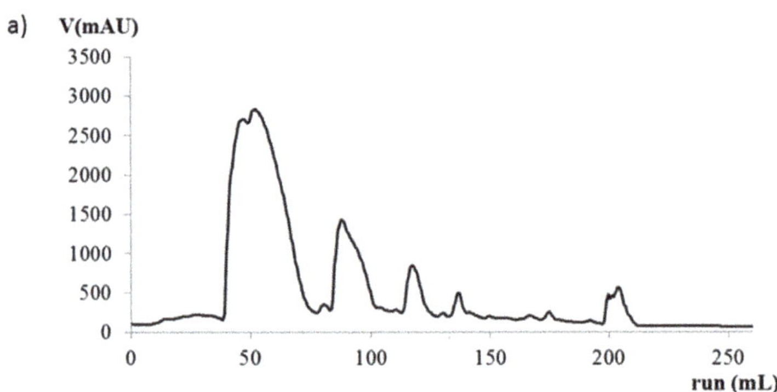

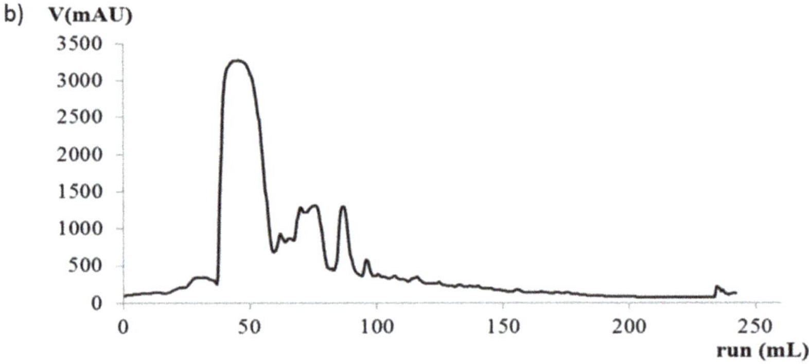

Figure 4. FPLCprofile of HCAs elution with (**a**) methanol and (**b**) ethanol as solvent.

Then, the methanol protocol was used in ten successive sample loadings, which obtained a repetitive elution profile. The first peak was assigned to CFA by HPLC analyses of its fractions. All the fractions that contained CFA at the minimal purity of 98% have been collected and freeze-dried. The final amount of crystallized CFA was 82 mg, which means a potential of 93.48 mg of compound obtained from 1 kg of fresh grapes if a verjuice yield of 57% *v/m* is considered. As previously demonstrated [41], high-speed counter-current chromatography (HSCCC) leads to the recovery of high pure CA-derivates, i.e., 97.0% for CFA, 97.2% CUA, and 90.4% for fertaric acid. The method here proposed achieves similar results in terms of CFA purity with a strong reduction of solvents and time; indeed, the HSCCC method permitted the isolation and recovery of 6 mg of caftaric acid, from 10 g of freeze-dried pomace, after a preliminary extraction that required 800 mL of methanol/0.1% HCl *v/v* and 400 mL of ethyl acetate followed by the compounds separation in about 120 mL of hexane/ethyl acetate/methanol/water 3:7:3:7 *v/v/v/v*/ 0.5% TFA plus 40 mL of ether/acetonitrile/n-butanol/water, 2:2:1:5, *v/v/v/v*/ 0.5% TFA, while in this new method, 8.2 mg of caftaric acid is obtained by the direct separation of 50 mL of verjuice in 70 mL of methanol 1:6 *v/v*/ 0.1% TFA.

3. Materials and Methods

3.1. Materials and Sample Preparation

Unripe grapes of five varieties, namely Pinot Noir (PN), Chardonnay (CH), Merlot (ME), Sangiovese (SG), and Glera (GL), were collected in the experimental vineyard of

"Scuola Enologica di Conegliano G.B. Cerletti" (Treviso, Italy) in four successive weeks of 2019, between stage 73 and 83 of the BBCH scale. Samples were promptly added with 0.2 g/kg of potassium metabisulphite and processed in a basket press. The obtained juice was centrifuged at 2000× *g* for 5 min, then vacuum filtered through 1.6 µm glass fiber filters (VWR, Milan, Italy) and kept frozen until HPLC analyses.

Additionally, unripe grape juice obtained from Rhine Riesling harvested at the véraison stage in 2020 was used for hydroxycinnamic acids recovery. Grape clusters were destemmed and washed before pressing with a small-scale stainless steel basket press. The basket press was loaded with berries in presence of 0.2 g/kg of potassium metabisulphite. The juice was centrifuged and filtered as described above and used for HPLC analyses and FPLC immediately. All reagents were analytical grade and were purchased from Sigma (Milan, Italy) unless otherwise stated. Chromatographic identification and quantification of caffeic acid, coumaric acid, and caftaric acid (CFA) were performed by the comparison of Riesling verjuice peaks with their commercial standard, while coutaric acid (CUA) was identified after juice enzymatic treatment. CFA standard curve was used for the quantification. The enzymatic treatment was performed using a commercial pectolytic enzyme with cinnamoyl esterase secondary activity. Verjuice (10 mL) was treated with 10 g/hL of enzyme and kept for 30 min at room temperature (25 °C) until the end of the reaction. Then, the sample was treated as described above before the injection.

3.2. Grape Degree of Maturation Parameters

Verjuice was immediately characterized by sugar content (SC) and total acidity (TA). Sugars were enzymatically determinaed by Hyperlab automatic multi-parametric analyzer (Steroglass, Perugia, Italy) by means of enzymatic kits, while titratable acidity was measured according to the official methods of wine analysis (Commission Regulation (EC) No1293/2005 of 5 August 2005 amending Regulation (EEC) No2676/90 determining Community methods for the analysis of wines).

3.3. HCAs Determination in High Performance Liquid Chromatography (HPLC)

Hydroxycinnamic acids (HCAs) separation was performed by C18 Lichrospher (4 × 250 mm, 5 µm, Agilent Technologies Italia, Milan, Italy) using a 1525 Binary Pump (Waters, Milan, Italy) equipped with 2487 Dual Band Absorbance Detector (Waters, Milan, Italy). Freshly prepared verjuice was centrifuged and filtered (0.2 µm), then it was injected (10 µL) and analyzed using the method proposed by Vanzo and colleagues [42] with modifications. Mobile phase was kept as proposed by the authors, while the flow rate was raised to 0.6 mL/min and the gradient was modified as follows: (A) Milli-Q water and 0.5% of formic acid *v/v* and (B) gradient-grade methanol and 2.0% of formic acid *v/v*. The gradient program was 0 min, 16% B; 7 min, 50% B; 8 min, 100% B; 8–12 min, 100% B; 13 min, 18% B; and 13–18 min, 18% B. The column temperature was kept at 40 °C. Hydroxycinnamic acids and esters were detected at the wavelength of 280 nm for purity determination and 330 nm for HCAs quantification; the peak areas were analyzed by software Breeze Version 3.3 (Waters, Milan, Italy).

3.4. HCAs Retrieve by Fast Protein Liquid Chromatography (FPLC)

Filtered Riesling verjuice (50 mL) was loaded onto a Bio Scale Column MT20 (15 × 113 mm, internal volume 20 mL, Bio-Rad Laboratories, Milan, Italy) packed with LiChrosorb RP-18 (Sigma-Aldrich, Milan, Italy) and connected to an FPLC (AKTA purifier 10). The column was previously equilibrated with deionized water with 0.1% trifluoroacetic acid (TFA) and, after the sample loading, the column was washed with the same buffer to remove unbound sample components. The target compounds were eluted with a gradient of methanol 0.1% TFA, which linearly achieved 30% in 100 min with a flow rate of 2 mL/min. Fractions of 3.5 mL were collected by means of a fraction collector. The elution was monitored by recording the signal at 280 nm, and the purity was checked by HPLC analysis. Fractions containing at least 98% of CFA were pooled together and freeze-dried by Heto cooling trap (Analitica De Mori, Milan, Italy).

3.5. Statistical Analyses

R software (R version 3.0.1) was used for statistical analysis. Differences were evaluated by one-way ANOVA and the Games–Howell post-hoc analyses. Variable relationships were tested using Pearson correlation. Statistical significance was attributed with p-value < 0.05.

4. Conclusions

Hydroxycinnamic acids and their derived ester gained new attention recently in light of their potential application as antioxidants and as bioactive molecules in food and cosmetic formulations.

Nowadays, hydroxycinnamic acids are mainly extracted form *Echinacea purpurea* roots, which are cultivated in an industrial plant set up with airlift bioreactors and require strictly controlled conditions of light, temperature, and nutrient availability, conditions that determine high cost of management. Nevertheless, other vegetables and herbs represent rich sources of HCAs and among them, grape berries.

In general, the extraction of HCAs from grapes' raw material, such as grape pomaces, faces the main problem of HCAs isolation from other phenolic compounds. On the other hand, verjuice polyphenols consist of a major part of hydroxycinnamic acids. This allows the reduction of costs and time for extraction and separation; the method here proposed demonstrates that a low-pressure separation procedure using fast protein liquid chromatography (FLPC) can be easily used to obtain high purity caftaric acid.

This work proposes the unripe grape juice as a new source of hydroxycinnamic acids, mainly represented by caftaric acid. This new approach gives two important technological advantages: the valorization of vineyard by-product in place of the installation of industrial plant for specific raw material production and the possibility of a more handy isolation of the target molecules. It should be underlined that this solution meets the general requirements of a new low-environmental impact alternative toward the reduction of solvents and the simplification of pure molecule recovery.

Author Contributions: Conceptualization: A.V. and S.V.; methodology: A.V. and S.V.; formal analysis: A.V. and V.V.; investigation: V.V.; resources: V.V. and S.V.; data curation: V.V.; writing—original draft preparation: V.V.; writing—review and editing: A.V. and S.V.; supervision: S.V.; funding acquisition: S.V. All authors have read and agreed to the published version of the manuscript.

Funding: This research received no external funding.

Informed Consent Statement: Not applicable.

Data Availability Statement: The data presented in this study are openly available in [repository name e.g., FigShare] at [doi], reference number [reference number].

Conflicts of Interest: The authors declare no conflict of interest.

References

1. Taofiq, O.; González-Paramás, A.M.; Barreiro, M.F.; Ferreira, I.C.F.R.; McPhee, D.J. Hydroxycinnamic acids and their derivatives: Cosmeceutical significance, challenges and future perspectives, a review. *Molecules* **2017**, *22*, 281. [CrossRef] [PubMed]
2. Yuan, Y.; Jiang, L. Application of Chicoric Acid Extract in Echinacea Purpurea. CN102726550, 17 October 2012.
3. Marchiosi, R.; dos Santos, W.D.; Constantin, R.P.; de Lima, R.B.; Soares, A.R.; Finger-Teixeira, A.; Mota, T.R.; de Oliveira, D.M.; de Pavia Foletto-Felipe, M.; Abrahão, J.; et al. Biosynthesis and metabolic actions of simple phenolic acids in plants. *Phytochem. Rev.* **2020**, *19*, 865–906. [CrossRef]
4. Teixeira, J.; Gaspar, A.; Garrido, E.M.; Garrido, J.; Borges, F. Hydroxycinnamic acid antioxidants: An electrochemical overview. *BioMed Res. Int.* **2013**, *2013*, 251754. [CrossRef] [PubMed]
5. Maas, M.; Petereit, F.; Hensel, A. Caffeic acid derivatives from *Eupatorium perfoliatum* L. *Molecules* **2009**, *14*, 36–45. [CrossRef] [PubMed]
6. Lee, J.; Scagel, C.F. Chicoric acid: Chemistry, distribution, and production. *Front. Chem.* **2013**, *1*, 40. [CrossRef]

7. Lima, A.; Oliveira, C.; Santos, C.; Campos, F.M.; Couto, J.A. Phenolic composition of monovarietal red wines regarding volatile phenols and its precursors. *Eur. Food Res. Technol.* **2018**, *244*, 1985–1994. [CrossRef]
8. Lin, S.D.; Sung, J.M.; Chen, C.L. Effect of drying and storage conditions on caffeic acid derivatives and total phenolics of *Echinacea purpurea* grown in Taiwan. *Food Chem.* **2011**, *125*, 226–231. [CrossRef]
9. Bahri, M.; Hance, P.; Grec, S.; Quillet, M.C.; Trotin, F.; Hilbert, J.L.; Hendriks, T. A "novel" protocol for the analysis of hydroxycinnamic acids in leaf tissue of chicory (*Cichorium intybus* L., Asteraceae). *Sci. World J.* **2012**, *2012*, 142983. [CrossRef]
10. Papetti, A.; Daglia, M.; Aceti, C.; Sordelli, B.; Spini, V.; Carazzone, C.; Gazzani, G. Hydroxycinnamic acid derivatives occurring in *Cichorium endivia* vegetables. *J. Pharm. Biomed. Anal.* **2008**, *48*, 472–476. [CrossRef]
11. Stojakowska, A.; Malarz, J.; Szewczyk, A.; Kisiel, W. Caffeic acid derivatives from a hairy root culture of *Lactuca virosa. Acta Physiol. Plant.* **2012**, *34*, 291–298. [CrossRef]
12. Takenaka, M.; Yan, X.; Ono, H.; Yoshida, M.; Nagata, T.; Nakanishi, T. Caffeic acid derivatives in the roots of yacon (*Smallanthus sonchifolius*). *J. Agric. Food Chem.* **2003**, *51*, 793–796. [CrossRef] [PubMed]
13. Silva, T.; Oliveira, C.; Borges, F. Caffeic acid derivatives, analogs and applications: A patent review (2009–2013). *Expert Opin. Ther. Pat.* **2014**, *24*, 1257–1270. [CrossRef]
14. Deng, Q.; Penner, M.H.; Zhao, Y. Chemical composition of dietary fiber and polyphenols of five different varieties of wine grape pomace skins. *Food Res. Int.* **2011**, *44*, 2712–2720. [CrossRef]
15. Cortés, A.; Moreira, M.T.; Feijoo, G. Integrated evaluation of wine lees valorization to produce value-added products. *Waste Manag.* **2019**, *95*, 70–77. [CrossRef]
16. De Bona, G.S.; Adrian, M.; Negrel, J.; Chiltz, A.; Klinguer, A.; Poinssot, B.; Héloir, M.C.; Angelini, E.; Vincenzi, S.; Bertazzon, N. Dual mode of action of grape cane extracts against *Botrytis cinerea. J. Agric. Food Chem.* **2019**, *67*, 5512–5520. [CrossRef]
17. Wang, Y.; He, Y.N.; He, L.; He, F.; Chen, W.; Duan, C.Q.; Wang, J. Changes in global aroma profiles of cabernet sauvignon in response to cluster thinning. *Food Res. Int.* **2019**, *122*, 56–65. [CrossRef]
18. Fanzone, M.; Zamora, F.; Jofré, V.; Assof, M.; Peña-Neira, Á. Phenolic composition of Malbec grape skins and seeds from Valle de Uco (Mendoza, Argentina) during ripening. effect of cluster thinning. *J. Agric. Food Chem.* **2011**, *59*, 6120–6136. [CrossRef] [PubMed]
19. Gil, M.; Esteruelas, M.; González, E.; Kontoudakis, N.; Jiménez, J.; Fort, F.; Canals, J.M.; Hermosín-Gutiérrez, I.; Zamora, F. Effect of two different treatments for reducing grape yield in *Vitis vinifera* cv Syrah on wine composition and quality: Berry thinning versus cluster thinning. *J. Agric. Food Chem.* **2013**, *61*, 4968–4978. [CrossRef]
20. The Commission of the European Communities. *Regulation (EU) No 1306/2013* **2013**, *L 347*, 671–854. Available online: https://eur-lex.europa.eu/legal-content/EN/LSU/?uri=celex:32013R1306 (accessed on 3 February 2021).
21. Salah Eddine, N.; Tlais, S.; Alkhatib, A.; Hamdan, R. Effect of four grape varieties on the physicochemical and sensory properties of unripe grape verjuice. *Int. J. Food Sci.* **2020**, *2020*, 1–7. [CrossRef]
22. Dupas de Matos, A.; Magli, M.; Marangon, M.; Curioni, A.; Pasini, G.; Vincenzi, S. Use of verjuice as an acidic salad seasoning ingredient: Evaluation by consumers' liking and Check-All-That-Apply. *Eur. Food Res. Technol.* **2018**, *244*, 2117–2125. [CrossRef]
23. Nasser, M.; Cheikh-Ali, H.; Hijazi, A.; Merah, O.; Al-Rekaby, A.E.A.N.; Awada, R. Phytochemical profile, antioxidant and antitumor activities of green grape juice. *Processes* **2020**, *8*, 507. [CrossRef]
24. Fia, G.; Bucalossi, G.; Gori, C.; Borghini, F.; Zanoni, B. Recovery of bioactive compounds from unripe red grapes (cv. Sangiovese) through a green extraction. *Foods* **2020**, *9*, 566. [CrossRef] [PubMed]
25. Honisch, C.; Osto, A.; Dupas de Matos, A.; Vincenzi, S.; Ruzza, P. Isolation of a tyrosinase inhibitor from unripe grapes juice: A spectrophotometric study. *Food Chem.* **2020**, *305*, 125506. [CrossRef] [PubMed]
26. Burin, V.M.; Ferreira-Lima, N.E.; Panceri, C.P.; Bordignon-Luiz, M.T. Bioactive compounds and antioxidant activity of *Vitis vinifera* and *Vitis labrusca* grapes: Evaluation of different extraction methods. *Microchem. J.* **2014**, *114*, 155–163. [CrossRef]
27. Conde, C.; Silva, P.; Fontes, N.; Dias, A.C.P.; Tavares, R.M.; Sousa, M.J.; Agasse, A.; Delrot, S.; Gerós, H. Biochemical changes throughout grape berry development and fruit and wine quality. *Food* **2007**, *1*, 1–22.
28. Wu, C.H.; Murthy, H.N.; Hahn, E.J.; Paek, K.Y. Large-scale cultivation of adventitious roots of *Echinacea purpurea* in airlift bioreactors for the production of chichoric acid, chlorogenic acid and caftaric acid. *Biotechnol. Lett.* **2007**, *29*, 1179–1182. [CrossRef]
29. Wu, C.H.; Murthy, H.N.; Hahn, E.J.; Paek, K.Y. Enhanced production of caftaric acid, chlorogenic acid and cichoric acid in suspension cultures of *Echinacea purpurea* by the manipulation of incubation temperature and photoperiod. *Biochem. Eng. J.* **2007**, *36*, 301–303. [CrossRef]
30. Peng, Y. The Method for Preparing Chicoric Acid from Echinacea purpurea. CN1587251A, 2 March 2005.
31. Chen, B.; Guo, J.; Xie, J. Method for Purifying Chicoric Acid and Monocaffeyltartaric Acid from Echinacea purpurea Extract. CN20081143072, 9 October 2008.
32. Koriem, K.M.M. Caftaric acid: An overview on its structure, daily consumption, bioavailability and pharmacological effects. *Biointerface Res. Appl. Chem.* **2020**, *10*, 5616–5623.
33. Sun, R.Z.; Cheng, G.; Li, Q.; He, Y.N.; Wang, Y.; Lan, Y.B.; Li, S.Y.; Zhu, Y.R.; Song, W.F.; Zhang, X.; et al. Light-induced variation in phenolic compounds in cabernet sauvignon grapes (*Vitis vinifera* L.) involves extensive transcriptome reprogramming of biosynthetic enzymes, transcription factors, and phytohormonal regulators. *Front. Plant. Sci.* **2017**, *8*, 1–18. [CrossRef]

34. Coelho, J.; Barros, L.; Dias, M.I.; Finimundy, T.C.; Amaral, J.S.; Alves, M.J.; Calhelha, R.C.; Santos, P.F.; Ferreira, I.C.F.R. *Echinacea purpurea* (L.) Moench: Chemical characterization and bioactivity of its extracts and fractions. *Pharmaceuticals* **2020**, *13*, 125. [CrossRef]
35. Buiarelli, F.; Coccioli, F.; Merolle, M.; Jasionowska, R.; Terracciano, A. Identification of hydroxycinnamic acid-tartaric acid esters in wine by HPLC-tandem mass spectrometry. *Food Chem.* **2010**, *123*, 827–833. [CrossRef]
36. Pour Nikfardjam, M.S. General and polyphenolic composition of unripe grape juice (verjus/verjuice) from various producers. *Mitteilungen Klosterneubg.* **2008**, *58*, 28–31.
37. Mulero, J.; Pardo, F.; Zafrilla, P. Antioxidant activity and phenolic composition of organic and conventional grapes and wines. *J. Food Compos. Anal.* **2010**, *23*, 569–574. [CrossRef]
38. Kammerer, D.; Claus, A.; Carle, R.; Schieber, A. Polyphenol screening of pomace from red and white grape varieties (*Vitis vinifera* L.) by HPLC-DAD-MS/MS. *J. Agric. Food Chem.* **2004**, *52*, 4360–4367. [CrossRef]
39. Lamuela-Raventós, R.M.; Waterhouse, A.L. A Direct HPLC separation of wine phenolics. *Am. J. Enol. Vitic.* **1994**, *45*, 1–5.
40. Ignat, I.; Volf, I.; Popa, V.I. A critical review of methods for characterisation of polyphenolic compounds in fruits and vegetables. *Food Chem.* **2011**, *126*, 1821–1835. [CrossRef] [PubMed]
41. Maier, T.; Sanzenbacher, S.; Kammerer, D.R.; Berardini, N.; Conrad, J.; Beifuss, U.; Carle, R.; Schieber, A. Isolation of hydroxycinnamoyltartaric acids from grape pomace by high-speed counter-current chromatography. *J. Chromatogr. A* **2006**, *1128*, 61–67. [CrossRef]
42. Pritchard, J.K.; Stephens, M.; Donnelly, P. Inference of population structure using multilocus genotype data. *Genetics* **2000**, *155*, 945–959. [PubMed]

Article

Color Stabilization of Apulian Red Wines through the Sequential Inoculation of *Starmerella bacillaris* and *Saccharomyces cerevisiae*

Matteo Velenosi [1,2,†], Pasquale Crupi [1,†], Rocco Perniola [1], Antonio Domenico Marsico [1], Antonella Salerno [1,2], Hervè Alexandre [3], Nicoletta Archidiacono [2], Mario Ventura [2,*] and Maria Francesca Cardone [1,*]

[1] Consiglio per la Ricerca in Agricoltura e L'analisi Dell'economia Agraria-Centro di Ricerca Viticoltura ed Enologia (CREA-VE), Turi, 148-70010 Bari, Via Casamassima, Italy; matteo.velenosi@tiscali.it (M.V.); pasquale.crupi@crea.gov.it (P.C.); rocco.perniola@crea.gov.it (R.P.); adomenico.marsico@crea.gov.it (A.D.M.); antonellasalerno10@gmail.com (A.S.)

[2] Dipartimento di Biologia, Università degli Studi di Bari Aldo Moro, 4-70124 Bari, Via Orabona, Italy; nicoletta.archidiacono@uniba.it

[3] UMRProcédésAlimentaires et Microbiologiques, Equipe VAlMiS (Vin, Aliment, Microbiologie, Stress), AgroSupDijon—Université de Bourgogne/Franche-Comté, IUVV, Rue Claude Ladrey, BP 27877, 21000 Dijon, France; rvalex@u-bourgogne.fr

[*] Correspondence: mario.ventura@uniba.it (M.V.); mariafrancesca.cardone@crea.gov.it (M.F.C.); Tel.: +0039-3494763122 (M.V.); +0039-3495743089 (M.F.C.)

[†] These authors contributed equally to this work.

Abstract: Mixed fermentation using *Starmerella bacillaris* and *Saccharomyces cerevisiae* has gained attention in recent years due to their ability to modulate the qualitative parameters of enological interest, such as the color intensity and stability of wine. In this study, three of the most important red Apulian varieties were fermented through two pure inoculations of *Saccharomyces cerevisiae* strains or the sequential inoculation of *Saccharomyces cerevisiae* after 48 h from *Starmerella bacillaris*. The evolution of anthocyanin profiles and chromatic characteristics were determined in the produced wines at draining off and after 18 months of bottle aging in order to assess the impact of the different fermentation protocols on the potential color stabilization and shelf-life. The chemical composition analysis showed titratable acidity and ethanol content exhibiting marked differences among wines after fermentation and aging. The 48 h inoculation delay produced wines with higher values of color intensity and color stability. This was ascribed to the increased presence of compounds, such as stable A-type vitisins and reddish/violet ethylidene-bridge flavonol-anthocyanin adducts, in the mixed fermentation. Our results proved that the sequential fermentation of *Starmerella bacillaris* and *Saccharomyces cerevisiae* could enhance the chromatic profile as well as the stability of the red wines, thus improving their organoleptic quality.

Keywords: HPLC-UV-ESI-MSn; free anthocyanins; co-pigmented anthocyanins; mixed fermentation; *starmerella bacillaris*; PCA

Citation: Velenosi, M.; Crupi, P.; Perniola, R.; Marsico, A.D.; Salerno, A.; Alexandre, H.; Archidiacono, N.; Ventura, M.; Cardone, M.F. Color Stabilization of Apulian Red Wines through the Sequential Inoculation of *Starmerella bacillaris* and *Saccharomyces cerevisiae*. *Molecules* **2021**, *26*, 907. https://doi.org/10.3390/molecules26040907

Academic Editor: Mirella Nardini

Received: 18 January 2021
Accepted: 5 February 2021
Published: 9 February 2021

Publisher's Note: MDPI stays neutral with regard to jurisdictional claims in published maps and institutional affiliations.

1. Introduction

Yeast metabolism, during the winemaking process, influences the wine organoleptic properties and, consequently, wine quality. It can directly or indirectly affect the content of several compounds related to both the aroma and color characteristics. Recent studies on mixed starter cultures have proved that the resulting wines differ significantly, concerning both their chemical composition and sensory characteristics. Different yeast species and the ratio of non-*Saccharomyces/Saccharomyces* yeasts determine the organoleptic properties of the final product, and therefore contribute differently to the improvement or depreciations of wine quality [1].

The color is the most important visual attribute of red wines [2], which strongly impresses consumers' purchasing preference [3]. Moreover, it influences the perception of other sensory properties, such as aroma and flavor. Therefore, winemakers have accustomed to adopting suitable practices that improve color extraction and enhance the stability of chromatic characteristics of wine over time [4]. The color of red wines is mainly due to anthocyanins, which are transferred from grape skins into wine throughout the maceration/fermentation process [5]. Whereas, the stability of color during wine aging is affected by the phenolic derivatives which stabilize anthocyanins through co-pigmentation reactions [6,7]. The types and concentrations of polyphenols in wine may depend on the grape variety, the degree of ripening [8], and the vine growing methods employed, specifically the pruning and training system [7,9]. The joining of additives (i.e., enzymes, yeasts, or tannins) during winemaking is also a determinant [4,10,11].

In this aspect, there has been growing interest in the use of non-*Saccharomyces* yeasts due to the positive impact some of their metabolites exert on wine quality [12,13]. Many authors have demonstrated that non-*Saccharomyces* yeasts have a protective effect on wine color [4,10,14]. Among these, *Starmerella bacillaris (S. bacillaris)* [15] has been considered one of the most promising non-*Saccharomyces* yeasts [16–18] (having strong fructophilicity, high tolerance to low temperatures, and ability to grow at an elevated sugar concentration) [19]. However, non-*Saccharomyces* yeasts possess low fermentation ability and cannot carry out the must fermentation alone, due to their ethanol sensitivity [20,21]. Consequently, their use in combination with selected *Saccharomyces cerevisiae (S. cerevisiae)* (Desm. Meyen 1838) strains is necessary for completing the fermentation and taking advantage of their unique features [22].

Recently, a meaningful knowledge has been accumulated about the importance of yeast inoculation density, timing, and combination of strains in improving the organoleptic properties of wines [16,23,24]. The use of *S. bacillaris* during winemaking has allowed increasing the must total acidity and enhancing the color intensity of wine [25,26]. Similarly, this yeast strain has led to a higher production of pyruvic acid, which is involved in the formation of stable pigments (i.e., vitisin A and B), compared to *Saccharomyces* [27]. Thereby, it could be hypothesized that a mixed fermentation (by employing both the yeasts, sequentially) works in improving the color intensity as well as the color stability of wine. This study aimed at comparing the anthocyanin profiles and chromatic characteristics of wines produced through two mono-*S. cerevisiae* fermentations (SCE16 and SCE138, respectively) or the sequential fermentation of *S. bacillaris* and *S. cerevisiae* SCE16/SCE138 inoculated 48 h later. The analyses were conducted on wines produced from the most important red Apulian varieties (Primitivo, Negramaro, and Aleatico) at draining off and after 18 months of bottle aging, to investigate the potential of the color stabilization and shelf life of these wines.

2. Results

2.1. Interaction between Saccharomyces Yeast Strains and Pilot Scale Fermentation

In order to evaluate the suitability of the three yeast strains in mixed fermentation, we first evaluated the phytotoxic activity towards each other both on plate and liquid culture assays.

In the experiment performed on the plate assay, the three yeast strains were able to grow independently of the previous growth of the other tested yeast strain on the cellophane disc. Furthermore, the growth curves of each yeast strain are similar regardless of the type of filtered supernatant added (Supplementary Figure S1). Likewise, no inhibition of growth was observed in the liquid culture assay combining two yeast strains together, both considering the interaction of *S. cerevisiae* strains or *S. bacillaris* with each of the *S. cerevisiae* strain. Taking into consideration the absence of any phytotoxic activity among the different combinations of yeast strains, we were able to test their effect on wine production in a mixed fermentation where the two *S. cerevisiae* strains (SCE16 and SCE138, 1:1) were added together 48 h after the inoculation of *S. bacillaris* (FA18), and compare this trial with mono-*saccharomyces* fermentation. Moreover, in order to assess the fermentation ability of the chosen yeast combination with respect to mono-fermentation and, in particular, to further verify the absence of any negative interaction in mixed fermentation among the yeast strains, fermentation kinetics were followed for each trial (Supplementary Figure S2). Mono inoculation SCE16 and SC138 showed a similar or equal consumption in sugar level in every variety considered, thus demonstrating the same fermentation capacity of the two *S. cerevisiae* strains. On the contrary, the mixed FA18 was characterized by a slow start, regarding the sugar consumption, reaching up to 7% in Primitivo, 8% in Negroamaro, and 13% in Aleatico. The higher delay we found in the Primitivo could be ascribed to the sugar concentration effect on the *S. bacillaris* activity, as previously described [28]. Indeed, the sugar and nitrogen composition of the grape must are key factors for the evolution of the alcoholic fermentation and the development of the yeasts [29,30]. Notably, also in the FA18 mixed fermentation, complete sugar consumption was reached around 4 days after the inoculum with the two *S. cerevisiae* strains, thus confirming the absence of a negative interaction between the yeast strains both considering *S. bacillaris* against *S. cerevisiae*, and between the two *S. cerevisiae* strains.

2.2. Basic Oenological Parameters and Chemical Composition

The chemical composition of Primitivo, Negramaro, and Aleatico wines produced by pure and mixed culture fermentation at draining off and after 18 months of bottle aging were listed in Table 1.

Table 1. Chemical analysis and polyphenolic indexes of Primitivo, Negramaro, and Aleatico at draining off (A) and after 18 months of bottle aging (B).

	Primitivo A			Primitivo B		
	SCE16	**SCE138**	**FA18**	**SCE16**	**SCE138**	**FA18**
CI	1.19 ± 0.10 [#]	1.04 ± 0.15	1.24 ± 0.05	0.92 ± 0.07	0.80 ± 0.12	0.96 ± 0.04
MA (mg/L)	176 ± 5ac	157 ± 11ab	148 ± 6b	121 ± 3c	109 ± 3d	102 ± 10d
TA (mg/L)	324 ± 16a	292 ± 20ab	294 ± 11ab	182 ± 17b	158 ± 16c	165 ± 12bc
TP (mg/L)	2390 ± 70a	2230 ± 60ab	2430 ± 80a	2081 ± 40b	1900 ± 100c	2110 ± 50b
pH	3.13 ± 0.03	3.21 ± 0.04	3.26 ± 0.05	3.39 ± 0.03	3.36 ± 0.04	3.39 ± 0.02
A (g/L)	7.81 ± 0.10	7.62 ± 0.15	7.67 ± 0.08	7.7 ± 0.2	7.50 ± 0.07	7.55 ± 0.07
ET (% *v/v*)	15.52 ± 0.12	14.9 ± 0.5	15.21 ± 0.12	15.31 ± 0.10	14.7 ± 0.4	15.0 ± 0.2
VA (g/L)	0.29 ± 0.04ab	0.25 ± 0.02c	0.28 ± 0.02b	0.23 ± 0.03c	0.24 ± 0.03c	0.32 ± 0.02a
H	0.283 ± 0.019c	0.24 ± 0.04d	0.256 ± 0.015cd	0.597 ± 0.003a	0.606 ± 0.006a	0.562 ± 0.009b
CEI	−2.6 ± 0.2c	−3.2 ± 0.7c	−2.99 ± 0.17c	−0.675 ± 0.008a	−0.650 ± 0.017a	−0.76 ± 0.02b
	Negramaro A			Negramaro B		
	SCE16	**SCE138**	**FA18**	**SCE16**	**SCE138**	**FA18**
CI	0.553 ± 0.018b	0.58 ± 0.03b	0.743 ± 0.019a	0.425 ± 0.014c	0.45 ± 0.02c	0.571 ± 0.015b
MA (mg/L)	117 ± 4a	107 ± 5a	98 ± 4b	97 ± 5b	72 ± 10c	69 ± 11c
TA (mg/L)	211 ± 3a	202 ± 2ab	189 ± 7b	120 ± 5c	110 ± 9cd	103 ± 6d
TP (mg/L)	2040 ± 90	1900 ± 70	1790 ± 100	2060 ± 70	1830 ± 90	1600 ± 100
pH	3.40 ± 0.05	3.30 ± 0.03	3.36 ± 0.03	3.56 ± 0.06	3.52 ± 0.04	3.54 ± 0.02
A (g/L)	6.07 ± 0.04b	6.10 ± 0.03ab	6.4 ± 0.03a	5.7 ± 0.03c	6.00 ± 0.02b	6.2 ± 0.03a
ET (% *v/v*)	12.3 ± 0.2	12.52 ± 0.12	12.33 ± 0.06	12.3 ± 0.3	12.43 ± 0.06	12.26 ± 0.13
VA (g/L)	0.21 ± 0.02b	0.23 ± 0.03b	0.23 ± 0.04b	0.20 ± 0.02b	0.21 ± 0.02b	0.38 ± 0.03a
H	0.435 ± 0.016d	0.437 ± 0.016d	0.412 ± 0.015c	0.73 ± 0.03a	0.717 ± 0.006a	0.694 ± 0.012b
CEI	−1.30 ± 0.08d	−1.29 ± 0.08d	−1.13 ± 0.06c	−0.37 ± 0.06b	−0.395 ± 0.012ab	−0.41 ± 0.03b
	Aleatico A			Aleatico B		
	SCE16	**SCE138**	**FA18**	**SCE16**	**SCE138**	**FA18**
CI	0.43 ± 0.05	0.40 ± 0.08	0.49 ± 0.03	0.33 ± 0.04	0.31 ± 0.06	0.39 ± 0.02
MA (mg/L)	95 ± 6a	97 ± 3a	85 ± 3b	70 ± 5bc	80 ± 7b	65 ± 4c
TA (mg/L)	157 ± 8a	165 ± 8a	152 ± 6ab	97 ± 9cb	103 ± 7b	92 ± 4c
TP (mg/L)	1700 ± 50a	1720 ± 20a	1600 ± 60b	1430 ± 50bc	1440 ± 80bc	1320 ± 40c
pH	3.26 ± 0.04	3.27 ± 0.06	3.22 ± 0.04	3.42 ± 0.02	3.43 ± 0.02	3.38 ± 0.04
A (g/L)	5.61 ± 0.02b	5.59 ± 0.10b	6.18 ± 0.04a	5.50 ± 0.14b	5.50 ± 0.02b	6.05 ± 0.07a
ET (% *v/v*)	12.0 ± 0.18	11.9 ± 0.03	11.7 ± 0.06	11.95 ± 0.13	11.83 ± 0.04	11.60 ± 0.02
VA (g/L)	0.22 ± 0.02b	0.21 ± 0.01b	0.26 ± 0.04a	0.17 ± 0.01c	0.18 ± 0.01c	0.24 ± 0.01ab
H	0.543 ± 0.017	0.527 ± 0.019	0.564 ± 0.016	0.760 ± 0.013	0.753 ± 0.018	0.78 ± 0.03
CEI	−0.84 ± 0.03	−0.90 ± 0.03	−0.806 ± 0.013	−0.22 ± 0.03	−0.23 ± 0.04	−0.26 ± 0.05

Each value was calculated as means of three independent replicates ± [#]standard deviation at $p < 0.05$. Different letters on the same line are significantly different at a 5% level (Tukey's HSD post-hoc test). CI: Color intensity; MA: Monomeric anthocyanins; TA: Total anthocyanins; TP: Total polyphenols; A: Total acidity; ET: Alcoholic degree; VA: Volatile acidity; H: Hue; CEI: Color evolution index.

Overall, the fermentation type factor influenced the titratable acidity (A) of the wines. Indeed, samples obtained by mixed fermentation generally contained more acids, in particular, Negramaro and Aleatico FA18 wines had a significantly higher A ($p < 0.01$). These differences (ranging from 0.25 to 0.57 g/L) cannot be imputed to the main organic acids (citric, malic, tartaric, and lactic acids) whose values did not significantly change in all the wines (Supplementary Table S1).

Furthermore, pH values were not affected by the different fermentation protocols at drying off (Table 1). Conversely, these findings may be due to the capability of *S. bacillaris* strains to relatively synthesize high concentrations of keto acids either during the early stages of fermentation from sugar metabolism or from the corresponding amino acids (alanine for pyruvic acid and glutamate for α-keto glutaric acid), as previously reported [27,31,32]. On the contrary, we revealed a significantly higher pH value in 18 months aged wines connected to the partial tartaric precipitation that happened during aging in the bottle. However, not surprisingly, the slight decrease of A (total acidity) during the wines aging could also be due to a series of maturation reactions involving pyruvic acid [7].

No significant difference in the alcoholic degree (% *v/v*) was registered between pure and mixed fermentation in all the samples (Table 1). Moreover, the volatile acidity was strongly influenced by the fermentation protocol and bottle aging, as well as by the interaction of the two factors ($p < 0.001$), even though all the wines contained <0.40 g/L (Table 1), which cannot be considered detrimental to the sensorial quality of wine as in agreement with literature data [33]. Furthermore, we analyzed the polyphenolic content and we found that monomeric anthocyanins (MA), total anthocyanins (TA), and total polyphenols (TP) values appeared significantly higher in SCE16 and SCE138 than in the FA18 samples, especially for Negramaro and Aleatico (Table 1). Moreover, we detected a decrease of phenolics after 18 months which, was generally more marked in FA18 than SCE16 and SCE138 wines (Table 1).

2.3. HPLC-MS Analysis of Anthocyanin Profile in the Wines

The color changes during wine maturation are usually attributed to anthocyanin polymerization reactions and the evolution of co-pigments resulting from interactions between anthocyanins and other compounds at the fermentation phase [34,35]. For these reasons, we investigated the anthocyanin profile of the wines by HPLC-MS analyses and the pigments, identified through their retention time (RT), molecular ion (M$^+$), and principal MS/MS fragments, as listed in Table 2.

Five mono glucoside anthocyanins, namely delphinidin (3), cyanidin (5), petunidin (6), peonidin (8), and malvidin (9), together with malvidin-3-*O*-acetylglucoside (15), malvidin-3-*O*-caffeoylglucoside (17), cyanidin-3-*O*-(*p*-coumaroyl)glucoside (17), peonidin-3-*O*-trans-(*p*-coumaroyl)glucoside (20), and malvidin-3-*O*-trans-(*p*-coumaroyl)glucoside (21) were revealed in all the samples. Whilst other acyl compounds, such as peonidin-3-*O*-acetylglucoside (14), petunidin-3-*O*-(*p*-coumaroyl)glucoside (18), and malvidin-3-*O*-cis-(*p*-coumaroyl)glucoside (18), also belonging to the group of free-anthocyanins directly extracted from grape skin [36,37], were not detected in Aleatico wines. Four compounds corresponding to carboxy-pyranoanthocyanins derived from the reaction between glucoside anthocyanins and pyruvic acid (A-type vitisins) were also identified (Table 2). In particular, petunidin (7) and malvidin (10) 3-*O*-glucoside pyruvate were present in all the samples, while peonidin (13) and malvidin (14) 3-*O*-(*p*-coumaroyl)glucoside pyruvate were absent in Aleatico wines. Two well resolved chromatographic peaks (11 and 12) referring to isobaric ions with similar MS/MS spectra were achieved for the species with [M]+ at m/z 809, which were identified as isomers of malvidin-3-*O*-glucoside-8-ethyl-(epi)catechin [38]. Then, other ethylidene-bridged flavanol anthocyanins, namely peonidin-3-*O*-(*p*-coumaroyl)glucoside-8-ethyl-(epi)catechin (19) and malvidin-3-*O*-(*p*-coumaroyl)glucoside-8-ethyl-(epi)catechin (23), were revealed in the wines (Table 2). With regards to vinyl-linked flavanol anthocyanins, also known as flavanol pyranoanthocyanins [33], malvidin-3-*O*-acetylglucoside-4-vinyl-(epi)catechin (16) and malvidin-3-*O*-glucoside-4-vinyl-(epi)catechin (22) were only detected in Primitivo and Negramaro wines, respectively. Finally, three flavanol-anthocyanins derivatives, having molecular ions and fragmentation patterns typical of (epi)-catechin-peonidin (1) or malvidin-3-*O*-glucoside (2 and 4) adducts [7] were found (Table 2).

Table 2. Chromatographic and mass spectral data of the identified anthocyanin compounds.

Peak	RT	Compound	[M]$^+$ (*m/z*)	MS/MS
1	9.896	(epi)-catechin-peonidin-3-*O*-glucoside	751	589, 463, 437
2	10.603	(epi)-catechin-malvidin-3-*O*-glucoside	781	619, 493, 467
3	11.356	delphinidin-3-*O*-glucoside	465	303
4	12.236	di(epi)catechin-malvidin-3-*O*-glucoside	1069	907, 781, 619
5	13.511	cyanidin-3-*O*-glucoside	449	287
6	14.975	petunidin-3-*O*-glucoside	476	317
7	16.458	petunidin-3-*O*-glucoside pyruvate	547	385
8	17.520	peonidin-3-*O*-glucoside	463	301
9	19.147	malvidin-3-*O*-glucoside	493	331
10	22.096	malvidin-3-*O*-glucoside pyruvate	561	399
11	28.755	malvidin-3-*O*-glucoside-8-ethyl-(epi)catechin	809	647,519,357
12	30.251	malvidin-3-*O*-glucoside-8-ethyl-(epi)catechin	809	647,519,357
13	31.197	peonidin-3-*O*-(*p*-coumaryl)-glucoside pyruvate	677	369
14	31.957	peonidin-3-*O*-acetylglucoside	505	301
14	31.957	malvidin-3-*O*-(*p*-coumaryl)-glucoside pyruvate	707	399
15	33.237	malvidin-3-*O*-acetylglucoside	535	331
16	34.843	malvidin-3-*O*-acetylglucoside-4-vinyl-(epi)catechin	847	643,491
17	35.622	malvidin-3-*O*-caffeoylglucoside	655	331
17	35.622	cyanidin-3-*O*-(*p*-coumaryl)-glucoside	595	287
18	36.914	petunidin-3-*O*-(*p*-coumaryl)-glucoside	625	317
18	36.914	malvidin-3-*O*-*cis*-(*p*-coumaryl)-glucoside	639	331
19	37.694	peonidin-3-*O*-(*p*-coumaryl)-glucoside-8-ethyl-(epi)catechin	925	635,617,327
20	39.841	peonidin-3-*O*-*trans*-(*p*-coumaryl)-glucoside	609	301
21	41.231	malvidin-3-*O*-*trans*-(*p*-coumaryl)-glucoside	639	331
22	42.355	malvidin-3-*O*-glucoside-4-vinyl-(epi)catechin	805	643,491
23	49.702	malvidin-3-*O*-(*p*-coumaroyl)-glucoside-8-ethyl-(epi)catechin	955	665,647,357

In order to investigate the influence of the fermentation type on the formation and evolution of anthocyanin derived pigments, involved in the color intensity and stability, PCA analyses were performed on Primitivo, Negramaro, and Aleatico wines at draining off and after 18 months of bottle storage. Moreover, the percentage content of the five different classes of pigments were compared among the wines at the two time-points of aging (Figures 1–3). Overall, the mixed fermentation protocol provoked the increasing synthesis of stable pigments in the wines during the vinification process. Indeed, at draining off, the FA18 samples appeared richer in pyranoanthocyanins and ethylidene-bridged compounds, whose content was also enhanced during the bottle aging, thus contributing to the intensity and stability of the color. This was in agreement with the effect of sequential inoculum (delay of 5 days) with *S. bacillaris* CZ1 in the production of wines with a higher level of A-type vitisins [39]. Regarding Primitivo at draining off (Figure 1a), FA18 was characterized by a higher content of vitisin A (10), but also reddish/violet ethylidene-bridged compounds

(11, 12, and 19) and bluish pigment (4). On the contrary, SCE16 (and less SCE138) showed greater amounts of free anthocyanins, especially the compounds 6, 9, 17, 20, and 21 together with pyruvic and vinyl derivatives (7 and 16, respectively). Having $\lambda_{max} > 530$ nm [31], the relative predominance of the compounds 4, 10, 11, 12, and 19 could partially explain the slightly higher CI in FA18 than SCE wines (Table 1).

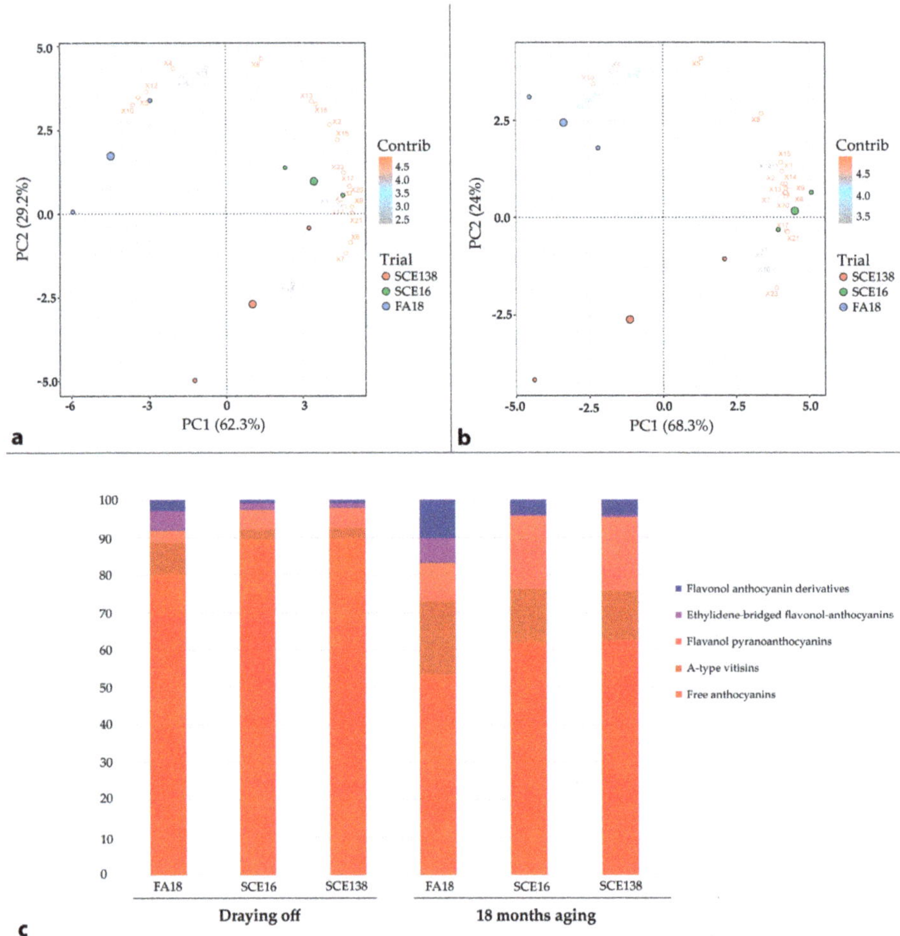

Figure 1. PCA—Primitivo. Principal component diagram of anthocynin-derived red pigments in Primitivo wines SCE138 (green point), SCE 16 (blue point), FA18 (red point), and distribution (percentage) calculated (**a**) at draining off and (**b**) after 18 months of bottle aging; in (**c**), we report the percentage of each pigment as measured by HLPC assays both at draining off and after 18 months of bottle aging. Variables correspond to peaks reported in Table 2.

Moreover, Negramaro FA18 wines at draining off (Figure 2a,c) were distinguished for having a higher content of stable pigments 4, 10, and 23, which positively affected their color intensity (Table 1). Whereas, SCE wines were separated on the sore plot since more correlated to the free anthocyanins 3, 5, 6, 9, 14, 15, 17, and 21 showing greater factor loadings (>|0.9|) on PC1 and PC2 (Figure 2a).

Finally, with regards to Aleatico, even though the use of *S. bacillaris* in winemaking partially enhanced the formation of stable conjugated forms (especially vitisin A 10 and compound 23) in wines at draining off (Figure 3a,c), this was not enough to intensify and stabilize the color. Indeed, there was no significant variation of CI, H, and CEI among the three wines (Table 1). Furthermore, SCE 18 month-old wines were less clearly separated from FA18 and their relative percentage of pigment families was very close (Figure 3b).

These findings, coupled with the highest H and CEI values in the aged samples (Table 1), indicated a similar and faster color change from red to orange tone and color loss [4]. A possible explanation for this behavior can be attributed to the very low content of anthocyanins (TA) and polyphenols (TP) extracted from grapes in Aleatico wines during both fermentation types.

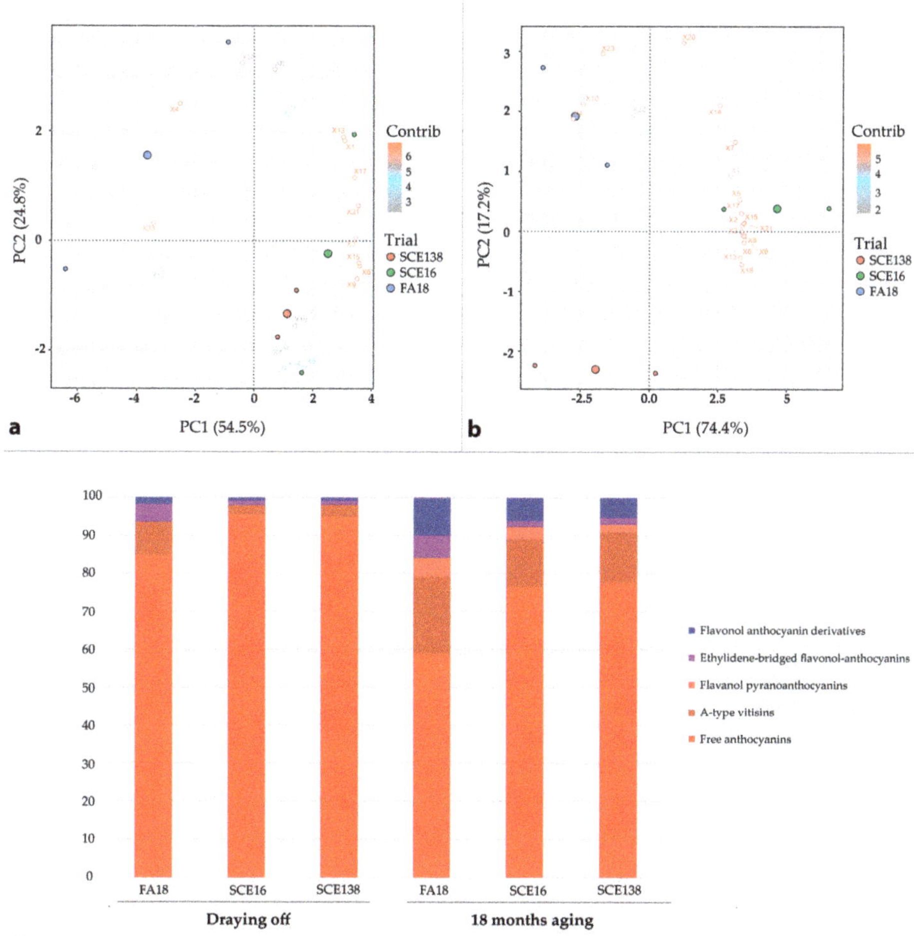

Figure 2. PCA—Negramaro. Principal component diagram of anthocynin-derived red pigments in Negramaro wines SCE138 (green point), SCE 16 (blue point), FA18 (red point), and distribution (percentage) calculated (**a**) at draining off and (**b**) after 18 months of bottle aging; in (**c**), we report the percentage of each pigment as measured by HLPC assays both at draining off and after 18 months of bottle aging. Variables correspond to peaks reported in Table 2.

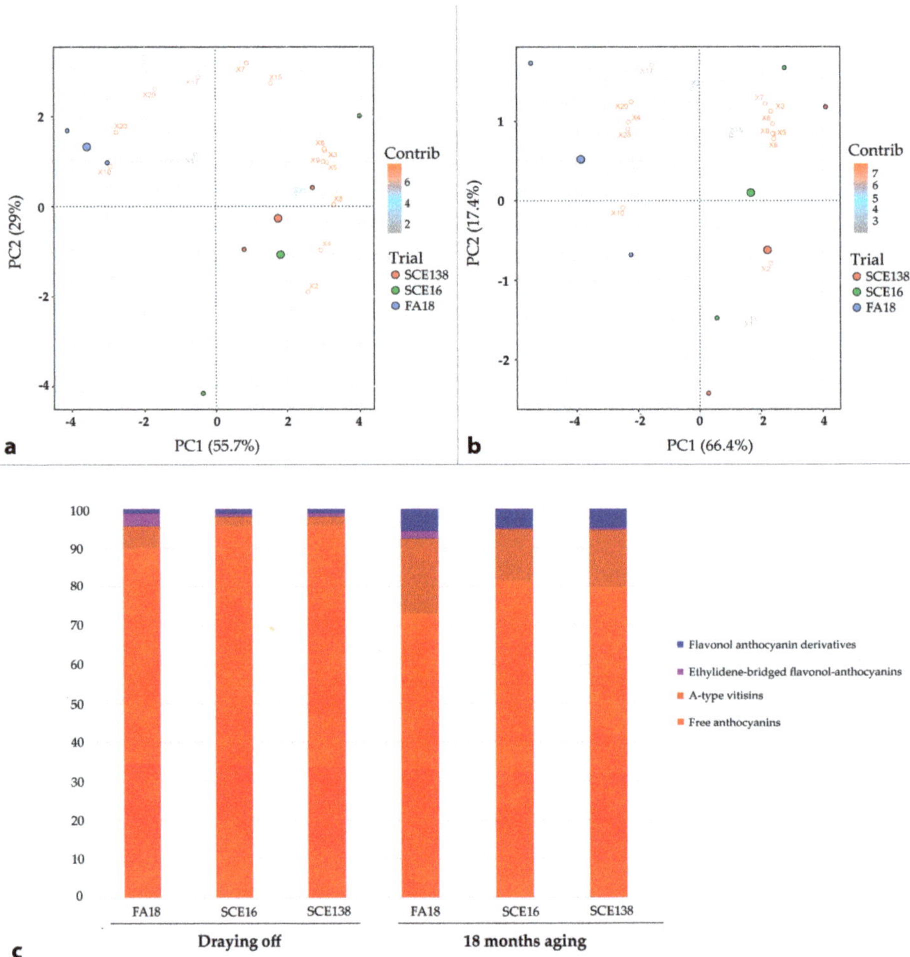

Figure 3. PCA—Aleatico. Principal component diagram of anthocynin-derived red pigments in Aleatico wines SCE138 (green point), SCE 16 (blue point), FA18 (red point), and distribution (percentage) calculated (**a**) at draining off and (**b**) after 18 months of bottle aging; in (**c**), we report the percentage of each pigment as measured by HLPC assays both at draining off and after 18 months of bottle aging. Variables correspond to peaks reported in Table 2.

3. Discussion

Wine is the result of a complex biochemical process, that starts with grape harvesting, continues with the alcoholic and malolactic fermentations, wine aging, and bottling [40]. In this process, the diversity and composition of the yeast micro-population may significantly contribute to the organoleptic characteristics of wine, and consequently, those known as terroir. Indeed, modern oenology is increasingly oriented today to the development of technologies and strategies that allow enhancing the typicity and the quality of autochthonous vines. In this regard, one of the most promising ways is the identification of yeasts which are used as a starter in innovative winemaking processes and allow improving the quality of wines. A combination of *S. bacillaris* and *S. cerevisiae* in a sequential fermentation has been described promising to satisfy the modern market and consumer preferences due to its peculiar characteristics [18].

In the present paper, we investigated how mixed fermentation combining the use of *S. bacillaris* with *S. cerevisiae* might influence the color and its stability during aging, one of the most important organoleptic characteristics in red wine, on three of the most typical and commercially important wines in the South of Italy, Primitivo, Negramaro, and Aleatico. In our trials, we first demonstrated that no killer effect exists of the *S. bacillaris* strain FA18 against the chosen *S. cerevisiae* strains (SCE16 and SCE138), thus confirming their suitability in mixed fermentation. Moreover, the kinetics of fermentation and chemical analysis demonstrated that the two *S. cerevisiae* strains have a similar fermentation capacity on all the three cultivars, thus confirming their suitability of combination in *S. bacillaris*.

Our results revealed that mixed fermentation influences both basic parameters and chemical compounds (i.e., pyranoanathocyanins) specifically related to the co-pigments formation and color stabilization. *S. bacillaris* has been described to affect the chemical composition of the musts and wines by producing various metabolites of enological interest [18].

Among these effects, the reduction of ethanol levels in wines has been described when *S. bacillaris* was used coupling with plus *S. cerevisiae* [16,18,25,26]. However, we did not find any variation in the alcoholic degree (% v/v) (Table 1). Indeed, no significant differences in the ethanol production have been described between mono-*Saccharomyces* and mixed fermentations with some *S. bacillaris* strains. On the contrary, the significant reduction in ethanol is shown when *S. cerevisiae* is added 48 to 72 h after the *S. bacillaris* inoculation, and oxygen is applied during the fermentation process in order to favor the respiration rather than fermentation [16]. Furthermore, the reduction in ethanol for the sequential fermentation is emphasized when the fermentation occurs in a synthetic must medium rather than the natural grape must [41].

Moreover, our data revealed that the fermentation type significantly affected ($p < 0.05$) MA, TA, and TP in the analyzed wines. Pure fermentations allowed a better extraction of anthocyanins and polyphenols as demonstrated by the significant higher value of MA, TA, and TP in SCE16 and SCE138 than in the FA18 samples, especially for Negramaro and Aleatico (Table 1). Despite the aforementioned non-variation of ethanol in our samples, it is known that mixed fermentation of *S. bacillaris* and *S. cerevisiae* leads to a slower development of ethanol in the early stages of winemaking [25,26], thus reducing the extraction of phenolic compounds during the skin maceration [42]. This could partially explain the reduction in phenolic compounds we observed in FA18. Moreover, we detected an even more evident decrease in MA and TA, as well as in TP during aging which in fact is due to the precipitation and degradation phenomena (both oxidative and reductive), that can involve the less stable and oxidizable forms of red wine (such as cyanidin-3-*O*-glucoside and peonidin-3-*O*-glucoside) already described in literature data [7,43].

Most relevant, substantial differences emerged among our wines considering several compounds playing a critical role in the wine color. Indeed, the evolution of wine color is influenced by a number of factors, such as the amount of tannin and acids, grape variety, alcohol and acetaldehyde concentrations, as well as the winemaking and storage conditions of wine [42,44,45]. In particular, the color changes during wine maturation are usually attributed to anthocyanin polymerization reactions and the evolution of co-pigments resulting from interactions between anthocyanins and other compounds at the fermentation phase and during aging [34,35].

Overall, our data highlighted that a 48 h sequential fermentation employing the FA18 *S. bacillaris* in Primitivo, Negramaro, and Aleatico enhances the synthesis of stable anthocyanin pigments, in particular, A-type vitisins and ethylidene-bridge flavonol-anthocyanin adducts, as well as their preservation after 18 months of aging in the bottle. The acidogenic nature of *S. bacillaris*, leading to a more consistent production of pyruvic and acetaldehyde during fermentation, would be responsible for the preferential synthesis of these compounds [25,46]. It is worth pointing out that Primitivo, Negramaro, and Aleatico grapes, used in winemaking, derived from minimal or no canopy management grown vineyards and, thus, were poorer in anthocyanins and polyphenols with respect to conventional conditions, as previously reported in literature [47]. This could motivate the lack of various pigments, such as pinotins, anthocyanin dimers, and trimers, as well as more different vinyl-linked and ethylidene-bridged compounds, compared to wines analyzed by direct injection [33] or after fractionation [7].

Notably, the pyranic structure of malvidin-3-*O*-glucoside pyruvate (10) is recognized as more resistant to the bleaching effect due to SO_2 than malvidinic free anthocyanins, thereby its presence in wine implies a greater red color stabilization [48]. Furthermore, this vitisin A is resistant to a pH increase [48] and oxidative degradation [49], as well as temperature changes [50]. It is worth noting that, although free anthocyanins more strongly decreased in FA18, mixed fermentation seemed to protect the wine from further non-oxidative degradation reactions. It was confirmed by the relative unstable ethyl linked anthocyanins (11, 12, and 19), whose percentage slightly increased during aging (Figure 1b,c), and the reduced color loss, as proved by the significant lower values of H and CEI than those found in SCE16 and SCE138 after 18 months in the bottle (Table 1). This would be a very important finding from a technological standpoint, since the use of *S. bacillaris* in tandem with *S. cerevisiae* could contribute to mitigate the often-reported rapid change of Primitivo color into orange hue compared to other international wines [7]. In addition, the significant lower values of H and CEI highlighted that Negramaro derived from the mixed inoculum of *S. bacillaris/S. cerevisiae* remained more stable in the color after bottle storage than SCEs (Table 1). This was corroborated by the most pronounced increase in vitisins, ethylidene-bridged pigments, and flavanol-anthocyanin adducts percentage in FA18 aged wines (Figure 2b,c). However, the remarked difference in the color stability of Negramaro wines was less evident respect to Primitivo ones (Table 1), maybe due to the different ethyl linked compounds prevailing in the former (i.e., malvidin-3-*O*-glucoside-8-ethyl-epicatechin) despite the latter samples (i.e., malvidin-3-*O*-glucoside-8-ethyl-epicatechin isomers and peonidin-3-*O*-pcoumaroyl-glucoside-8-ethyl-epicatechin), as well as their relative concentrations (Figure 2).

Notably, at our knowledge, this is the first evidence that mixed fermentation induced the production of ethylidene-bridge flavonol-anthocyanin adducts. Indeed, these adducts have been previously found unstable and intermediate products formed during winemaking and aging, also using different vinification procedures [51,52] or present only at a low concentration, in addition to their importance has been hypothesized [53]. As a matter of fact, these ethylidene linked pigments are associated to a color increase with a shift towards violet [54,55]. Moreover, these pigments undergo further polymerization phenomena, thus leading to an important reduction in astringency [37] that improve the organoleptic quality of the red wines.

4. Materials and Methods

4.1. Yeast Strains

Two *S. cerevisiae* strains and one *S. bacillaris* strain available at the I.U.V.V.—Institut Universitaire de la Vigne et du Vin Jules Guyot of Dijon (France) were inoculated in red vinification experiments. The two *S. cerevisiae* strains were isolated from 'Savignin Jura' and were coded SCE16 and SCE138, while the *S. bacillaris* strain was isolated from 'Pinot noir' in Burgundy and identified as FA18. These strains were previously isolated in Burgundy, characterized, and then selected based on their oenological performances [24,56]. The 5.8S ITS rDNA sequencing confirmed the pure culture condition of these strains and the correct identity of these species [57].

4.2. Grape Varieties and Vineyard Conduction

The experiments were carried out in 2017 on three important Apulian *Vitis vinifera* L. red grape varieties: Primitivo, Negramaro, and Aleatico, chosen as used for the most important enological production in Apulia Region, Southern Italy. They were cultivated in an experimental vineyard of the CREA-VE, located in the area around Rutigliano (Bari), Apulia Region, Southern Italy. The vineyards are composed of 13-year-old vines, grafted34

E.M., trained on Gobelet Alberello, and pruned with four spurs of two buds. Plants are planted 1.5 m between rows and 1.0 m in the row. All the vines were cultivated without water supply, chemical inputs, and canopy management. Samples of 130 kg per each variety were hand-harvested at the same time in mid-October, at technical maturity [58]. At harvest, the total soluble solids (TSS) content, A, and pH were as follows: Aleatico: TSS 19.8 °Brix, A 5.9 g/L, pH 3.40; Negramaro: TSS 21 °Brix, A 6.9 g/L, pH 3.38; Primitivo: TSS 25 °Brix, A 7.1 g/L, pH 3.42. The grapes were hand-picked in small pierced plastic crates and immediately crushed and destemmed. After crushing and destemming, 4 g/hL of potassium metabisulphite (the equivalent of 20 mg/L of SO_2) was added in the unpasteurized must. Organic and inorganic nitrogen sources were added, as described in the laboratory scale protocol of Nisiotou et al. [59]. The obtained must were directly processed for winemaking.

4.3. Interaction between Saccharomyces Yeast Strains

In order to test the killer action between the three yeast strains we performed two experiments.

Experiment 1: Cellophane agar layer technique [60]. Sterilized disc of 90 mm diameter of cellophane was laid aseptically over the solidified Yeast Peptone Dextrose Agar (YPDA) medium in culture plates. The plates were laid overnight to allow the excess moisture to evaporate. In addition, 10 μL (at the concentrations of 1.0×10^7 CFU/mL) of each yeast strain (*S. cerevisiae* SCE16, *S. cerevisiae* SCE138, and *S. bacillaris* FA18) were uniformly distributed on the cellophane disc. For each yeast, six plates were produced. Moreover, 10 μL of sterilized YPDA without yeast were used as a control on nine different plates. After 48 h of incubation at 25 °C, the cellophane disc with and without yeast was removed from the plates. On the first three plates previously covered with the cellophane disc with the *S. cerevisiae* strain SCE16, 10 μL (at the concentrations of 1.0×10^7 CFU/mL) of the *S. cerevisiae* strain SCE138 were uniformly distributed and on the other three plates, 10 μL (at the concentrations of 1.0×10^7 CFU/mL) of the *S. bacillaris* strain FA18. The same procedure was used for the six plates covered with the cellophane agar with the saccharomyces strain SCE138 and for the six plates covered with the cellophane disc with the *S. bacillaris* strain FA18. On the plates used as a control, the three yeast strains were uniformly distributed. After incubation at 25 °C for 48 h, the growth of each yeast strain was recorded.

Experiment 2: Growth in liquid media. Each yeast strain was grown in a tube containing liquid YPD for 24 h at 25 °C. Cells were removed by a double centrifugation at 7240 g for 5 min and the supernatant was filtered through a syringe filter (0.22 μm pore size). In addition, 7.5 mL of the filtered supernatant of the *S. cerevisiae* strain SCE138 were placed in two sterilized tubes and added with 7.5 mL of liquid YPD containing the *S. cerevisiae* strain SCE16 or liquid YPD containing the *S. bacillaris* strain FA18, both at the concentration of 1.0×10^6 CFU/mL. The same procedure was followed using the filtered supernatant of SCE16 and FA18 added with liquid YPD containing living cells of other yeast strains. The tubes were placed on an orbital shaker at 25 °C for 48 h. After 18, 24, 42, and 48 h, two aliquots of 1 mL each were aseptically withdrawn from each tube. Using a spectrophotometer (Thermo Scientific NanoDrop2000) the growth of yeast cultures was monitored by measuring the optical density (OD) at 600 nm. For each aliquot, five replicates/lectures have been performed and the average values were used to plot the growth curve of each yeast strain in the presence of the filtered supernatant of another yeast strain.

4.4. Pilot Scale Fermentation Procedure

Pilot scale vinification trials of 20 kg (equal solid/liquid ratio in each trial) were conducted in stainless steel fermenters. The must obtained, corresponding to 18 Lt from each sample of single variety were fermented separately following a standard red winemaking procedure and three independent replicates for each trial were finally carried out.

Each trial was as follows: (i) Mono-SCE16 inoculation (SCE16), (ii) mono-SCE138 inoculation (SCE138), and (iii) a mixed fermentation where the two *S. cerevisiae* (SCE16 and SCE138, 1:1) were added together 48 h after the inoculation of *S. bacillaris* (FA18). Each yeast strain was inoculated at a starting concentration of about 5×10^6 CFU/mL. The possible lack of nutrients was avoided through a standard addition of nitrogen nutrients and enzymatic cofactors into the fermenting juice (20 g/hL of organic nitrogen). This was applied when sugar consumption reached 50 gr/Lt in the mono-SCE16 and -SCE138, while it was implemented in the mixed fermentations (FA18) after 48 h, when the two *S. cerevisiae* strains were inoculated, thus to enhance the *Saccharomyces* metabolic activities, avoiding nutrients depletion and preventing *Saccharomyces* growth arrest.

The fermentation proceeded at a constant temperature of 25 ± 0.5 °C, performing manual pushing down of the pomace cap three times a day during the first half of the fermentation and two times a day until the end. Fermentation kinetics were measured, checking the level of sugar consumption (°Babo), utilizing a standard hydrometer. Macerations and fermentations were considered ended when residual sugar levels, measured with a hydrometer (Babo Klosterneuburg Mostimeter), reached 0 °Babo (8–10 days). The complete fermented must was pressed (up to 2–3 bar) and kept in the cellar for 2 days before storage in a refrigerated room (4–5 °C) to allow the residual solid parts (solid lees) to settle down. The wines were racked after a week to remove the solid lees. Consequently, wines were poured in 0.75 L glass bottles, supplemented with potassium metabisulphite to achieve a final concentration of 80 mg/L of total SO_2. The wines were stored at a constant temperature of 15 °C and analyzed at draining off and after 18 months of aging to assess the color stabilization and the variation of chromatic characteristics.

4.5. Chemical Analysis

A chemical analysis on wine and must was performed according to the EEC regulation 2676/90, as reported by the International Organization for Vine and Wine (OIV, 2018: https://www.oiv.int/en/technical-standards-and-documents/methods-of-analysis/ compendium-of-international-methods-of-analysis-of-wines-and-musts-2-vol (accessed on 24 April 2020)). Titratable acidity, A (g/L of tartaric acid) was measured following OIV MA-AS313-01 R2015 par.5.3;pH: OIV MA-AS313-15 R2011; volatile acidity, VA (g/L of acetic acid): OIV MA-AS313-02 R2015; alcoholic degree, ET (% *v/v*): OIV MA-AS312-01A R2016 par. 4C. The wine color was assessed by the Glories chromatic parameters [61]: Color intensity (CI) was calculated as the sum of absorbance ($\lambda_{420} + \lambda_{520} + \lambda_{620}$ nm); hue (H) was defined as the ratio $\lambda_{420}/\lambda_{520}$ nm, while the color evolution index (CEI) was calculated as $(\lambda_{420} - \lambda_{520}$ nm$)/\lambda_{420}$ nm.

4.6. Phenolic Indexes

Total polyphenols (TP), total anthocyanins (TA), and monomeric anthocyanins (MA) were measured spectrophotometrically to assess the phenolic wine composition and the overall chromatic characteristics.

TP was determined following the method suggested by Waterhouse et al. [62]. From each sample, 20 μL were collected in separate cuvettes, and mixed with 1.58 mL water and 100 μL of Folin-Ciocalteu reagent. After 5 min, 300 μL Na2CO3 10% were added and the solution was shacked. The absorbance of each solution was read at λ_{750} nm against a blank after waiting for 2 h at 20 °C. A calibration curve (R2 = 0.9264) was set with a polyphenolic concentration between 0–3000 mg/L of gallic acid, considering the effective range of the assay. Results were reported as mg/L of gallic acid equivalents (GAE).

TA was determined as already reported [38]. Briefly, the samples were diluted in a solution consisting of 70/30/1 (*v/v/v*) ethanol/water/HCl. The relative absorbance for each sample was measured at λ_{max} of 540 nm. The total anthocyanin content was expressed as mg/L of malvidin-3-*O*-glucoside equivalents.

Finally, MA was measured by the spectrophotometric determination reported by Lee et al. [63]. Briefly, all the dilutions were performed in 50 mL volumetric flasks. At the beginning, the appropriate dilution factor by diluting the test portion with a pH 1.0 buffer was determined until absorbance at λ_{max} of 520 nm was within the linear range (between 0.2 and 1.4 AU). Using the appropriate dilution factor, two dilutions of each test sample, either for pH 1.0 (potassium chloride, 0.0025 M) or pH 4.5 (sodium acetate, 0.4 M) buffers were prepared. Hence, the determination proceeded through pH 1.0 and 4.5 buffer dilutions of the samples, reading them both at λ_{max} of 520 and 700 nm. The measure at 700 nm was considered a wine haze correction of the reading at 520 nm. The content of anthocyanin pigments was expressed as mg/L of cyanidin-3-*O*-gluoside equivalents.

4.7. Anthocyanin Profile Determined by HPLC-DAD-MS

An HPLC 1100 equipped with a DAD and XCT-trap Plus mass detector (Agilent Technologies, Palo Alto, CA., USA) coupled with an ESI interface was used. The reversed stationary phase employed was a Zorbax C18 5 μm (250 × 4.6 mm i.d., Agilent Technologies) with a pre-column Gemini C18 5 μm (4 × 2 mm i.d., Phenomenex, Castel Maggiore, Bologna, Italy). The following gradient system was used with water/formic acid (90:10, *v*/*v*) (solvent A) and acetonitrile (solvent B): 0 min, 95% A— 5% B; 10 min, 87% A —13% B; 20 min, 85% A—15% B; 30 min, 78% A—22% B; 50 min 78% A—22% B; 55 min 5% A— 95% B; stop time at 70 min. Finally, the column was re-equilibrated with the initial solvent mixture for 15 min. The flow was maintained at 0.7 mL/min; the sample injection was 5 μL. Wine samples were filtered (0.2 μm RC syringe filters, Phenomenex) before the HPLC analysis. The diode array detection was between 250 and 650 nm, and absorbance was recorded at 520 nm. The positive electrospray mode was used for ionization of the molecules with capillary voltage at 4000 V and skimmer voltage at 30 V. The nebulizer pressure was 40 psi and the nitrogen flow rate was 9 L/min. The temperature of drying gas was 350 °C. The monitored mass range was from m/z 100 to 1200.

Free and co-pigmented anthocyanins were identified by matching the chromatographic elution order, molecular ions, and MS/MS fragments with those reported in the literature [7]. Semi-quantitation was performed using extracted ion chromatograms (EIC): For each compound, the EIC at the corresponding molecular ion was obtained and the relevant peak was integrated (Supplementary Table S2). Subsequently, peak areas were summed with respect to the type of pigment to calculate the percentage content of the different classes determined in the wines.

4.8. Organic Acids Determination by HPLC-UV

An HPLC 1100 equipped with a VWD detector (Agilent Technologies, Palo Alto, CA, USA) was used. The reversed stationary phase employed was a Synergy Hydro-RP-80A 5 μm (250 × 4.6 mm i.d., Phenomenex, Castel Maggiore, Bologna, Italy) with a pre-column Gemini C18 5 μm (4 × 2 mm i.d., Phenomenex, Castel Maggiore, Bologna, Italy). The separation was conducted in an isocratic mode using water/orthophosphoric acid (0.1%) as the mobile phase. The flow was maintained at 0.7 mL/min and sample injection was 5 μL. Wine samples were 2-folds diluted and filtered (0.2 μm RC syringe filters, Phenomenex) before the HPLC analysis. Absorbance was recorded at 210 nm.

4.9. Statistical Analysis

Data were analyzed using the R package software (version 3.4.0). Specifically, after testing their normal distribution by the Mardia test, a two-way multivariate analysis of variance (MANOVA) was performed on the chemical composition data in order to evaluate the effect of the factors fermentation type and aging, whose significance was discussed in the text. Tukey's HSD post-hoc test was used to separate the means ($p < 0.05$) when the interaction between the factors was significant (Table 1). Furthermore, the principal component analysis (PCA) of the dataset was performed on semi-quantified HPLC-anthocyanin profiles of each wine at draining off and after 18 months in the bottle to explore qualitative differences. In the PCA, only the first two components were considered accounting for more than 80% of the total variance explained.

5. Conclusions

In conclusion, the presented results highlighted that the use of *S. bacillaris* in tandem with *S. cerevisiae* has positively contributed to the evolution and stability of the wine color during the aging process. Although preliminary, our data are a further step that highlight the applicative technological potential of mixed fermentations with *S. bacillaris*. [18]. Indeed, our results support the importance of mixed fermentations to enhance the organoleptic characteristics (such as color intensity and stability) and shelf-life of wines that belong to the winemaking tradition. In particular, mixed fermentation with *S. bacillaris* might represent a valuable technological tool for mitigating the often reported rapid change of the color of some mono-varietal wines (such as Primitivo) towards an orange-brown hue. Moreover, we highlighted new clues on the impact of individual components produced in the presence of different starters on the final wine quality. This is a small pilot scale fermentation trial, but as a future perspective, the possibility of testing mixed cultures on different musts while also studying more in-depth yeast interactions, offer the opportunity to evaluate their benefits and limitations in order to select the best starters capable of fully enhancing the qualities of the resulting wines.

Supplementary Materials: The following are available online. Table S1: Concentration of organic acids in the studied wines; Table S2: Quantities of the identified anthocyanins into wines; Figure S1: Cell concentration of each yeast strain grown in liquid YPD amended with filtered supernatant obtained from the growth of each yeast strain on YPD. Data are the mean value of two replicates and five lectures/replicates. The time in hours is reported on abscises, while the ordinate axes reported the 10 logarithms of the number of cells per mL; Figure S2: Fermentation kinetics of Primitivo, Negroamaro, and Aleatico in pilot scale conditions: The days of fermentation are reported on abscises, while °Babo is reported on the ordinates. SCE16 and SCE138: Mono *S. cerevisiae* fermentations; FA18: Mixed fermentation of *S. bacillaris* FA18 and the two *S. cerevisiae* strains (48 h of delay). Red arrows indicate the addition of the two *S.cerevisiae* strains.

Author Contributions: Investigation, validation, methodology, software, data curation, writing—original draft, M.V. (Matteo Velenosi) and P.C.; formal analysis, software, validation, R.P. and A.D.M.; formal analysis on the yeast used in the pilot scale fermentation, A.S.; validation, data curation, writing—original draft, H.A.; supervision and resources, reviewing, N.A.; supervision, resources, data curation, validation, writing—reviewing and editing, M.V. (Mario Ventura) and M.F.C. All authors have read and agreed to the published version of the manuscript.

Funding: This research received no external funding from any specific grant from funding agencies in the public, commercial, or not-for-profit sectors. The experiments were funded by local funds from the Consiglio per la ricerca in agricoltura e l'analisi dell'economia agraria-Centro di ricerca Viticoltura ed Enologia (CREA-VE) and the University of Bari Aldo Moro. Matteo Velenosi performed the experiments in the UMR ProcédésAlimentaires et Microbiologiques, EquipeVAlMiS (Vin, Aliment, Microbiologie, Stress), AgroSup Dijon—Université de Bourgogne thanks to GlobalDoc thesis grant.

Informed Consent Statement: Informed consent was obtained from all the subjects involved in the study.

Data Availability Statement: The data presented in this study are available on request from the corresponding author.

Conflicts of Interest: The authors declare no conflict of interest.

Sample Availability: Samples are available from the authors.

References

1. Capozzi, V.; Garofalo, C.; Chiriatti, M.A.; Grieco, F.; Spano, G. Microbial terroir and food innovation: The case of yeast biodiversity in wine. *Microbiol. Res.* **2015**, *181*, 75–83. [CrossRef]
2. Hernández, B.; Sáenz, C.; Alberdi, C.; Alfonso, S.; Diñeiro, J.M. Colour Evolution of Rosé Wines after Bottling. *S. Afr. J. Enol. Vitic.* **2016**, *32*, 42–50. [CrossRef]
3. Salinas, M.R.; Garijo, J.; Pardo, F.; Zalacain, A.; Alonso, G.L. Color, polyphenol, and aroma compounds in rosé wines after prefermentative maceration and enzymatic treatments. *Am. J. Enol. Vitic.* **2003**, *54*, 195–202.

4. Benucci, I.; Cerreti, M.; Liburdi, K.; Nardi, T.; Vagnoli, P.; Ortiz-Julien, A.; Esti, M. Pre-fermentative cold maceration in presence of non- Saccharomyces strains: Evolution of chromatic characteristics of Sangiovese red wine elaborated by sequential inoculation. *Food Res. Int.* **2018**, *107*, 257–266. [CrossRef]

5. Canals, R.; Llaudy, M.C.; Valls, J.; Canals, A.J.M.; Zamora, F. Influence of Ethanol Concentration on the Extraction of Color and Phenolic Compounds from the Skin and Seeds of Tempranillo Grapes at Different Stages of Ripening. *J. Agric. Food Chem.* **2005**, *53*, 4019–4025. [CrossRef]

6. Torchio, F.; Segade, S.R.; Gerbi, V.; Cagnasso, E.; Rolle, L. Changes in chromatic characteristics and phenolic composition during winemaking and shelf-life of two types of red sweet sparkling wines. *Food Res. Int.* **2011**, *44*, 729–738. [CrossRef]

7. Dipalmo, T.; Crupi, P.; Pati, S.; Clodoveo, M.L.; Di Luccia, A. Studying the evolution of anthocyanin-derived pigments in a typical red wine of Southern Italy to assess its resistance to aging. *LWT* **2016**, *71*, 1–9. [CrossRef]

8. Pérez-Magariño, S.; González-Sanjosé, M.L. Evolution of Flavanols, Anthocyanins, and Their Derivatives during the Aging of Red Wines Elaborated from Grapes Harvested at Different Stages of Ripening. *J. Agric. Food Chem.* **2004**, *52*, 1181–1189. [CrossRef]

9. Jackson, D.I.; Lombard, B.P. Environmental and management practices affecting grape composition and wine quality: A review. *Am. J. Enol. Vitic.* **1993**, *44*, 409–430.

10. Pérez-Lamela, C.; García-Falcón, M.; Simal-Gandara, J.; Orriols-Fernández, I. Influence of grape variety, vine system and enological treatments on the colour stability of young red wines. *Food Chem.* **2007**, *101*, 601–606. [CrossRef]

11. Asenstorfer, R.E.; Markides, A.J.; Iland, P.G.; Jones, G.P. Formation of vitisin A during red wine fermentation and maturation. *Aust. J. Grape Wine Res.* **2003**, *9*, 40–46. [CrossRef]

12. Jolly, N.; Augustyn, O.; Pretorius, I. The Role and Use of Non-Saccharomyces Yeasts in Wine Production. *S. Afr. J. Enol. Vitic.* **2017**, *27*, 15–39. [CrossRef]

13. Loira, I.; Morata, A.; Comuzzo, P.; Callejo, M.J.; González, C.; Calderón, F.; Suárez-Lepe, J.A. Use of Schizosaccharomyces pombe and Torulaspora delbrueckii strains in mixed and sequential fermentations to improve red wine sensory quality. *Food Res. Int.* **2015**, *76*, 325–333. [CrossRef] [PubMed]

14. Ortega, A.F.; Lopez-Toledano, A.; Mayen, M.; Merida, J.; Medina, M. Changes in Color and Phenolic Compounds during Oxidative Aging of Sherry White Wine. *J. Food Sci.* **2003**, *68*, 2461–2468. [CrossRef]

15. Duarte, F.; Pimentel, N.H.; Teixeira, A.; Fonseca, A. Saccharomyces bacillaris is not a synonym of Candida stellata: Reinstatement as Starmerella bacillaris comb. nov. *Antonie van Leeuwenhoek* **2012**, *102*, 653–658. [CrossRef]

16. Englezos, V.; Rantsiou, K.; Cravero, F.; Torchio, F.; Ortiz-Julien, A.; Gerbi, V.; Rolle, L.; Cocolin, L. Starmerella bacillaris and Saccharomyces cerevisiae mixed fermentations to reduce ethanol content in wine. *Appl. Microbiol. Biotechnol.* **2016**, *100*, 5515–5526. [CrossRef] [PubMed]

17. Junior, W.J.F.L.; Nadai, C.; Crepalde, L.T.; De Oliveira, V.S.; De Matos, A.D.; Giacomini, A.; Corich, V. Potential use of Starmerella bacillaris as fermentation starter for the production of low-alcohol beverages obtained from unripe grapes. *Int. J. Food Microbiol.* **2019**, *303*, 1–8. [CrossRef]

18. Englezos, V.; Giacosa, S.; Rantsiou, K.; Rolle, L.; Cocolin, L. Starmerella bacillaris in winemaking: Opportunities and risks. *Curr. Opin. Food Sci.* **2017**, *17*, 30–35. [CrossRef]

19. Sipiczki, M. Candida zemplinina sp. nov., an osmotolerant and psychrotolerant yeast that ferments sweet botrytized wines. *Int. J. Syst. Evol. Microbiol.* **2003**, *53*, 2079–2083. [CrossRef]

20. Torija, M.-J.; Rozès, N.; Poblet, M.; Guillamón, J.M.; Mas, A. Yeast population dynamics in spontaneous fermentations: Comparison between two different wine-producing areas over a period of three years. *Antonie van Leeuwenhoek* **2001**, *79*, 345–352. [CrossRef]

21. Romancino, D.P.; Di Maio, S.; Muriella, R.; Oliva, D. Analysis of non-Saccharomycesyeast populations isolated from grape musts from Sicily (Italy). *J. Appl. Microbiol.* **2008**, *105*, 2248–2254. [CrossRef]

22. Fleet, G.H. Wine yeasts for the future. *FEMS Yeast Res.* **2008**, *8*, 979–995. [CrossRef]

23. Comitini, F.; Gobbi, M.; Domizio, P.; Romani, C.; Lencioni, L.; Mannazzu, I.; Ciani, M. Selected non-Saccharomyces wine yeasts in controlled multistarter fermentations with Saccharomyces cerevisiae. *Food Microbiol.* **2011**, *28*, 873–882. [CrossRef]

24. Sadoudi, M.; Tourdot-Maréchal, R.; Rousseaux, S.; Steyer, D.; Gallardo-Chacón, J.-J.; Ballester, J.; Vichi, S.; Guérin-Schneider, R.; Caixach, J.; Alexandre, H. Yeast–yeast interactions revealed by aromatic profile analysis of Sauvignon Blanc wine fermented by single or co-culture of non-Saccharomyces and Saccharomyces yeasts. *Food Microbiol.* **2012**, *32*, 243–253. [CrossRef] [PubMed]

25. Englezos, V.; Rantsiou, K.; Cravero, F.; Torchio, F.; Giacosa, S.; Ortiz-Julien, A.; Gerbi, V.; Rolle, L.; Cocolin, L. Volatile profiles and chromatic characteristics of red wines produced with Starmerella bacillaris and Saccharomyces cerevisiae. *Food Res. Int.* **2018**, *109*, 298–309. [CrossRef]

26. Englezos, V.; Cocolin, L.; Rantsiou, K.; Ortiz-Julien, A.; Bloem, A.; Dequin, S.; Camarasa, C. Specific Phenotypic Traits ofStarmerella bacillarisRelated to Nitrogen Source Consumption and Central Carbon Metabolite Production during Wine Fermentation. *Appl. Environ. Microbiol.* **2018**, *84*. [CrossRef] [PubMed]

27. Magyar, I.; Nyitrai-Sárdy, D.; Leskó, A.; Pomázi, A.; Kállay, M. Anaerobic organic acid metabolism of Candida zemplinina in comparison with Saccharomyces wine yeasts. *Int. J. Food Microbiol.* **2014**, *178*, 1–6. [CrossRef]

28. Lleixà, J.; Manzano, M.; Mas, A.; Portillo, M.C. Saccharomyces and non-Saccharomyces Competition during Microvinification under Different Sugar and Nitrogen Conditions. *Front. Microbiol.* **2016**, *7*, 1959. [CrossRef]

29. Bell, S.-J.; Henschke, P.A. Implications of nitrogen nutrition for grapes, fermentation and wine. *Aust. J. Grape Wine Res.* **2005**, *11*, 242–295. [CrossRef]

30. Medina, K.; Boido, E.; Dellacassa, E.; Carrau, F. Growth of non-Saccharomyces yeasts affects nutrient availability for Saccharomyces cerevisiae during wine fermentation. *Int. J. Food Microbiol.* **2012**, *157*, 245–250. [CrossRef] [PubMed]

31. Hazelwood, L.A.; Daran, J.-M.; Van Maris, A.J.A.; Pronk, J.T.; Dickinson, J.R. The Ehrlich Pathway for Fusel Alcohol Production: A Century of Research on Saccharomyces cerevisiae Metabolism. *Appl. Environ. Microbiol.* **2008**, *74*, 2259–2266. [CrossRef] [PubMed]

32. Morata, A.; Loira, I.; Heras, J.M.; Callejo, M.J.; Tesfaye, W.; González, C.; Suárez-Lepe, J.A.; Barrado, A.D.M. Yeast influence on the formation of stable pigments in red winemaking. *Food Chem.* **2016**, *197*, 686–691. [CrossRef]

33. Pati, S.; Crupi, P.; Savastano, M.L.; Benucci, I.; Esti, M. Evolution of phenolic and volatile compounds during bottle storage of a white wine without added sulfite. *J. Sci. Food Agric.* **2020**, *100*, 775–784. [CrossRef]

34. Mateus, N.; Freitas, V. Evolution and Stability of Anthocyanin-Derived Pigments during Port Wine Aging. *J. Agric. Food Chem.* **2001**, *49*, 5217–5222. [CrossRef]

35. Mateus, N.; Silva, A.M.S.; Vercauteren, J.; Freitas, V. Occurrence of anthocyanin-derived pigments in red wines. *J. Agric. Food Chem.* **2001**, *49*, 4836–4840. [CrossRef] [PubMed]

36. He, F.; Liang, N.-N.; Mu, L.; Pan, Q.-H.; Wang, J.; Reeves, M.J.; Duan, C.-Q. Anthocyanins and Their Variation in Red Wines I. Monomeric Anthocyanins and Their Color Expression. *Molecules* **2012**, *17*, 1571–1601. [CrossRef] [PubMed]

37. He, F.; Liang, N.-N.; Mu, L.; Pan, Q.-H.; Wang, J.; Reeves, M.J.; Duan, C.-Q. Anthocyanins and Their Variation in Red Wines II. Anthocyanin Derived Pigments and Their Color Evolution. *Molecules* **2012**, *17*, 1483–1519. [CrossRef] [PubMed]

38. Coletta, A.; Trani, A.; Faccia, M.; Punzi, R.; Dipalmo, T.; Crupi, P.; Antonacci, D.; Gambacorta, G. Influence of viticultural practices and winemaking technologies on phenolic composition and sensory characteristics of Negroamaro red wines. *Int. J. Food Sci. Technol.* **2013**, *48*, 2215–2227. [CrossRef]

39. Romboli, Y.; Mangani, S.; Buscioni, G.; Granchi, L.; Vincenzini, M. Effect of Saccharomyces cerevisiae and Candida zemplinina on quercetin, vitisin A and hydroxytyrosol contents in Sangiovese wines. *World J. Microbiol. Biotechnol.* **2015**, *31*, 1137–1145. [CrossRef]

40. Romano, P. Function of yeast species and strains in wine flavour. *Int. J. Food Microbiol.* **2003**, *86*, 169–180. [CrossRef]

41. Lemos, W.J.; Bovo, B.; Nadai, C.; Crosato, G.; Carlot, M.; Favaron, F.; Giacomini, A.; Corich, V. Biocontrol Ability and Action Mechanism of Starmerella bacillaris (Synonym Candida zemplinina) Isolated from Wine Musts against Gray Mold Disease Agent Botrytis cinerea on Grape and Their Effects on Alcoholic Fermentation. *Front. Microbiol.* **2016**, *7*, 1249. [CrossRef] [PubMed]

42. Ribéreau-Gayon, P.; Dubourdieu, D.; Donèche, B.; Lonvaud, A. Traité d'oenologie. In *Microbiologie du vin Vinifications*; Dunod: Paris, France, 2017.

43. Canuti, V.; Puccioni, S.; Giovani, G.; Salmi, M.; Rosi, I.; Bertuccioli, M. ffect of oenotannin addition on the composition of sangiovese wines from grapes with different characteristics. *Am. J. Enol. Vitic.* **2012**, *63*, 220–231. [CrossRef]

44. Suárez-Lepe, J.; Morata, A. New trends in yeast selection for winemaking. *Trends Food Sci. Technol.* **2012**, *23*, 39–50. [CrossRef]

45. Božič, J.T.; Butinar, L.; Albreht, A.; Vovk, I.; Korte, D.; Vodopivec, B.M. The impact of Saccharomyces and non-Saccharomyces yeasts on wine colour: A laboratory study of vinylphenolic pyranoanthocyanin formation and anthocyanin cell wall adsorption. *LWT* **2020**, *123*. [CrossRef]

46. Fulcrand, H.; Dueñas, M.; Salas, E.; Cheynier, V. Phenolic reactions during winemaking and aging. *Am. J. Enol. Vitic.* **2006**, *57*, 289–297.

47. Di Profio, F.; Reynolds, A.G.; Kasimos, A.; Lopes, P.; Marques, J.; Lino, J.; Coelho, J.; Alves, C.; Roseira, I.; Mendes, A.; et al. Canopy Management and Enzyme Impacts on Merlot, Cabernet franc, and Cabernet Sauvignon. II. Wine Composition and Quality. *Am. J. Enol. Vitic.* **2011**, *62*, 152–168. [CrossRef]

48. Alcalde-Eon, C.; Escribano-Bailón, M.T.; Santos-Buelga, C.; Rivas-Gonzalo, J.C. Changes in the detailed pigment composition of red wine during maturity and ageing. *Anal. Chim. Acta* **2006**, *563*, 238–254. [CrossRef]

49. Bakker, J.; Timberlake, C.F. Isolation, Identification, and Characterization of New Color-Stable Anthocyanins Occurring in Some Red Wines. *J. Agric. Food Chem.* **1997**, *45*, 35–43. [CrossRef]

50. Sarni-Manchado, P.; Fulcrand, H.; Souquet, J.-M.; Cheynier, V.; Moutounet, M. Stability and Color of Unreported Wine Anthocyanin-derived Pigments. *J. Food Sci.* **1996**, *61*, 938–941. [CrossRef]

51. Rivas-Gonzalo, J.C.; Bravo-Haro, S.; Santos-Buelga, C. Detection of Compounds Formed through the Reaction of Malvidin 3-Monoglucoside and Catechin in the Presence of Acetaldehyde. *J. Agric. Food Chem.* **1995**, *43*, 1444–1449. [CrossRef]

52. Escribano-Bailon, T.; Alvarez-Garcia, M.; Rivas-Gonzalo, J.; Heredia, F.J.; Santos-Buelga, C. Color and stability of pig-ments derived from the acetaldehyde-mediated condensation between malvidin 3-O-glucoside and (+)-catechin. *J. Agric. Food Chem.* **2001**, *49*, 1213–1217. [CrossRef]

53. Drinkine, J.; Lopes, P.; Kennedy, J.A.; Teissedre, P.-L.; Saucier, C. Ethylidene-Bridged Flavan-3-ols in Red Wine and Correlation with Wine Age. *J. Agric. Food Chem.* **2007**, *55*, 6292–6299. [CrossRef] [PubMed]

54. Dallas, C.; Ricardo-Da-Silva, A.J.M.; Laureano, O. Products Formed in Model Wine Solutions Involving Anthocyanins, Procyanidin B2, and Acetaldehyde. *J. Agric. Food Chem.* **1996**, *44*, 2402–2407. [CrossRef]

55. Es-Safi, N.-E.; Le Guernevé, C.; Cheynier, V.; Moutounet, M. New phenolic compounds formed by evolution of (+)-catechin and glyoxylic acid in hydroalcoholic solution and their implication in color changes of grape-derived foods. *J. Agric. Food Chem.* **2000**, *48*, 4233–4240. [CrossRef]

56. Gobert, A.; Tourdot-Maréchal, R.; Morge, C.; Sparrow, C.; Liu, Y.; Quintanilla-Casas, B.; Vichi, S.; Alexandre, H. Non-Saccharomyces Yeasts Nitrogen Source Preferences: Impact on Sequential Fermentation and Wine Volatile Compounds Profile. *Front. Microbiol.* **2017**, *8*, 2175. [CrossRef]
57. Diguta, C.; Vincent, B.; Guilloux-Benatier, M.; Alexandre, H.; Rousseaux, S. PCR ITS-RFLP: A useful method for identifying filamentous fungi isolates on grapes. *Food Microbiol.* **2011**, *28*, 1145–1154. [CrossRef] [PubMed]
58. Jones, G.V.; Davis, R.E. Climate influences on grapevine phenology, grape composition, and wine production and quality for Bordeaux, France. *Am. J. Enol. Vitic.* **2000**, *51*, 249–261.
59. Nisiotou, A.; Sgouros, G.; Mallouchos, A.; Nisiotis, C.-S.; Michaelidis, C.; Tassou, C.; Banilas, G. The use of indigenous Saccharomyces cerevisiae and Starmerella bacillaris strains as a tool to create chemical complexity in local wines. *Food Res. Int.* **2018**, *111*, 498–508. [CrossRef]
60. Gibbs, J.N. A Study of the Epiphytic Growth Habit of Fomes annosus. *Ann. Bot.* **1967**, *31*, 755–774. [CrossRef]
61. Glories, Y. La couleur des vins rouges. 2e partie: Mesure, origine et interprétation. *OENO ONE* **1984**, *18*, 253. [CrossRef]
62. Waterhouse, A.L. Determination of Total Phenolics by Folin-Ciocalteu Colorimetry. In *Current Protocols in Food Analytical Chemistry*; Wiley & Sons: Hoboken, NJ, USA, 2002.
63. Lee, J.M.; Durst, R.W.; Wrolstad, R.E. Determination of total monomeric anthocyanin pigment content of fruit juices, beverages, natural colorants, and wines by the pH differential method: Collaborative study. *J. AOAC Int.* **2006**, *88*, 1269–1278. [CrossRef]

Review

The Role of Bioactive Phenolic Compounds on the Impact of Beer on Health

Roberto Ambra *, Gianni Pastore and Sabrina Lucchetti

Council for Agricultural Research and Economics, Research Centre for Food and Nutrition, 00178 Rome, Italy; giovanni.pastore@crea.gov.it (G.P.); sabrina.lucchetti@crea.gov.it (S.L.)
* Correspondence: roberto.ambra@crea.gov.it; Tel.: +39-06-51594-570

Abstract: This review reports recent knowledge on the role of ingredients (barley, hop and yeasts), including genetic factors, on the final yield of phenolic compounds in beer, and how these molecules generally affect resulting beer attributes, focusing mainly on new attempts at the enrichment of beer phenols, with fruits or cereals other than barley. An entire section is dedicated to health-related effects, analyzing the degree up to which studies, investigating phenols-related health effects of beer, have appropriately considered the contribution of alcohol (pure or spirits) intake. For such purpose, we searched Scopus.com for any kind of experimental model (in vitro, animal, human observational or intervention) using beer and considering phenols. Overall, data reported so far support the existence of the somehow additive or synergistic effects of phenols and ethanol present in beer. However, findings are inconclusive and thus deserve further animal and human studies.

Keywords: beer; phenols; alcohol; health

Citation: Roberto Ambra, Gianni Pastore and Sabrina Lucchetti The Role of Bioactive Phenolic Compounds on the Impact of Beer on Health. *Molecules* **2021**, 26, 486. https://doi.org/10.3390/molecules26020486

Academic Editor: Nardini Mirella; Chemat Farid

Received: 30 November 2020
Accepted: 15 January 2021
Published: 18 January 2021

Publisher's Note: MDPI stays neutral with regard to jurisdictional claims in published maps and institutional affiliations.

1. Introduction

Beer is a natural drink and historical evidences indicate a common use since ancient times also for medical and religious purposes [1]. Antique recipes prove widespread production back to 5000 years ago [2]. Beer is actually the most consumed alcoholic beverage in the EU and annual per capita consumption (L/year) has sharply increased in the Czech Republic (141 L), US (50–80 L) and France (33 L) [3]. Such a level of consumption has led some research to focus on the nutritional appropriateness of beer, merely considering health aspects like, for example, the intake of minerals [4] or the ability to prevent dysbiosis [5], properties also present in other beverages. Unfortunately, like wine, beer naturally contains ethanol, a well-known toxic and carcinogenic molecule [6].

Nonetheless, characteristic of beer is the high content in phenolic compounds, which are the focus of this review. The consumption of polyphenol-rich foods, like beer, is a well-accepted factor involved in the prevention of oxidative stress-associated diseases [7]. Traditionally, beer is obtained from as little as four basic ingredients: barley, hop, yeast and water. The first two ingredients naturally contain phenolics, however during beer production, these molecules undergo chemical modifications and new molecules are formed, influencing both the yield and final characteristics of a beer. Aroma, flavors, taste, astringency, body and fullness are the result of the metabolic activity of microbes on raw materials, and scientific evidences suggesting that they are all influenced by phenol content are summarized here. Moreover, this review focusses more deeply on most recent advances on the role of phenolic compounds on affecting human health status, considering how seriously researchers have tackled the effects of alcohol.

2. Main and Minor Beer Phenols

The polyphenolic composition of beers is considered as one of the quality indicators of beer processing and marketing [8]. In fact, the type and quantity of phenols influence taste, aroma and color, but also colloidal and foam stability, shortening beer's shelf-life taste (see Section 4, "Phenols and beer attributes"). Several different groups of phenolic compounds have been reported in beer, the main ones being phenolic acids and tannins, and flavones and flavonols [9]. Because of its high concentration, also thanks to high producing yeasts (see Section 5, "The role of barley, yeast and hop genetics on beer phenols"), the simple phenolic alcohol tyrosol is one of the main phenols looked at in beer, present also in alcohol-free beers [10]. Concentration is so high in certain beers, reaching that of red wine [11], that authors have hypothesized that tyrosol could represent an indirect source, through biotransformation, of the more biologically active hydroxytyrosol [12] (see Section 6, "Phenols-related health effects of beer consumption"). In alcoholic beers, both phenols possibly protect yeast from the stress generated by high levels of ethanol, a phenomenon that has been demonstrated for wine's resveratrol [13], indicating that phenols not only undergo changes during brewing, but they also direct it. Accordingly, non-alcoholic beers normally have lower phenolic content [14], supporting the existence of a correlation between phenols and alcohol concentrations. Among minor phenols, those derived from barley, for example alkylresorcinols, are a group of phenolic lipids for which in vitro antioxidant and antigenotoxic [15] and in vivo diet-induced obesity-suppressing [16] activities have been reported. Even if contribution to alkylresorcinols dietary intake appears not significant, higher amounts were reported in stout beer [11]. Other quantitatively minor phenols derived from hop, for example, xanthohumol and other prenylated flavonoids, contribute significantly to beer flavor and aromas and have antibacterial, anti-inflammatory and antioxidant properties, and phytoestrogen activity [17,18]. Prenylflavonoids are of particular interest for beer as, on the one hand, no other food sources other than hop are known and, on the other hand, they are present regardless of the fermentation method, ale or lager, even if higher concentrations were found in stout and India Pale Ale styles [11].

Despite the fact that prenylated flavonoids can last for 10 years in beer stored at room temperature [19], monophenols and flavonoids show a temperature- and time-dependent decay in beer [20,21]. This phenomenon was initially studied using radioactive isotopes that revealed that almost 65% of molecules belonging to the tannin fraction go through oxidation [22]. Later, other evidences supported the role of oxidation in the time-dependent decay of phenols in beer, also demonstrating the role of the intrinsic haze-forming ability of some phenols [23]. Meanwhile, acetaldehyde was also involved in haze formation, because of its ability to polymerize polyphenols and compromise beer's flavanols level [24]. A resolutive approach to this problem could come from the implementation of dry-conservation. It was recently reported that production of microparticles from beer through high-temperature (up to 180 °C) spray-drying, used for the development of functional food with a specific heath objective, yielded a well-accepted beverage, in terms of appearance, taste and color, that kept, up to the entire period of dry-conservation (180 days), the initial amount of total phenols (measured using the Folin–Ciocalteu method) [25]. Even if no qualitative indication of phenols was reported, the study supports the validity of spray-drying in the production of non-alcoholic, high-phenols, beer-flavored beverages (see Section 9, "Phenols in non-alcoholic and isotonic beers").

3. Phenols' Fate during Malting and Brewing

As mentioned in the introduction, beer content in phenols depends on the type of barley and hops used for production. Even if hops contain a huge amount of phenols (up to 4% of dry matter) compared to barley (up to 0.1%), on average, four fifths of beer's phenols come from malt or other mashed cereals, because of their significantly higher starting amount [26]. Phenols undergo both quantitative and qualitative changes during seed germination and brewing processes [27] (Figure 1). The germination of barley seed, i.e., malting, has been studied deeply and is preceded by seed hydration (steeping), during which phenolic content decreases due to leaching, and followed by seed-drying (kilning), during which the improved crumbliness of the grain enhances the enzymatic release of bound phenolic acids. Kilning can be performed at different temperatures, for example in special malts brewing in order to bring desirable flavors and colors [28]. At temperatures lower than 80 °C, kilning normally induces an increase in the amount of water-soluble total phenolic compounds [29], thanks to a Maillard-enzymatic release of phenols in the matrix [30] and to increased friability and extraction from the grain [31]. According to Leitao and colleagues, total phenolic content of barley (whose antioxidant contribution is mostly for ferulic and sinapic acids) increases four-fold during the transition to malt. Even if final yields depend on the malting procedures, the amount of phenolic compounds present in malt is inversely correlated with the degree of steeping and positively influenced by the germination temperature [32]. More recently, Koren and coworkers reported a 3- to 5-fold increase in the amount of total polyphenols during malting in six barley varieties, independently from the initial amounts [33].

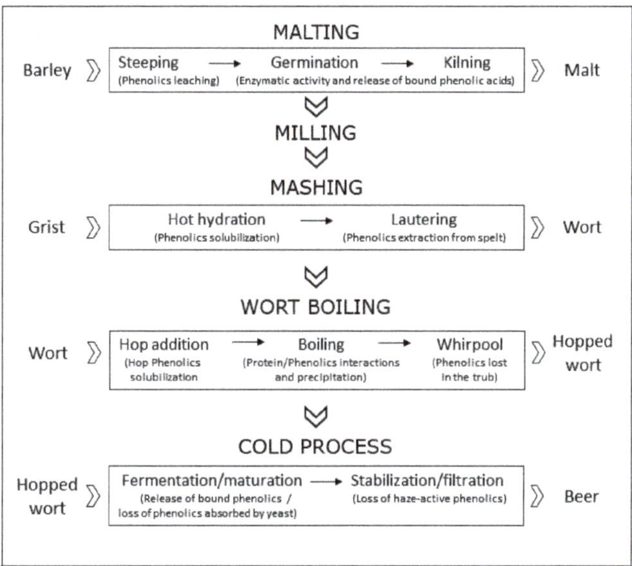

Figure 1. Phenolic compounds' fate during the phases of malting and brewing processes: in the phase of mashing, after an initial decrease, total phenolics amount increases 3- to 5-fold; afterwards, phenolics continue to increase throughout mashing and during hop addition, but dramatically decrease during wort boiling, whirpool, fermentation, maturation, stabilization and filtration, so that, during the entire brewing process, about 60% of the malt phenolic content is lost.

The amount of polyphenols reached in malt then significantly falls during brewing steps, depending on the protocol adopted, with a higher decrease for malt milled in wet conditions [34]. Enzymatic and non-enzymatic solubilization of phenols take place during the first step of mashing (hot hydration), and both are influenced by temperature and time, as well as the separation of wort, during which extraction of phenolic-rich spelt material occurs [9]. A successive increase of total phenolic compounds occurs in the wort separation (lautering) due to the extraction from spelt materials. Brewing is fundamentally ascribable to the metabolic activity of a fermentable carbohydrate source in the absence of oxygen, yielding alcohol and carbon dioxide. Fermentation is normally performed at fixed temperature but can be pushed at higher or lower temperatures. Hops, which were formerly included in the brewing process mainly for their preserving properties, are then added and wort boiling is started. Hops addition actually has several advantages, improving not only the bitter taste and astringency but giving protection to beer brewing yeasts, thanks to its antibacterial activity, against Gram-positive bacteria, and lowering pH to 4–4.2 [35]. During boiling, hop polyphenols are released and polymerization reactions with proteins occur, yielding precipitated complexes, responsible for the formation of chill haze, that are then lost in the successive whirpool process and during the final filtration and stabilization. Final processes are critical for polyphenols and include fermentation, warm rest, chill-lagering filtration and clarification [36]. During brewing, around 60% of the malt phenolic content is lost. Decay affects all phenolic compounds, excepting *p*-hydroxybenzoic acid and sinapic acid, whose concentration increases by even four-fold [31]. However, different brewing processes can deeply influence total phenolic compounds, for example bock beers are normally three times richer than dealcoholized beer, with intermediate and decreasing quantities for abbey, ale, wheat, pilsner and lager beers [36]. Recent data also indicate that beer's content in phenols is associated with the production scale. In fact, the lesser characterized craft beers (unpasteurized and unfiltered) [37], whose production scale is limited by law in several countries (200,000 hL/year in Italy), exhibit higher total phenolic compounds' values compared to large-scale beers [38], mainly thanks to the lack of filtration. Finally, the phenolic content of beer is affected negatively by higher temperature pasteurization treatments [39].

4. Phenols and Beer Attributes

The ability of phenols to influence beer taste has been well known since the early 1960s, when the so-called "sunlight flavor" was ascribed mainly to humulone and lupulone addition after beer fermentation [40]. Phenols' ability to interfere with aroma, instead, was noticed around forty years ago, thanks to a *S. cerevisiae* "killer strain" producing a clove-like aroma [41]. Later, a study clarified that presence of the main phenolic flavors relies on yeasts capability to decarboxylate or reduce phenolic acids: 4-vinylguaiacol and 4-vinylphenol from *S. cerevisiae* and 4-ethylguaiacol and 4-ethylphenol from *Brettanomyces* sp. [42]. More recent data indicate that the ability of phenols to selectively characterize beer's flavors relies on their chemical transformations. For example, thermal decarboxylation of ferulic acid to 4-vinyl guaiacol, occurring during wort boiling and during fermentation, induces a three-orders-of-magnitude increase in its flavor threshold [43]. Unfortunately, some metabolic reactions have side effects, like that involving cinnamic acid and yielding the toxicologically relevant styrene [44]. Moreover, higher concentrations of monophenol can turn spicy or vanilla-like sweet flavor notes to unpleasant medicinal-like flavors [45]. A recent deep analysis of the association between metabolites and sensory characteristics using two-way orthogonal partial least squares indicates that isoferulic acid affects beer's fruity sensory attributes [46], suggesting the possibility to predict to some extent the formation of specific flavors.

With respect to aroma, phenols' protecting properties were found almost 25 years ago: phenols were found to prevent the formation of off-flavors, before and during malting, and the phenomenon was ascribed to their antioxidant activity in barley and malt [47]. More recently, some specific monophenols that confer the typical aroma of some popular beers were identified [48] and recently reviewed [49]. Worthy of interest are Czech beers whose distribution of individual phenolic compounds, that has been brought back to the origin of raw materials and the technology used for processing, is so unique that they have been proposed for authenticity analysis [50,51]. With respect to color, after high-affinity selective removing of tannins, Dadic and Van Gheluwe observed a severe discoloration of beer, demonstrating for the first time the correlation between phenols and beer color [52]. The involvement of monoflavanols' oxidation on beer color was further demonstrated by the recovery of oxidized molecules in polyethylene terephthalate bottle-stored beer [20]. More recently, several works have clearly demonstrated the relationship between phenols and beer color, both in small- and large-scale brewed beers [38].

Barley seeds' phenolic acids, flavonoids and proanthocyanidins influence quality indexes like viscosity, diastatic power and nitrogen content [53], and have an impact on beer turbidity [54], taste, bitterness and aroma [55]. With regard to hop, which was antiquely added in beer especially for its pleasant aroma and bitterness, brewing trials indicate that hop phenols can selectively reduce flavor deterioration during storage [56], specifically the sunstruck off-flavor that is formed in beer upon light exposure [57]. More recent data indicate a temporal effect. In fact, later addition of hop, just before the end of wort boiling, significantly increases phenolic content [58]. Astringency, bitterness and fullness, which are affected by the boiling time [39], have been linked to different hop phenols fractions [59,60].

5. The Role of Barley, Yeast and Hop Genetics on Beer Phenols

The yield in phenols of a beer necessarily depends on the genetic background of its raw ingredients, and differences were reported in barley grain [61], hop [62] and yeast [63]. Unfortunately, domestication of barley and hop has reduced phenols' diversity. Nevertheless, total polyphenol content could be linked to specific quantitative trait loci in barley [64] and some specific combinations of phenols in barley can still be attributed to different genotypes. For example, the ratio between barley's main phenolic acids, ferulic acid and *p*-coumaric acid, is genetically determined and combinations can also influence key agronomic traits, such as hull adherence and grain color [65], through functionally related genes [53]. Studies combining genetics and environment on wild barley cultivars, that show a wider genetic diversity in agronomic traits and abiotic stress tolerance, identified some genes involved in phenol accumulation in barley seeds. Such studies are of special relevance as they can give a picture of the loss of genetic variation due to domestication and provide information for the set-up of breeding applications for phenols-related beer improvement. For example, a network analysis of gene expression and secondary metabolites, induced by the well-known stressor drought [66] in developing grains from several different Tibetan wild barley cultivars, recently allowed the identification of genes whose manipulation is believed to help the development of cultivars with specific contents of phenolic compounds [67]. Less data is available for a role of the genetic background on hop phenols. For example, a significant cultivar-dependent role has been recently reported for 2-phenylethyl glucoside [68], but the relevance on final quantities recoverable in beer is still lacking.

The ability of yeasts to adapt to different chemical (sugar, nitrogen) and physical (temperature, pH, oxygen, sulfur dioxide) properties resides in the great genetic diversity that has been exploited by the beer industry, i.e., for the development of strains with distinct flavor profiles. The production of different metabolites, like volatile phenols, is the direct consequence of human influence through wine and beer production. A first evidence testifying the role of the genetic background of yeasts in beer phenols came from the observation, at the beginning of the twentieth century, of volatile "ethereal substances" in English stock ales, during fermentation by Brettanomyces [69]. *Brettanomyces bruxellensis*, the first microorganism to be patented for beer production, was also involved in the spoilage of draught beer [70] and in the clove off-flavor (the ethylphenol 4-vinylguaiacol) [71] but, after being reported together with *Lactobacillus vini* as a contaminant in several ethanol-producing plants [72], was finally isolated from a number of fermented beverages and food, from cider to olives [73]. Spoilage depends on a still not fully identified gene pathway that involves two phenylacrylic acid decarboxylase (PAD) enzymes [74]. Ethylphenols production has been related to strain-dependent PAD amino acid sequence variability [75]. Thanks to their ability to convert ferulic acid to 4-vinylguaiacol, yeasts are believed to have a stronger impact on phenols than thermal processing steps [76]. Yeasts also have a fundamental impact in barrel beer ageing. Barrel-aged beers are sensorially enriched beers obtained by storage of already fermented beers in wood casks or by fermentation of beer's wort directly in wood barrels. Such processes mainly occur because of the spontaneous growth of microbes present in breweries' atmosphere and in barrels [77]. During this fermentative incubation, a bi-directional exchange of different molecules occurs from wood and beer: some beer's molecules are retained by the wood while others are released from wood to the beverage. *Dekkera bruxellensis*, another spoilage-related microbe in wine, is considered the main contributor to the aroma of aged beers, through its ability to convert hydroxycinnamic acids to volatile phenols, and has several advantages, from high ethanol yield to low pH tolerance [78]. Its spontaneous growth is accompanied by some enzymatic activities that transform wort composition and yield the final chemical and sensory profiles of aged beer.

Aiming at finding optimal conditions for accelerating wort transformations, research is focused at finding optimal chemical conditions to produce beers with specific and preferred bacterial metabolites, normally avoiding those from non-Saccharomyces species, in multi-starter cultures. For such purpose, Coelho and coworkers recently found that low glucose or high ethanol conditions favor the yield of *D. bruxellensis*-related metabolites over *S. cerevisiae* ones [79]. Ethanol-resistance and increased dominance towards other *S. cerevisiae* strains were also reported on mixed starter fermentations for the high polyphenols-producing *S. cerevisiae* var. *boulardii* strain [80]. A recent deep genomes/phenomes analysis involving 157 industrial *S. cerevisiae* strains [81] reported that production of 4-vinylguaiacol relies on specific genetic variants able to ferment maltotriose [81]. More recently, next-generation sequencing allowed the identification of a Brettanomyces strain void of phenolic off-flavors, limiting economic losses during production [82], a problem that was bypassed in *S. cerevisiae* by the selection of strains with inactivated alleles and/or functional copies [83]. Worth mentioning is a recent work that, seeking to explain different adaptive abilities, profiled microsatellite markers and ploidy-states of 1488 isolates coming from niches dispersed all over the world [84].

6. Phenols-Related Health Effects of Beer Consumption

While the serious damages of high alcohol intake are known, the effects of moderate consumption of alcoholic beverages are still a source of heated debate. Moderate beer consumption is believed to be associated with protective cardiovascular function and reduction in the development of neurodegenerative disease. Moreover, there is no evidence that moderate beer consumption can stimulate cancer. Nevertheless, alcohol consumption can become a problem for people at high risk of developing alcohol-related cancer or for those affected by cardiomyopathy, cardiac arrhythmia, depression, liver and pancreatic diseases, and is not recommended for children, adolescents, pregnant women and frail people at risk of alcoholism [85]. Anyway, beer, like wine, contains the already mentioned substances with indubitable protective capacities, not merely anti-inflammatory and antioxidant, as demonstrated by huge in vitro work on single substances [86]. However, the ambitious objective in studying the effects of beer consumption on human health is to analyze it in toto and, in order to understand the single contribution of phenols and alcohol, parallel experiments with similar doses of an equivalent non-alcoholic beer and of alcohol alone are essential. For example, Karatzi and coworkers [87] reported that both non-alcoholic and alcoholic beers improved some arterial biomarkers (reduced aortic stiffness and increased pulse pressure amplification), but the effects were also similar in a parallel vodka intervention, containing the same amount of ethanol as the alcoholic beer. However, as some other effects (wave reflections reduction) were higher in the alcoholic beer intervention compared to alcohol alone (vodka), and the endothelial function was significantly improved only after beer consumption, the authors concluded that the non-alcoholic and the alcoholic fractions of beer could have additive or synergistic effects [87].

We thus thought to analyze the fraction of similar publications that considered, in the search of the health effects of beer containing phenols, also the effects of the presence of alcohol. For such purpose, we used in Scopus.com the search string TITLE-ABS-KEY (beer AND (phenol OR polyphenol OR flavonoid) AND (observational OR administration OR consumption OR drinking OR prospective OR intervention OR crossover OR trial)) AND (LIMIT-TO (DOCTYPE, "ar")). The search was performed on October 2020 and returned 161 documents, including 31 reviews (even if they were already excluded by the string search), 7 not pertinent articles, 9 studies merely evaluating phenols' population intakes, 51 chemical-only reports (papers reporting chemical analyses of phenols of commercial or improved beers) and 22 reports using only single phenols in in vitro or in vivo models. For the remaining 41 (minus one not available even by the authors themselves [88]), experimental models, parameters tested and main findings are summarized in the next section and sorted chronologically by the most recent, in Tables 1–3, about in vitro and animal models, human intervention and human observational, highlighting the use of alcohol alone (spirits, eventually vodka or gin), as well as non-alcoholic beer.

6.1. In Vitro and Animal Experiments

As demonstrated by in vitro cancer cell models (Table 1), several cancer types are sensitive to the antiproliferative action of some beer components, including ethanol. For example, epithelial cells' viability was reduced in a similar way by beer or an equivalent amount of ethanol [89]. Unfortunately, the authors did not test an alcohol-free beer. Using single molecules or a matrix containing all beer components, Machado and coworkers showed that phenols' activities are synergic [90]. Unfortunately, in this case, ethanol was not tested. Similarly, a total extract obtained from dark beer conferred higher protection to rat C6 glioma and human SH-SY5Y neuroblastoma cells against an oxidant stressor challenge (hydrogen peroxide) compared total extracts obtained from non-alcoholic and lager beers [91]. Again, neither the phenolic compounds of beers nor an alcoholic reconstituted extract were tested.

Table 1. In vitro and animal studies.

Experimental Model	Tested Parameters	Observations	Non-Alcoholic Beer	Alcoholic Beer	Ethanol	References
in vitro, rat C6 glioma and human SH-SY5Y neuroblastoma cells, treated with total extracts from dark, non-alcoholic or lager beers	cell viability and adenosine receptors gene expression and protein levels following oxidant stressor (hydrogen peroxide) challenge	alcoholic dark beer extract conferred higher protection compared to lager or non-alcoholic beer extracts	yes	no	no	[91]
animal, 36 prepuberal Wistar rats fed with beer or ethanol (both 10%) or water for 2/4 weeks	plasma reproductive hormones, cleaved caspase-3 immunolocalization and neuronal nitric oxide synthase level in Leydig cells	beer decreased sex hormones compared to ethanol or water rats and inhibited ethanol-induced increase of cleaved caspase-3	no	yes	yes	[92]
animal, 70 male Wistar rats, with monocrotaline-induced pulmonary arterial hypertension, fed with xanthohumol-fortified beer or ethanol (both 5.2%) for 4 weeks	cardiopulmonary exercise testing and hemodynamic recordings, analysis of pulmonary vascular remodeling and cardiac function	xanthohumol-fortified beer attenuated pharmacologically induced pulmonary vascular remodeling and improved cardiac function, compared to ethanol rats	no	yes	yes	[93]
animal, 40 male Wistar rats, with aluminium nitrate-induced inflammatory status, fed with low alcoholic-beer (0.9%) or hops or silicons for 3 months	animal behavior and brain antioxidant and anti-inflammatory status	non-alcoholic beer, but also silicon and hops alone, prevented aluminum-induced inflammation and neurodegenerative effects	yes	no	no	[94]
animal, 30 male Wistar rats, with streptozotocin-induced diabetes, fed with alcoholic beer or xanthohumol-enriched or 5% ethanol for 5 weeks	hepatic glucolipid metabolism, levels lipogenic enzymes and glucose transporter 2	alcoholic beer enriched with xanthohumol (but not normal beer nor ethanol) prevented the streptozotocin-induced liver catabolic state alterations	no	yes	yes	[95]

Table 1. *Cont.*

Experimental Model	Tested Parameters	Observations	Non-Alcoholic Beer	Alcoholic Beer	Ethanol	References
animal, 30 male Wistar rats, with skin induced wound healing and streptozotocin-induced diabetes, fed with alcoholic beer or xanthohumol-enriched or 5% ethanol for 5 weeks	effects on wound healing, through evaluation of angiogenesis, inflammation and oxidative stress modulation	alcoholic and xanthohumol-enriched beers respectively, prevented and reversed the alcohol-induced markers of inflammation, oxidative stress and angiogenesis	no	yes	yes	[96]
animal, 24 male Wistar rats, with skin induced wound healing, fed with xanthohumol-fortified alcoholic beer or 5% ethanol for 4 weeks	angiogenesis and inflammation markers (serum vascular endothelial growth factor levels, N-acetylglucosaminidase activity, Interleukin-1 β concentration)	alcoholic and xanthohumol-enriched beer respectively, prevented and reversed the alcohol-induced markers of inflammation, oxidative stress and angiogenesis	no	yes	yes	[97]
in vitro, MKN-28 gastric epithelial cells, treated with different alcoholic beverages, at a similar ethanol concentration	tetrazolium (MTT) assay at 30, 60 and 120 min	alcoholic beer reduced cell viability like ethanol, while red wine, even dealcoholated, protected	no	yes	yes	[89]
animal, 32 spontaneously hypertensive and 32 normotensive Wister rats, fed intragastrically with lyophilized beer for 10 days	aminooxyacetic acid-induced γ-aminobutyric acid (GABA) accumulation in hypothalamus and pons-medulla	lyophilized beer decreased GABA accumulation	yes	no	no	[98]
animal, 36 male Wistar rats fed (4 weeks) with lyophilized, polyphenol-free, beer or white wine	plasma lipids and lipid peroxides	polyphenol-free beer (not polyphenol-free wine) significantly decreased lipids and lipid peroxides	no	yes	no	[99]
animal, 60 Wistar rats fed (4 weeks) with alcoholic (4%) or lyophilized beer	plasma lipids and lipid peroxides	both alcoholic and lyophilized beers similarly decreased lipids and lipid peroxides	yes	yes	no	[100]

Wistar rats were used in several experiments with beer. One publication reported that both administration of alcoholic (4%) or lyophilized beer for 4 weeks had low, but statistically significant, beneficial effects on plasma lipidemic and antioxidant markers (total cholesterol, low-density lipoprotein (LDL) cholesterol, triglycerides and lipid peroxides), however alcohol alone was not tested and the authors themselves concluded that minimal effects observed could rely on relatively low alcoholic content of beer [100]. Next, using only a polyphenol-free beer, the same group concluded that lipid effects had to be ascribed to beer proteins, as long as effects were absent in rats fed with polyphenol-free wine [99]. In rats with skin incision-induced wound healing, feeding for 4 weeks with alcoholic beer prevented alcohol-induced markers of inflammation, oxidative stress and angiogenesis [97]. Notably, when beer was enriched with 10 mg of xanthohumol, effects were even more ameliorated. Similar results were obtained using animals with streptozotocin-induced diabetes [96]. On the same streptozotocin-induced diabetes model, hepatic glucolipid metabolism, lipogenic enzymes and glucose transporter 2 levels were tested after 5 weeks of administration of xanthohumol-enriched alcoholic beer for 5 weeks [95]. Interestingly, beer prevented all the streptozotocin-induced liver catabolic state alterations tested (fibrosis, apoptosis, glycogen depletion, GLUT2 upregulation, lipogenesis reduction) and the effect was not observed in rats fed with normal beer. The authors also tested the effect of ethanol alone but, in none of these last three works were an alcohol-free beer, nor xanthohumol alone, tested, thus it is impossible to distinguish neither the effect of beer components nor of the polyphenol itself. Furthermore, in vitro and in vivo work on xanthohumol metabolites (isoxanthohumol and 8-prenylnaringenin) previously indicated opposite effects on angiogenesis and inflammation processes (pro-angiogenetic for 8-prenylnaringenin and anti-angiogenic and anti-inflammatory for the other two) [101]. Nevertheless, a xanthohumol-fortified alcoholic beer was used again to demonstrate attenuated pharmacologically induced pulmonary vascular remodeling and improved cardiac function [93]. Also, in this case, even if effects were absent in rats fed only with ethanol, no rats were tested with an alcohol-free beer. It is noteworthy that the authors could identify the involvement of extracellular signal-regulated kinase1/2, phosphatidylinositol 3-kinase/protein kinase B and VEGF receptor 2 in the protective properties of beer towards pulmonary arterial hypertension [93]. In a prepubertal rat model, beer with 10% alcohol significantly decreased, after 4 weeks, the levels of sex hormones, compared to ethanol- or water-fed rats [92]. Again, even if authors concluded that beer inhibited the ethanol-induced increase of cleaved caspase-3 in Leydig cells, a non-alcoholic beer was not tested. In addition to the works recovered using the Scopus.com search string and mentioned in Table 1, worthy of mention are experiments showing that alcoholic-free beer can decrease the aminooxyacetic acid-induced GABA accumulation in hypertensive animals [98], and prevent brain inflammation and neurodegenerative effects induced by aluminum nitrate [94]. However, while as expected hops administration alone had a beer-overlapping positive effects to some extent, so did silicon administration, reinforcing the need for an appropriate set-up of experimental models.

6.2. Role of Alcohol on Phenols' Metabolism and Beer Antioxidant and Anti-Inflammatory Properties, and on Cardiovascular-Related Effects

Phenolic acids' absorption, previously reported both in low-alcohol [102] and alcoholic beer [103], is impaired by ethanol removal from beer [104]. The opposite effect of alcohol has been reported for tyrosol metabolization to hydroxytyrosol following beer consumption, as mentioned above. In particular, the administration of a single dose of 250 mL of blonde beer was associated to higher urinary recovery of tyrosol, whilst an identical dose of alcohol-free beer yielded higher urinary recovery of hydroxytyrosol [12]. However, as alcohol consumption proportionally increases hydroxytyrosol excretion through dopamine metabolism [105], hydroxytyrosol bioavailability is hardly attributable only to beer phenols.

Among first beer intervention studies (Table 2), there is an almost-perfectly set-up randomized acute administration of either 4.5% alcoholic beer ($n = 14$), or dealcoholized beer or 4.5% water solution of ethanol ($n = 7$), for the evaluation of the contribution of beer's alcohol [104]. Results demonstrated that a significant increase in plasma antioxidant capacity (TRAP) could be obtained only following alcoholic beer administration. Unfortunately, no crossover intervention was performed, and the effects were studied only in a temporally limited manner. In another similar, but a crossover, acute intervention of beer or wine (or vodka for the evaluation of the contribution of alcohol) inhibition of oxidative stress induced (by 100% normobaric O_2 breathing) was tested [106]. Analysis of stiffness 3 h after administration showed that only wine prevented oxygen-induced oxidative stress, possibly because of the higher content of polyphenols compared to beer i.e., 2.6 g/L vs. 0.4 g/L gallic acid equivalents (GAE) [106]. No one can say if such a low phenols amount in an equivalent alcohol-free beer could have produced the effects observed with wine. Daily supplementation of breastfeeding mothers ($n = 30$) with 660 mL of non-alcoholic beer was associated with an improvement of mothers' plasma and breastmilk antioxidant capacities, assessed 30 days postpartum, compared to control non-supplemented mothers [107]. For obvious reasons, an alcoholic beer was not tested. Administration of alcohol-free beer

(500 mL) for 45 days to postmenopausal women ($n = 29$) was associated with a reduction of several indicators of early protein oxidation, especially reducing cholesterol levels in subjects with higher than 240 mg/dL [108], supporting the usefulness of long-term alcohol-free beer consumption in fighting low-grade chronic inflammation and preventing metabolic disorders. As an alcoholic beer was not tested, one might speculate that alcohol can abolish the beneficial effect. However, previous work that used a crossover intervention trial (healthy drinkers, $n = 27$) to switch consumption of beers with similar phenolic content (310–330 mg/L) for 4 weeks, from low (0.9%) to high (4.9%) alcohol and vice versa, indicates that while the switch to low-alcohol did not change in vitro LDL oxidizability, the opposite switch did [109]. On the other hand, only non-alcoholic beer daily consumption for one week (17 healthy females, 330 mL) was associated to an increase in the urinary antioxidant capacity, as measured by Trolox equivalents [110], contradicting the results of the study reported at the beginning of this paragraph.

Table 2. Intervention studies (n, subjects' number; y, age (years)).

Experimental Model	Tested Parameters	Observations	Non-Alcoholic Beer	Alcoholic Beer	Ethanol	References
intervention trial (healthy adults), 30 days, 355 mL beer/day with (4.9%, $n = 33$, 21–55 y) or without alcohol (0.5%, $n = 35$, 21–53 y)	microbiota composition, fasting blood serum glucose, β-cell function	both beer interventions increased microbiota diversity, but only non-alcoholic beer increased heathier diversity and β-cells function and decreased fasting blood serum glucose	yes	yes	no	[111]
controlled clinical trial (healthy adults, $n = 20$, 18–45 y, single blind, randomized, crossover), single dose of beer (250 mL), with (4.5 or 8,5%) or without (0%) alcohol	urinary tyrosol (TYR) and hydroxytyrosol (HT)	non-alcoholic beer intervention increased HT recovery (and reduced TYR recovery) compared to alcoholic beer	yes	yes	no	[12]
intervention controlled trial (high cardiovascular risk males, $n = 33$, 55–75 y, open, randomized, crossover), 4 weeks, daily: 660 mL beer (1029 mg polyphenols and 30 g ethanol) or 990 mL non-alcoholic beer (1243 mg polyphenols and <1 g ethanol) or 100 mL gin (30 g ethanol)	urinary metabolomics	both beer intervention increased to similar extent urine excretion of hop α-acids and fermentation products, compared to gin intervention	yes	yes	yes	[112]
intervention controlled trial (high cardiovascular risk males, $n = 33$, 55–75 y, open, randomized, crossover), 4 weeks, daily: 660 mL beer (1029 mg polyphenols and 30 g ethanol) or 990 mL non-alcoholic beer (1243 mg polyphenols and <1 g ethanol) or 100 mL gin (30 g ethanol)	atherosclerotic and inflammation plasma biomarkers and peripheral blood mononuclear cells immunophenotyping	only non-alcoholic beer intervention reduced leukocyte adhesion molecules and inflammatory biomarkers, but alcoholic beer and gin interventions improved plasma lipid and atherosclerosis inflammatory markers	yes	yes	yes	[113]

Table 2. *Cont.*

Experimental Model	Tested Parameters	Observations	Non-Alcoholic Beer	Alcoholic Beer	Ethanol	References
intervention controlled trial (high cardiovascular risk males, $n = 33$, 55–75 y, open, randomized, crossover), 4 weeks, daily: 660 mL beer (1029 mg polyphenols and 30 g ethanol) or 990 mL non-alcoholic beer (1243 mg polyphenols and <1 g ethanol) or 100 mL gin (30 g ethanol)	number of circulating endothelial progenitor cells (EPC)	8-fold and 5-fold increases of EPC number respectively in alcoholic and non-alcoholic beer interventions and statistically not significant 5-fold decrease in gin administration	yes	yes	yes	[114]
intervention controlled trial (high cardiovascular risk males, $n = 33$, 55–75 y, open, randomized, crossover), 4 weeks, daily: 660 mL beer (1029 mg polyphenols and 30 g ethanol) or 990 mL non-alcoholic beer (1243 mg polyphenols and <1 g ethanol) or 100 mL gin (30 g ethanol)	urinary isoxanthohumol	beer administrations (not gin) induced similar excretion of urinary isoxanthohumol	yes	yes	yes	[115]
intervention trial (stressed healthy females, $n = 17$, 40.9 ± 10.5 y, randomized, crossover), 2 weeks 330 mL beer/day, first week non-alcoholic, second week alcoholic	antioxidant capacity in urine	non-alcoholic beer administration induced higher antioxidant capacity compared to alcoholic beer one	yes	yes	no	[110]
intervention trial (healthy males $n = 17$, 28.5 ± 5.2 y, randomized, single-blind, crossover), single dose (800 mL) beer (48 mg polyphenols and 20 g ethanol) or non-alcoholic beer (48 mg polyphenols) or vodka (20 g ethanol)	endothelial function, aortic stiffness, pressure wave reflections and aortic pressure	non-alcoholic and alcoholic beer interventions improved (similarly) arterial biomarkers but the effects were observed also for the vodka intervention alcoholic beer intervention improved wave reflections reduction better than vodka intervention	yes	yes	yes	[87]
intervention trial (postpartum breastfeeding-mother-infants dyads), 30 days 660 mL/day non-alcoholic beer ($n = 30$, 30 ± 5 y) or not ($n = 30$, 31 ± 3 y)	breastmilk, plasma and urine oxidative status	non-alcoholic beer increased breastmilk and plasma antioxidant capacities	yes	no	no	[107]
intervention trial (healthy male marathon runners, double-blind), 5 weeks (from 3 before to 2 after marathon) 1.0–1.5 L non-alcoholic beer ($n = 142$, 36–51 y) or control beverage without polyphenols ($n = 135$, 35–49)	blood inflammatory markers and upper respiratory tract illness (URTI) rates	non-alcoholic beer intervention reduced after-run blood inflammatory markers and URTI rates, compared to the polyphenols-free beverage	yes	no	no	[116]

Table 2. *Cont.*

Experimental Model	Tested Parameters	Observations	Non-Alcoholic Beer	Alcoholic Beer	Ethanol	References
intervention trial (healthy males, $n = 10$, 21–29 y, randomized, single-blind, crossover), single dose (7 mL/kg body wt) alcoholic beer (0.4 g/L GAE polyphenols and 0.32 g ethanol/kg body wt) or vodka (0.32 g ethanol/kg body wt)	plasma lipid peroxides, uric acid concentration and arterial stiffness following 100% O_2 breathing-oxidative stress	alcoholic beer intervention protected against oxygen-induced increase in arterial stiffness but so did vodka	no	yes	yes	[106]
intervention (post-menopausal healthy females, $n = 29$, 64.5 ± 5.3 y, longitudinal), 45 days 500 mL alcoholic-free beer/day	lipid profile and plasma inflammatory markers	alcoholic-free beer intervention improved lipid profile and plasma inflammatory markers	yes	no	no	[108]
controlled clinical trial (hypercholesterolemic non-drinker males, $n = 42$, 43–71 y, randomized, single-blind), 30 days, daily: 330 mL 5.4% beer (20 g alcohol and 510 mg polyphenols) or water (containing beer mineral)	coronary atherosclerosis plasma markers	alcoholic beer intervention improved coronary atherosclerosis plasma markers compared to control administration water	no	yes	no	[117]
intervention (healthy adults, $n = 10$, 25–45 y, randomized), single dose (500 mL) 4.5% alcoholic beer	phenolic acids plasma metabolites	alcoholic beer intervention demonstrates absorption and metabolism of phenolic acids to glucuronide and sulfate conjugates	no	yes	no	[103]
intervention (healthy normotensive drinking men, $n = 28$, 20–65 y, randomized, crossover), 4 weeks, daily: 1125 mL 4.6% beer (41 g alcohol) or 375 mL 13% red wine 2023 mg/L polyphenols) or 375 mL dealcoholized red wine (2094 mg/L polyphenols)	blood pressure and vascular function following brachial artery flow-mediated and glyceryl trinitrate-mediated dilatation	alcoholic beer (but also wine) increased awake systolic blood pressure and asleep heart rate	no	yes	no	[118]
intervention (healthy adults, 25–45 y, randomized no crossover), single dose (500 mL): 4.5% alcoholic ($n = 14$) or dealcoholized beer or 4.5% ethanol ($n = 7$)	total plasma antioxidant status	alcoholic beer administration improved higher plasma antioxidant capacity compared to the dealcoholized one, thanks to higher absorption of phenolic acids	yes	yes	yes	[104]
intervention (healthy males, $n = 5$, 23–40 y), single dose (4 L) low-alcohol (1%) beer	urinary ferulic and its glucuronide	beer administration demonstrates bioavailability of ferulic acid	yes	no	no	[102]
intervention trial (healthy male drinkers, $n = 27$, 49.2 ± 2.3 y, randomized, crossover), 4 weeks, daily 375 mL: 4.9% or 0.9% beer (similar phenolic content 310–330 mg/L)	LDL in vitro oxidizability and characterization	switch from low to high alcoholic beer intervention increased LDL oxidizability	yes	yes	no	[109]

One observational study (1604 subjects of the IMMIDIET (Dietary Habit Profile in European Communities with Different Risk of Myocardial Infarction: the Impact of Migration as a Model of Gene-Environment Interaction) study, 26–65 years, see Table 3) supports a somewhat interfering property of alcohol on non-alcoholic components of beer. In fact, adjustment of beer intake for alcohol content broke the association between beer consumption and higher plasma and red blood cell omega 3 fatty acids [119]. In the overweight or class 1 obese healthy subjects, the daily consumption of alcoholic beer (but not of alcoholic-free beer with similar amount of total phenols) for four weeks raised HDL levels in subjects with low LDL-lipid profile and facilitated cholesterol efflux from macrophages, without affecting body mass index (BMI), liver and kidney functions, potentially reducing the risk of vessels occlusion by cholesterol deposition [120]. As the consumption of alcohol alone was not tested, it is not possible to exclude that the effects could be at least partially ascribable to alcohol. Similarly, in a crossover study of 28 daily healthy nonsmoking normotensive men consuming alcoholic beer (1125 mL; 41 g alcohol) for 4 weeks, an increase of the awake systolic blood pressure and the asleep heart rate was reported, however the effects were identical in men consuming red wine containing the same amount of alcohol [118], and an alcohol-free beer was not tested. Similarly, analysis of stiffness, 3 h after administration of alcoholic beer or vodka, showed that both protected against oxygen-induced increase in arterial stiffness, making the authors conclude that the observation was probably due to a central vasodilatatory effect of alcohol itself [106]. Again, Gorinstein and coworkers found that alcoholic beer consumption (330 mL daily, containing 510 mg of polyphenols and 20 g of alcohol for 30 days) ameliorated markers of coronary atherosclerosis of hypercholesterolemic in non-drinker males (n = 42, 43–71 years) during recovery from coronary bypass surgery [117]. Unfortunately, the control group of the randomized single-blind trial had only water "with minerals of beer", making it impossible to ascribe effects to either phenols or to alcohol. In a double-blind intervention of healthy male runners (n = 277), daily consumption of non-alcoholic beer, for 3 weeks before and 2 weeks after a marathon, reduced interleukin-6 immediately after the race, total blood leukocyte counts immediately and 24 h after the race and post-marathon incidence of upper respiratory tract illness [116]. However, like for breastfeeding mothers mentioned above, alcoholic beer was not tested, we guess for similar obvious reasons. Also, other observational studies (Table 3) suffer from this limitation. For example, a significant inverse association between beer consumption (and not for coffee, nuts, tea, olive oil and red or white wine) and hypertension was found by means of food frequency questionnaires submitted to 2044 adults [121], however neither the consumption of alcohol-free beer nor the contribution of pure alcoholic beverages were evaluated.

Table 3. Observational studies (n, subjects' number; y, age (years)).

Experimental Model	Tested Parameters	Observations	Non-Alcoholic Beer	Alcoholic Beer	Ethanol	References
observational (ALMICROBHOL adults n = 78, 25–50 y), alcoholics BCQ	microbiota composition (16S rRNA sequencing) and short chain fatty acid profile in fecal samples	higher butyric acid concentration and gut microbial diversity in consumers vs. non-consumers of beer	no	yes	no	[122]
observational (TwinsUK females n = 916, 16–98 y), alcoholics FFQ	microbiota composition in fecal samples (16S rRNA sequencing)	no association between beer (nor all alcohols except wine) consumption and gut microbial diversity	no	yes	yes	[123]
observational (MEAL Southern Italy adults, n = 2044, >18 y), phenolics FFQ	hypertension (arterial blood pressure measurement)	inverse association between beer consumption and hypertension	no	yes	no	[121]

Table 3. *Cont.*

Experimental Model	Tested Parameters	Observations	Non-Alcoholic Beer	Alcoholic Beer	Ethanol	References
observational prospective (2002–2003 CMHS Californian males n = 84,170, 45–69 y), alcoholics FFQ	prostate cancer registries (Surveillance Epidemiology and End Result)	no association between beer (nor wine nor liquor) consumption and prostate cancer	no	yes	yes	[124]
observational cross-sectional (IMMIDIET Italy–Belgium–UK female–male pairs, n = 1604, 26–65 y), alcoholics FFQ	plasma and red blood cell omega–3 fatty acids	no association between beer consumption and plasma or red blood cell omega 3 fatty acids (reduced for wine)	no	yes	no	[119]
observational 34 year prospective (PPSWG Sweden females, n = 1462; 38, 46, 50, 54, 60 y), alcoholics BCQ	dementia (neuropsychiatric years-repeated examinations)	direct association between beer (or wine) consumption and longevity and reduced dementia risk (compared to subjects consuming only spirits)	no	yes	yes	[125]
observational (over 10 years) case-control matched leukoplakia subjects (n = 187 + 187, 40–65 y), alcoholics FFQ	leukoplakia (clinical examination and biopsy)	no significant association between moderate beer drinking and leukoplakia risk (increased for spirit and reduced for wine)	no	yes	yes	[126]
observational case-control prospective (1987–2004 IWHS) diabetes postmenopausal females (n = 35,816; 55–69 y), flavonoids FFQ	self-reported diabetes	inverse association between beer (or other alcoholic beverages including liquor) consumption and diabetes risk	no	yes	yes	[127]
observational oral cancer mortality rate (20 Nations male 2002 age-standardized), national mean alcoholic beverage consumption	oral cancer mortality rates (International Agency for Research on Cancer)	no association between beer (nor wine, but association for spirits) consumption alone and oral cancer risk	no	yes	yes	[128]
observational case-control matched (1993–1996 King County, WA) prostate cancer subjects (n = 753 + 703; 40–64 y), alcoholics BCQ	prostate cancer registry (Seattle Puget Sound Surveillance Epidemiology and End Results Cancer Registry), histological confirmation	no association for beer consumption (nor liquor but association for wine) and prostate cancer risk	no	yes	yes	[129]
observational case-control matched prospective (1980–1993 Québec) child acute lymphoblastic leukemia (n = 491 + 491; 0–9 y), parents alcoholics BCQ	child acute lymphoblastic leukemia hospital diagnosis	inverse association between mothers' beer (but not spirits) consumption and child acute lymphoblastic leukemia	no	yes	yes	[130]

FFQ, food frequency questionnaires; BCQ, beverage consumption questionnaires; ALMICROBHOL, Effects of Alcohol Consumption on Gut Microbiota Composition in Adults; TwinsUK, UK Adult Twin Registry; MEAL, Mediterranean healthy Eating, Ageing, and Lifestyle; CMHS, California Men's Health Study; IMMIDIET, Dietary Habit Profile in European Communities with Different Risk of Myocardial Infarction: the Impact of Migration as a Model of Gene-Environment Interaction; PPSWG, The Prospective Study of Women in Gothenburg; IWHS, Iowa Women's Health Study.

An open, randomized, crossover, finely set-up controlled intervention trial of 33 high-cardiovascular risk males drinking daily, for 4 weeks, a non-alcoholic beer (containing

1243 mg of total polyphenols) or an alcoholic beer (containing 1209 mg of total polyphenol and 30 g of ethanol), was repeatedly used (apparently with the same composition of subjects) by a group of Spanish researchers during the last 6 years, to investigate the possible synergistic effects of beer polyphenols and alcohol, using as control an administration of gin (containing 30 g of ethanol). Firstly, in an attempt to use urinary isoxanthohumol as a marker of beer consumption, a similar amount of the metabolite was recovered following non-alcoholic or alcoholic beer consumption, and no excretion was found following gin administration [115]. Notably, group differences in a female sub-population were found, but only an alcoholic beer was tested. Next, they looked for circulating endothelial progenitor cells (EPC) and reported that non-alcoholic beer consumption increased the number of circulating EPCs by 5 units, while in the alcoholic beer group, the increase was 8-fold. However, even if observations were not statistically significant, alcohol alone (gin) induced a 5-fold decrease in the number of circulating EPCs [114], suggesting the existence of some influencing, maybe genetic, factors. Then, they reported that only non-alcoholic beer consumption reduced leukocyte adhesion molecules and inflammatory biomarkers (decreased homocysteine and increased serum folic acid) [113], suggesting a possible antagonistic effect between alcohol and the non-alcoholic fraction of beer. Importantly, the alcoholic beer improved other plasma lipid and inflammation markers (high-density lipoprotein cholesterol, apolipoproteins A1 and A2, and adiponectin) and decreased fibrinogen and interleukin 5, but the effects were ascribed to alcohol as identical effects were observed following administration of a gin dose containing the same amount of alcohol (30 g). Finally, the group of Spanish researchers applied liquid chromatography-coupled Linear Trap Quadropole-Orbitrap mass spectrometry to discover the urinary metabolites produced in the intervention study. Increased urine excretion of hop α-acids and fermentation products were found following beer consumption with respect to the gin administration, but differences were slight and not completely reported [112].

6.3. Role of Alcohol on Phenols-Related Effects of Beer on Cancer

A case-control association study (over 14 years) of child acute lymphoblastic leukemia (n = 491 + 491) found an inverse relation with maternal moderate consumption (self-reported) of beer (and wine, but not spirits), making authors suggest a protective effect of flavonoids [130]. However, a positive relation was reported also for fathers, which is difficult to explain and minimizes the observation's reliability. In a similar matched case-control study of drinking/smoking habits (over 10 years) of leukoplakia patients (n = 187 + 187; 40–65 years), while a role of regular wine consumption was associated with a decreased probability of disease occurrence (compared to that of spirit drinking that was associated to increased risk), no significant effect for moderate beer drinking was found [126]. The authors concluded that weaker effects of beer were probably due to the different composition in substances synergistically or antagonistically, i.e., polyphenols, interacting with ethanol [126]. Nonetheless, using a 20-country wide one-year (2002) evaluation of alcoholic beverages consumption and total deaths for oral cancer, the same authors estimated a lower risk for beer (and wine) consumers compared to heavy alcohol consumption from spirits [128]. Similarly, the consumption of beer (nor liquor) could not be associated with prostate cancer risk, in a population case-control study taking into account the self-reported alcohol consumption (n = 753 + 703; 40–64 years), even if the same authors reported a reduced relative risk associated with increasing level of red wine consumption [129]. More recently, lack of association with prostate cancer was reported for beer consumption (but also for wine and liquor) in a bigger prospective study (n = 84,170; 45–69 years) [124].

6.4. Role of Alcohol on Phenols-Related Effects of Beer on the Microbiota

According to a relationship between microbiota, host genes and diet [131], recent work investigated the possibility that alcohol-free beer, acting at the level of gut microbiota, could prevent the metabolic syndrome (MS). In fact, occurrence of MS can be promoted by gut microbiota dysbiosis, through low-grade inflammation and alteration of lipid metabolism. Gut microbiota dysbiosis can in turn be induced by alteration of the relative abundance of bacterial families [132]. Thus, a daily administration for one month of 355 mL of non-alcoholic beer was found associated to a decrease in fasting blood serum glucose and an increase in functional β-cells only, and the effect was not observed with an alcoholic beer containing a similar amount of phenolic compounds [111]. Moreover, the authors observed an enrichment of the microbiota diversity, also with the alcoholic beer, but only alcohol-free beer consumption was associated to a specific microbiota diversity with healthier function, suggesting that alcohol inhibited the positive effects of beer. As β-diversity was observed only after 30 days of treatment, the authors hypothesized that the effect on gut microbiota could depend on polyphenols and phenolic acids [111]. Similar results were obtained in an observational study, especially for higher butyric acid concentration in consumers versus non-consumers of beer [122], but no estimation of phenols intake was performed, nor were consumption of alcohol-free beer nor spirits-only drinkers recorded. On the other hand, another observational study on the microbiota of 916 UK female twins found association only for wine drinkers but not for beer (nor all other alcohols) [123], but also in this case, the consumption of alcohol-free beer was not considered. Figure 2 summarizes gut microbiota changes after beer consumption.

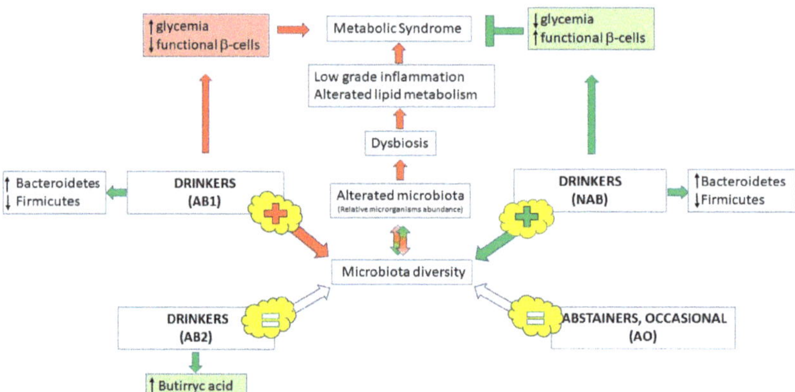

Figure 2. Schematic representation of relationship between beer, gut microbiota and metabolic syndrome. Phenolic compounds contained in non-alcoholic beer have a positive effect on the microbiota dysbiosis, one of the main causes of metabolic syndrome, but the effect is prevented by alcohol presence. Non-alcoholic beer consumption also determines a positive modification of some parameters typical of metabolic syndrome such as glycemia and the β-cells' function (AB1, drinkers of 355 mL/day of alcoholic beer; NAB, drinkers of 355 mL/day of non-alcoholic beer [111]). On the other hand, moderate beer consumption can increase the production of butyric acid, a fundamental molecule produced by the microbiota and useful for its healthy implications (AB2, drinkers of 200–600 mL/day; AO, abstainers or occasional consumers of <1.5 alcohol g/day [122]).

6.5. Role of Alcohol on Other Phenols-Related Effects of Beer

In an observational follow-up prospective study (34 years) of the association between alcoholic beverage consumption (using repeated surveys) and dementia (*n* = 1462 women, 38–60 years), beer consumption was associated to reduced dementia risk compared to subjects consuming only spirits [125]. Unfortunately, the consumption of alcohol-free beer was not taken into account. Moreover, in a case-control study of the association, in postmenopausal women (*n* = 35,816; 55–69 years), between specific self-reported drinking/smoking habits (over 20 years) and diabetes, a reduction of risk was observed for moderate consumption of either beer, red or white wines, but also for liquor, making the authors disprove the hypothesis that flavonoids could protect from diabetes onset [127].

7. Fruit-Based Enrichment of Beer Phenols

Beer is considered a promising beverage in the context of functional foods, which are food items with, in theory, health benefits, due to the enrichment with specific ingredients or bioactive compounds. Beer has high market opportunities because of an already high acceptancy of new organoleptic characteristics, due to widespread and previous diffusion of craft beers. Several ingredients have been added such as wheat, corn, rice and fruits. The phenolic profiles of several commercialized beers enriched with ingredients have already been reported and reviewed [8,133]. Studies agree that fruits' refermentation and maturation within beer production is associated to a significant increase of flavors and bioactive compounds supporting benefits of fruit contribution to beer's consumer acceptance. Both qualitative and quantitative increases in phenols have been reported in beers enriched with whole fruits during fermentation and works mainly focused on the role of the technological processes applied. However, rarely did a study report more than one fruit supplement. An exception is a recent report that compared individual phenols amounts in commercial beers enriched with cherry, raspberry, peach, apricot, grape, plum, orange or apple, and respective contribution to the antioxidant activity [134]. Importantly, this work demonstrates that fruit beers may be enriched with bioactive compounds (catechin, rutin, myricetin, quercetin and resveratrol) that are undetectable in conventional beers at identical extraction conditions and indicates enrichment with peels to be very promising because of the highest amount polyphenols and flavonoids content and antioxidant activity. Notably, resveratrol was found in beers enriched with all fruits except one (plum), with the highest level being measured in grape beer [134].

Other recent beer-added ingredients are quince fruit, mango, sweet potato and olive leaves. Because of organoleptic characteristics, quince fruit is specifically appreciated as a processed food. Many studies have shown that quince fruit lends itself as an affordable and good source of phenolic acids and flavonoids; in particular, in vitro assays have shown that phenols are the main compounds responsible for fruit's hydrophilic antioxidant activity [135]. Quince fruit phenols have been extensively studied [136] and recent data indicate that the addition of different quince cultivars, with different sensory attributes or antioxidant content, can selectively modulate the final content in specific phenols and related sensory descriptors attributes [137]. The addition of quince increased the total polyphenol content, the total hydroxicinnamic acids, concentration of main volatile compounds related with fruity sensory descriptors, and led to higher intensities of floral and fruity sensory attributes [137]. The addition of mango fruit, naturally reach in phenols [138], yielded beers with higher polyphenol content and aroma than traditional beer, especially if the fruit was homogenized before addition, on the condition that no thermal treatment was performed [139].

The addition of dried flakes of sweet potato, naturally rich in phenols [140], before beer brewing increased both total phenols (about 10%) and flavonoids (about 20%) content without changing physicochemical and sensory parameters of beer, which also benefited from an important increase in β-carotene [141].

Also, dried olives or resulting extracts, that contain not only common phenols, but also the olive tree family-exclusive secoiridoids [142], were added to beer, and the resulting beer had positive flavor and aroma, but low colloidal stability and showed increasing haze formation during storage due to very high polyphenols content [143]. Similar increase in colloidal haze was reported also for beers with added omija fruits, questioning the validity of increasing the phenolic content of beers too much [144]. Beers enriched with lignans from wood chips or extracts displayed excessive bitter taste and unusual resin aroma, indicating the need for technological approaches to avoid significant changes to the characteristics of beer. A possible solution could come from the use of hot water as a unique solvent, already applied for the removal of resins from wood chips or lignan extracts from the knots of spruce trees (*Picea abies*), a strategy that yielded beer with as much as 100 mg/L of lignans.

8. Cereal-Based Enrichment of Beer Phenols

Apart from barley, other malted cereals have been used since antiquity for the development of fermented beverages, in a somehow geographical way, for example, rice in India [145], millet in Nigeria [146], sorghum in South Africa [147] and Corn in Mexico [148]. Regarding the latter, a pulque-fermented drink known as "Sendechó" was antiquely prepared by the Mazahuas population in the Valley of Mexico, using chili and pigmented corn varieties with high content of phenolic compounds, mainly anthocyanins [149], which are completely absent in barley beer. In the attempt of developing a beer with traditional ingredients, pulque was substituted with hop and brewer's yeast in an ale fermentation process performed with guajillo chili and blue corn malt. The result was a beer with total polyphenols concentration up to 560 mg GAE/L and of total anthocyanins up to 19.4 mg cyaniding-3-glucoside/L [150]. More recently, the same laboratory obtained blue or red corn malt blended beers with even higher total phenols amount (up to 849.5 mg GAE/L) and identified anthocyanins responsible for the final color yield of red and blue corn beers (pelargonidin-3-glucoside and cyanidin-3-glucoside) [151]. Authors also identified, in corn beers both previously reported and unreported, volatile phenols conferring desirable aromas to beers. Such results are promising with respect to previous reports of lower content of phenols in corn-added beers [152]. Nevertheless, to our knowledge, no consumer acceptability of such beers has been evaluated, and this aspect is crucial especially for the high content of phenols that can contribute to high spicy perception and for astringency of anthocyanins [151]. Among other gluten-free beers, those obtained from oat [153], sorghum [147], teff [154], millet [146], buckwheat [155] and quinoa [156] are in theory valid alternatives in terms of phenols considering the grain natural content, even if very little is available on the phenolic content of such beers, i.e., only a sum of aromatic alcohols was reported for millet [157]. A noteworthy emerging exception is represented by indigenous beer-like fermented beverages "ikigage" [158], "burukutu" and "pito" [159], drinks for which 4-vinylphenols quantities have been reported. Another exception is that of rice-based alcoholic beverages of Assam, India, with total polyphenol content up to 631.33 mg GAE/L [160].

One frequent issue of non-barley cereals is the low diastatic power, that traps phenols, making necessary the combination with other cereals or the addition of exogenous enzymes [161]. Addition to the mashing process of recombinant ferulic acid esterase [162] was recently proven as a valid remedy also for the low amount of the desirable phenol 4-vinylguaiacol (derived from ferulic acid by enzymic decarboxylation), a common issue of top-fermented wheat beer [163]. More recently, the strategy was further implemented by producing yeasts expressing bacterial ferulic acid decarboxylase [164].

9. Phenols in Non-Alcoholic and Isotonic Beers

Driving laws and a healthier lifestyle have increased the popularity of non-alcoholic beers. In order to not exceed the limit of 0.5% (v/v) alcohol or to produce beer with a limited alcohol content, two approaches are exploited. The first one consists in limiting the fermentation process, and hence the alcohol production, using low-alcohol yeasts or producing a wort with low degrees Plato and low diastatic power in order to obtain more dextrins than fermentable sugars. The alternative approach involves physical methods to remove the alcohol at the end of brewing, for example by vacuum evaporation or reverse osmosis treatments. Unfortunately, limiting the fermentation process can bring about inadequate conversion of wort to beer and, on the other hand, physical methods for alcohol removal can deteriorate beer composition [165]. Osmotic distillation using a membrane contactor was recently shown to be able to maintain the total phenols content in a low-alcohol top-fermented beer [166]. On the other hand, using the fermentation interruption approach, De Fusco and coworkers recently obtained a low-alcohol isotonic beer with an amount of total phenolic compounds similar to that of Pilsen beer and sport drinks [167]. Isotonic beers are an improvement on low-alcohol beers with similar rehydration potential of sports drinks [168] (beverages with specific osmolality and carbohydrate content [169]) and with the advantage of containing bioactive molecules. Notably, De Fusco and coworkers found that fermentation interruption did not significantly affect total phenols level [167]. Nevertheless, experiments are needed to test the shelf-life of low-alcohol isotonic beers and specifically to test if phenols' antimicrobial activity is adequate in such low-alcoholic and carbohydrate-containing beverages [170].

10. Future Directions and Conclusions

Here, we attempted to review the more recent findings on beer phenols and their role in human health. Particular attention was dedicated to the role of genetic factors and to the enrichment with phenolic compounds by cereals different from barley or fruits naturally rich in phenols. In this respect, it would be interesting to investigate to what extent fruit addition also increases the alcoholic content, which has health and consumer acceptance consequences that are not negligible. One other interesting question regards the huge amount of debris produced by beer production, especially in terms of phenolic compounds (1% in by-product spent grain [171]) that could be recycled for beer enrichment itself. Recent reports indicate that the recovery of phenols can be improved using the fungi *Rhizopus oligosporus* as a fermenting organism [172].

Even if the Scopus.com search string we used is arbitrary and may not entirely represent the research on health effects of beer ascribable to the presence of phenolic compounds, less than of 25% (40 out of 161) of entries we retrieved were reports on the in vivo (human or animal) or in vitro effects of phenolic compounds within in toto beer. Most of the research is focused on evaluating the effects of single phenolic compounds of beer, which, however, can give rise to partial conclusions that need further experiments performed in physiological conditions. For example, as beer contains a mean amount of xanthohumol around 0.2 mg/L, what is the rationale for supplementing volunteers with an enriched drink containing a daily dose of 12 mg of xanthohumol [173], corresponding to the amount found in 60 L of beer? Indeed, several techniques have been used, starting from almost 20 years ago and featuring a patented addition of an enriched hop product [174], in order to increase the amount of this compound in beer up to 10 mg/L [175–177]. However, even if studies with enriched beers are helpful for assessing the metabolic fate of phenolic compounds, in order to correctly evaluate healthy effects of beer consumption, researchers should consider, besides the side effects of alcohol (where ethically possible), also those possibly due to yet uncharacterized molecules, i.e., those resulting from the addition of the enriched hop product, and those due to a non-physiological consumption of a single flavonoid for which pro-apoptotic effects are already known [178]. In fact, the only health claims authorized for phenolics by the European Food Safety Authority regard, at the moment, olive oil hydroxytyrosol and cocoa flavanols, with high daily amounts (5 and 200 mg, respectively) that can, however, be easily consumed in the context of a balanced diet [179,180]. Among retrieved reports, only six investigated the effects of phenols in the presence and absence of alcohol, thus also considering the effects of alcohol alone. Actually, four publications belong to the same Spanish research group [112–115], and thus probably refer to the same and unique small population of 33 high-cardiovascular-risk males.

In conclusion, studies applying a parallel administration of non-alcoholic beer or/and alcohol alone, in both animal and human intervention studies, support the existence of somehow interfering effects of phenols and ethanol. However, in order to better highlight additive or synergistic effects, further correctly set-up human interventional crossover or observational, or at least animal, studies are required. From this point of view, even insect models could deserve more attention. In fact, using *D. melanogaster* fed with a total beer extract, Merinas-Amo and colleagues were able to demonstrate the synergic interaction between different molecules contained in beer [181].

Author Contributions: Scopus.com investigation, R.A.; writing—review and editing, R.A., G.P. and S.L. All authors have read and agreed to the published version of the manuscript.

Funding: This research received no external funding.

Institutional Review Board Statement: Not applicable.

Informed Consent Statement: Not applicable.

Data Availability Statement: Not applicable.

Conflicts of Interest: The authors declare no conflict of interest.

References

1. Homan, M.M. Beer and Its Drinkers: An Ancient Near Eastern Love Story. *Near East. Archaeol.* **2004**, *67*, 84–95. [CrossRef]
2. Wang, J.; Liu, L.; Ball, T.; Yu, L.; Li, Y.; Xing, F. Revealing a 5,000-y-old beer recipe in China. *Proc. Natl. Acad. Sci. USA* **2016**, 201601465. [CrossRef]
3. Brewers. Contribution made by Beer to the European Economy. Available online: https://brewersofeurope.org/uploads/mycms-files/documents/publications/2020/contribution-made-by-beer-to-EU-economy-2020.pdf (accessed on 1 October 2020).
4. Osorio-Paz, I.; Brunauer, R.; Alavez, S. Beer and its non-alcoholic compounds in health and disease. *Crit. Rev. Food Sci. Nutr.* **2019**, 1–14. [CrossRef]
5. Cardona, F.; Andrés-Lacueva, C.; Tulipani, S.; Tinahones, F.J.; Queipo-Ortuño, M.I. Benefits of polyphenols on gut microbiota and implications in human health. *J. Nutr. Biochem.* **2013**, *24*, 1415–1422. [CrossRef]
6. Palmer, K. IARC Monographs on the Evaluation of Carcinogenic Risks to Humans. Volume 98: Painting, Firefighting and Shiftwork. International Agency for Research on Cancer. *Occup. Med. (Chic. Ill.)* **2011**, *61*, 521–522. [CrossRef]
7. Scalbert, A.; Williamson, G. Dietary Intake and Bioavailability of Polyphenols. *J. Nutr.* **2000**, *130*, 2073S–2085S. [CrossRef]
8. Mitić, S.S.; Paunović, D.Đ.; Pavlović, A.N.; Tošić, S.B.; Stojković, M.B.; Mitić, M.N. Phenolic Profiles and Total Antioxidant Capacity of Marketed Beers in Serbia. *Int. J. Food Prop.* **2014**, *17*, 908–922. [CrossRef]
9. Wannenmacher, J.; Gastl, M.; Becker, T. Phenolic Substances in Beer: Structural Diversity, Reactive Potential and Relevance for Brewing Process and Beer Quality. *Compr. Rev. Food Sci. Food Saf.* **2018**, *17*, 953–988. [CrossRef] [PubMed]
10. Bartolomé, B.; Peña-Neira, A.; Gómez-Cordovés, C. Phenolics and related substances in alcohol-free beers. *Eur. Food Res. Technol.* **2000**, *210*, 419–423. [CrossRef]
11. Boronat, A.; Soldevila-Domenech, N.; Rodríguez-Morató, J.; Martínez-Huélamo, M.; Lamuela-Raventós, R.M.; de la Torre, R. Beer Phenolic Composition of Simple Phenols, Prenylated Flavonoids and Alkylresorcinols. *Molecules* **2020**, *25*, 2582. [CrossRef]
12. Soldevila-Domenech, N.; Boronat, A.; Mateus, J.; Diaz-Pellicer, P.; Matilla, I.; Pérez-Otero, M.; Aldea-Perona, A.; De La Torre, R. Generation of the antioxidant hydroxytyrosol from tyrosol present in beer and red wine in a randomized clinical trial. *Nutrients* **2019**. [CrossRef]
13. Gharwalova, L.; Sigler, K.; Dolezalova, J.; Masak, J.; Rezanka, T.; Kolouchova, I. Resveratrol suppresses ethanol stress in winery and bottom brewery yeast by affecting superoxide dismutase, lipid peroxidation and fatty acid profile. *World J. Microbiol. Biotechnol.* **2017**, *33*, 205. [CrossRef]
14. Moura-Nunes, N.; Brito, T.C.; Da Fonseca, N.D.; De Aguiar, P.F.; Monteiro, M.; Perrone, D.; Torres, A.G. Phenolic compounds of Brazilian beers from different types and styles and application of chemometrics for modeling antioxidant capacity. *Food Chem.* **2016**. [CrossRef]
15. Parikka, K.; Rowland, I.R.; Welch, R.W.; Wähälä, K. In vitro antioxidant activity and antigenotoxicity of 5-n-alkylresorcinols. *J. Agric. Food Chem.* **2006**. [CrossRef]
16. Oishi, K.; Yamamoto, S.; Itoh, N.; Nakao, R.; Yasumoto, Y.; Tanaka, K.; Kikuchi, Y.; Fukudome, S.-I.; Okita, K.; Takano-Ishikawa, Y. Wheat alkylresorcinols suppress high-fat, high-sucrose diet-induced obesity and glucose intolerance by increasing insulin sensitivity and cholesterol excretion in male mice. *J. Nutr.* **2015**. [CrossRef]
17. Chen, X.; Mukwaya, E.; Wong, M.-S.; Zhang, Y. A systematic review on biological activities of prenylated flavonoids. *Pharm. Biol.* **2014**, *52*, 655–660. [CrossRef]
18. Štulíková, K.; Karabín, M.; Nešpor, J.; Dostálek, P. Therapeutic Perspectives of 8-Prenylnaringenin, a Potent Phytoestrogen from Hops. *Molecules* **2018**, *23*, 660. [CrossRef]
19. Intelmann, D.; Haseleu, G.; Dunkel, A.; Lagemann, A.; Stephan, A.; Hofmann, T. Comprehensive sensomics analysis of hop-derived bitter compounds during storage of beer. *J. Agric. Food Chem.* **2011**. [CrossRef]
20. Callemien, D.; Collin, S. Involvement of flavanoids in beer color instability during storage. *J. Agric. Food Chem.* **2007**. [CrossRef]
21. Heuberger, A.L.; Broeckling, C.D.; Lewis, M.R.; Salazar, L.; Bouckaert, P.; Prenni, J.E. Metabolomic profiling of beer reveals effect of temperature on non-volatile small molecules during short-term storage. *Food Chem.* **2012**. [CrossRef]
22. Owades, J.L.; Jakovac, J. Study of Beer Oxidation with O18. *Proc. Annu. Meet. Am. Soc. Brew. Chem.* **1966**, *24*, 180–183. [CrossRef]
23. Asano, K.; Ohtsu, K.; Shinagawa, K.; Hashimoto, N. Affinity of proanthocyanidins and their oxidation products for haze-forming proteins of beerand the formation of chill haze. *Agric. Biol. Chem.* **1984**. [CrossRef]
24. Delcour, J.A.; Dondeyne, P.; Trousdale, E.K.; Singleton, V.L. The reactions between polyphenols and aldehydes and the influence of acetaldehyde on haze formation in beer. *J. Inst. Brew.* **1982**. [CrossRef]
25. Maia, P.D.D.S.; dos Santos Baião, D.; da Silva, V.P.F.; Lemos Miguel, M.A.; Quirino Lacerda, E.C.; de Araújo Calado, V.M.; da Silva Carneiro, C.; Finotelli, P.V.; Pierucci, A.P.T.R. Microencapsulation of a craft beer, nutritional composition, antioxidant stability, and drink acceptance. *LWT* **2020**, *133*, 110104. [CrossRef]

26. Callemien, D.; Collin, S. Use of RP-HPLC-ESI(–)-MS/MS to Differentiate Various Proanthocyanidin Isomers in Lager Beer Extracts. *J. Am. Soc. Brew. Chem.* **2008**, *66*, 109–115. [CrossRef]

27. Corso, M.; Perreau, F.; Mouille, G.; Lepiniec, L. Specialized phenolic compounds in seeds: Structures, functions, and regulations. *Plant. Sci.* **2020**, *296*, 110471. [CrossRef]

28. Gretenhart, K.E. Specialty malts. *Tech. Q. Master Brew. Assoc. Am.* **1997**, *34*, 102–106.

29. Lu, J.; Zhao, H.; Chen, J.; Fan, W.; Dong, J.; Kong, W.; Sun, J.; Cao, Y.; Cai, G. Evolution of phenolic compounds and antioxidant activity during malting. *J. Agric. Food Chem.* **2007**. [CrossRef]

30. Maillard, M.-N.; Berset, C. Evolution of Antioxidant Activity during Kilning: Role of Insoluble Bound Phenolic Acids of Barley and Malt. *J. Agric. Food Chem.* **1995**, *43*, 1789–1793. [CrossRef]

31. Leitao, C.; Marchioni, E.; Bergaentzlé, M.; Zhao, M.; Didierjean, L.; Miesch, L.; Holder, E.; Miesch, M.; Ennahar, S. Fate of polyphenols and antioxidant activity of barley throughout malting and brewing. *J. Cereal Sci.* **2012**. [CrossRef]

32. Munoz-Insa, A.; Gastl, M.; Becker, T. Influence of malting on the protein composition of triticale (× Triticosecale Wittmack) "trigold.". *Cereal Chem.* **2016**. [CrossRef]

33. Koren, D.; Kun, S.; Hegyesné Vecseri, B.; Kun-Farkas, G. Study of antioxidant activity during the malting and brewing process. *J. Food Sci. Technol.* **2019**. [CrossRef]

34. Szwajgier, D. Dry and wet milling of malt. a preliminary study comparing fermentable sugar, total protein, total phenolics and the ferulic acid content in non-hopped worts. *J. Inst. Brew.* **2011**. [CrossRef]

35. Defernez, M.; Foxall, R.J.; O'Malley, C.J.; Montague, G.; Ring, S.M.; Kemsley, E.K. Modelling beer fermentation variability. *J. Food Eng.* **2007**, *83*, 167–172. [CrossRef]

36. Piazzon, A.; Forte, M.; Nardini, M. Characterization of phenolics content and antioxidant activity of different beer types. *J. Agric. Food Chem.* **2010**. [CrossRef]

37. Cheiran, K.P.; Raimundo, V.P.; Manfroi, V.; Anzanello, M.J.; Kahmann, A.; Rodrigues, E.; Frazzon, J. Simultaneous identification of low-molecular weight phenolic and nitrogen compounds in craft beers by HPLC-ESI-MS/MS. *Food Chem.* **2019**. [CrossRef]

38. Bertuzzi, T.; Mulazzi, A.; Rastelli, S.; Donadini, G.; Rossi, F.; Spigno, G. Targeted healthy compounds in small and large-scale brewed beers. *Food Chem.* **2020**. [CrossRef]

39. Zhao, H. Effects of Processing Stages on the Profile of Phenolic Compounds in Beer. In *Processing and Impact on Active Components in Food*; Academic Press: Cambridge, MA, USA, 2015; ISBN 9780124047099.

40. Obata, Y.; Koshika, M. Studies on the Sunlight Flavor of Beer. *Bull. Agric. Chem. Soc. Jpn.* **1960**, *24*, 644–646. [CrossRef]

41. Taylor, R.; Kirsop, B.H. Occurrence of a killer strain of Saccharomyces cerevisiae in a batch fermentation plant. *J. Inst. Brew.* **1979**, *85*, 325. [CrossRef]

42. Vanbeneden, N.; Van Roey, T.; Willems, F.; Delvaux, F.; Delvaux, F.R. Release of phenolic flavour precursors during wort production: Influence of process parameters and grist composition on ferulic acid release during brewing. *Food Chem.* **2008**. [CrossRef]

43. Vanbeneden, N.; Gils, F.; Delvaux, F.; Delvaux, F.R. Formation of 4-vinyl and 4-ethyl derivatives from hydroxycinnamic acids: Occurrence of volatile phenolic flavour compounds in beer and distribution of Pad1-activity among brewing yeasts. *Food Chem.* **2008**. [CrossRef]

44. Schwarz, K.J.; Stübner, R.; Methner, F.J. Formation of styrene dependent on fermentation management during wheat beer production. *Food Chem.* **2012**. [CrossRef] [PubMed]

45. Sterckx, F.L.; Missiaen, J.; Saison, D.; Delvaux, F.R. Contribution of monophenols to beer flavour based on flavour thresholds, interactions and recombination experiments. *Food Chem.* **2011**. [CrossRef] [PubMed]

46. Bettenhausen, H.M.; Barr, L.; Broeckling, C.D.; Chaparro, J.M.; Holbrook, C.; Sedin, D.; Heuberger, A.L. Influence of malt source on beer chemistry, flavor, and flavor stability. *Food Res. Int.* **2018**. [CrossRef]

47. Maillard, M.N.; Soum, M.H.; Boivin, P.; Berset, C. Antioxidant activity of barley and malt: Relationship with phenolic content. *LWT Food Sci. Technol.* **1996**. [CrossRef]

48. Langos, D.; Granvogl, M.; Schieberle, P. Characterization of the key aroma compounds in two Bavarian wheat beers by means of the sensomics approach. *J. Agric. Food Chem.* **2013**. [CrossRef]

49. Lentz, M. The impact of simple phenolic compounds on beer aroma and flavor. *Fermentation* **2018**, *4*, 20. [CrossRef]

50. Marova, I.; Parilova, K.; Friedl, Z.; Obruca, S.; Duronova, K. Analysis of phenolic compounds in lager beers of different origin: A contribution to potential determination of the authenticity of Czech beer. *Chromatographia* **2011**. [CrossRef]

51. Olšovská, J.; Čejka, P.; Sigler, K.; Hönigová, V. The phenomenon of Czech beer: A review. *Czech. J. Food Sci.* **2014**, *32*, 309–319. [CrossRef]

52. Dadic, M.; Van Gheluwe, G.E.A. Role of polyphenols and non-volatiles in beer quality. *Master Brew. Assoc. Am. Tech. Q.* **1973**, *10*, 69–73.

53. Cai, S.; Han, Z.; Huang, Y.; Chen, Z.-H.; Zhang, G.; Dai, F. Genetic Diversity of Individual Phenolic Acids in Barley and Their Correlation with Barley Malt Quality. *J. Agric. Food Chem.* **2015**, *63*, 7051–7057. [CrossRef] [PubMed]

54. Ye, L.; Huang, Y.; Hu, H.; Dai, F.; Zhang, G. Identification of QTLs associated with haze active proteins in barley. *Euphytica* **2015**. [CrossRef]

55. Han, Z.; Zhang, J.; Cai, S.; Chen, X.; Quan, X.; Zhang, G. Association mapping for total polyphenol content, total flavonoid content and antioxidant activity in barley. *BMC Genomics* **2018**. [CrossRef] [PubMed]

56. Mikyška, A.; Hrabák, M.; Hašková, D.; Šrogl, J. The role of malt and hop polyphenols in beer quality, flavour and haze stability. *J. Inst. Brew.* **2002**. [CrossRef]

57. Munoz-Insa, A.; Gastl, M.; Becker, T. Use of Polyphenol-Rich Hop Products to Reduce Sunstruck Flavor in Beer. *J. Am. Soc. Brew. Chem.* **2015**, *73*, 228–235. [CrossRef]

58. Wannenmacher, J.; Cotterchio, C.; Schlumberger, M.; Reuber, V.; Gastl, M.; Becker, T. Technological influence on sensory stability and antioxidant activity of beers measured by ORAC and FRAP. *J. Sci. Food Agric.* **2019**. [CrossRef]

59. Goiris, K.; Jaskula-Goiris, B.; Syryn, E.; Van Opstaele, F.; De Rouck, G.; Aerts, G.; De Cooman, L. The flavoring potential of hop polyphenols in beer. *J. Am. Soc. Brew. Chem.* **2014**. [CrossRef]

60. Bober, A.; Liashenko, M.; Protsenko, L.; Slobodyanyuk, N.; Matseiko, L.; Yashchuk, N.; Gunko, S.; Mushtruk, M. Biochemical composition of the hops and quality of the finished beer. *Potravin. Slovak J. Food Sci.* **2020**. [CrossRef]

61. Griffiths, D.W.; Welch, R.W. Genotypic and environmental variation in the total phenol and flavanol contents of barley grain. *J. Sci. Food Agric.* **1982**. [CrossRef]

62. Forster, A.; Beck, B.; Schmidt, R.; Jansen, C.; Mellenthin, A. On the composition of low molecular polyphenols in different varieties of hops and from two growing areas. *Monatsschr. Brauwiss.* **2002**, *55*, 98–108.

63. Thurston, P.A.; Tubb, R.S. Screening yeast strains for the ability to produce phenolic off-flavours: A simple method for determining phenols in wort and beer. *J. Inst. Brew.* **1981**. [CrossRef]

64. Mohammadi, M.; Endelman, J.B.; Nair, S.; Chao, S.; Jones, S.S.; Muehlbauer, G.J.; Ullrich, S.E.; Baik, B.-K.; Wise, M.L.; Smith, K.P. Association mapping of grain hardness, polyphenol oxidase, total phenolics, amylose content, and β-glucan in US barley breeding germplasm. *Mol. Breed.* **2014**, *34*, 1229–1243. [CrossRef]

65. Holtekjølen, A.K.; Kinitz, C.; Knutsen, S.H. Flavanol and Bound Phenolic Acid Contents in Different Barley Varieties. *J. Agric. Food Chem.* **2006**, *54*, 2253–2260. [CrossRef] [PubMed]

66. Ahmed, I.M.; Cao, F.; Han, Y.; Nadira, U.A.; Zhang, G.; Wu, F. Differential changes in grain ultrastructure, amylase, protein and amino acid profiles between Tibetan wild and cultivated barleys under drought and salinity alone and combined stress. *Food Chem.* **2013**, *141*, 2743–2750. [CrossRef] [PubMed]

67. Han, Z.; Ahsan, M.; Adil, M.F.; Chen, X.; Nazir, M.M.; Shamsi, I.H.; Zeng, F.; Zhang, G. Identification of the gene network modules highly associated with the synthesis of phenolics compounds in barley by transcriptome and metabolome analysis. *Food Chem.* **2020**, *323*, 126862. [CrossRef] [PubMed]

68. Morcol, T.B.; Negrin, A.; Matthews, P.D.; Kennelly, E.J. Hop (Humulus lupulus L.) terroir has large effect on a glycosylated green leaf volatile but not on other aroma glycosides. *Food Chem.* **2020**. [CrossRef] [PubMed]

69. Claussen, N.H. On a Method for the Application of Hansen's Pure Yeast System in the Manufacturing of Well-Conditioned English Stock Beers. *J. Inst. Brew.* **1904**, *10*, 308–331. [CrossRef]

70. Wiles, A.E. Studies of some yeasts causing spoilage of draught beer. *J. Inst. Brew.* **1950**. [CrossRef]

71. Chatonnet, P.; Dubourdie, D.; Boidron, J.-N.; Pons, M. The origin of ethylphenols in wines. *J. Sci. Food Agric.* **1992**. [CrossRef]

72. Passoth, V.; Blomqvist, J.; Schnürer, J. Dekkera bruxellensis and Lactobacillus vini form a stable ethanol-producing consortium in a commercial alcohol production process. *Appl. Environ. Microbiol.* **2007**, *73*, 4354–4356. [CrossRef]

73. Steensels, J.; Daenen, L.; Malcorps, P.; Derdelinckx, G.; Verachtert, H.; Verstrepen, K.J. Brettanomyces yeasts—From spoilage organisms to valuable contributors to industrial fermentations. *Int. J. Food Microbiol.* **2015**, *206*, 24–38. [CrossRef]

74. González, C.; Godoy, L.; Ganga, M.A. Identification of a second PAD1 in Brettanomyces bruxellensis LAMAP2480. *Antonie Leeuwenhoek Int. J. Gen. Mol. Microbiol.* **2017**. [CrossRef]

75. Lentz, M.; Harris, C. Analysis of Growth Inhibition and Metabolism of Hydroxycinnamic Acids by Brewing and Spoilage Strains of Brettanomyces Yeast. *Foods* **2015**. [CrossRef]

76. Coghe, S.; Benoot, K.; Delvaux, F.; Vanderhaegen, B.; Delvaux, F.R. Ferulic Acid Release and 4-Vinylguaiacol Formation during Brewing and Fermentation: Indications for Feruloyl Esterase Activity in Saccharomyces cerevisiae. *J. Agric. Food Chem.* **2004**. [CrossRef]

77. Spitaels, F.; Wieme, A.D.; Janssens, M.; Aerts, M.; Daniel, H.-M.; Van Landschoot, A.; De Vuyst, L.; Vandamme, P. The Microbial Diversity of Traditional Spontaneously Fermented Lambic Beer. *PLoS ONE* **2014**, *9*, e95384. [CrossRef]

78. Serra Colomer, M.; Funch, B.; Forster, J. The raise of Brettanomyces yeast species for beer production. *Curr. Opin. Biotechnol.* **2019**, *56*, 30–35. [CrossRef]

79. Coelho, E.; Azevedo, M.; Teixeira, J.A.; Tavares, T.; Oliveira, J.M.; Domingues, L. Evaluation of multi-starter S. cerevisiae/ D. bruxellensis cultures for mimicking and accelerating transformations occurring during barrel ageing of beer. *Food Chem.* **2020**, *323*, 126826. [CrossRef]

80. Capece, A.; Romaniello, R.; Pietrafesa, A.; Siesto, G.; Pietrafesa, R.; Zambuto, M.; Romano, P. Use of Saccharomyces cerevisiae var. boulardii in co-fermentations with S. cerevisiae for the production of craft beers with potential healthy value-added. *Int. J. Food Microbiol.* **2018**. [CrossRef]

81. Gallone, B.; Steensels, J.; Prahl, T.; Soriaga, L.; Saels, V.; Herrera-Malaver, B.; Merlevede, A.; Roncoroni, M.; Voordeckers, K.; Miraglia, L.; et al. Domestication and Divergence of Saccharomyces cerevisiae Beer Yeasts. *Cell* **2016**, *166*, 1397–1410.e16. [CrossRef]

82. Colomer, M.S.; Chailyan, A.; Fennessy, R.T.; Olsson, K.F.; Johnsen, L.; Solodovnikova, N.; Forster, J. Assessing Population Diversity of Brettanomyces Yeast Species and Identification of Strains for Brewing Applications. *Front. Microbiol.* **2020**, *11*, 637. [CrossRef]

83. Langdon, Q.K.; Peris, D.; Baker, E.P.; Opulente, D.A.; Nguyen, H.-V.; Bond, U.; Gonçalves, P.; Sampaio, J.P.; Libkind, D.; Hittinger, C.T. Fermentation innovation through complex hybridization of wild and domesticated yeasts. *Nat. Ecol. Evol.* **2019**, *3*, 1576–1586. [CrossRef] [PubMed]

84. Avramova, M.; Cibrario, A.; Peltier, E.; Coton, M.; Coton, E.; Schacherer, J.; Spano, G.; Capozzi, V.; Blaiotta, G.; Salin, F.; et al. Brettanomyces bruxellensis population survey reveals a diploid-triploid complex structured according to substrate of isolation and geographical distribution. *Sci. Rep.* **2018**, *8*, 4136. [CrossRef] [PubMed]

85. De Gaetano, G.; Costanzo, S.; Di Castelnuovo, A.; Badimon, L.; Bejko, D.; Alkerwi, A.; Chiva-Blanch, G.; Estruch, R.; La Vecchia, C.; Panico, S.; et al. Effects of moderate beer consumption on health and disease: A consensus document. *Nutr. Metab. Cardiovasc. Dis.* **2016**, *26*, 443–467. [CrossRef] [PubMed]

86. Dai, J.; Mumper, R.J. Plant phenolics: Extraction, analysis and their antioxidant and anticancer properties. *Molecules* **2010**, *15*, 7313–7352. [CrossRef] [PubMed]

87. Karatzi, K.; Rontoyanni, V.G.; Protogerou, A.D.; Georgoulia, A.; Xenos, K.; Chrysou, J.; Sfikakis, P.P.; Sidossis, L.S. Acute effects of beer on endothelial function and hemodynamics: Asingle-blind, crossover study in healthy volunteers. *Nutrition* **2013**. [CrossRef] [PubMed]

88. De Miranda, R.C.; Paiva, E.S.; Suter Correia Cadena, S.M.; Brandt, A.P.; Vilela, R.M. Polyphenol-rich foods alleviate pain and ameliorate quality of life in fibromyalgic women. *Int. J. Vitam. Nutr. Res.* **2017**. [CrossRef]

89. Loguercio, C.; Tuccillo, C.; Federico, A.; Fogliano, V.; Del Vecchio Blanco, C.; Romano, M. Alcoholic beverages and gastric epithelial cell viability: Effect on oxidative stress-induced damage. *J. Physiol. Pharmacol.* **2009**, *60*, 87–92.

90. Machado, J.C.; Faria, M.A.; Melo, A.; Ferreira, I.M.P.L.V.O. Antiproliferative effect of beer and hop compounds against human colorectal adenocarcinome Caco-2 cells. *J. Funct. Foods* **2017**. [CrossRef]

91. Alonso-Andrés, P.; Martín, M.; Albasanz, J.É.L. Modulation of adenosine receptors and antioxidative effect of beer extracts in in vitro models. *Nutrients* **2019**, *11*, 1258. [CrossRef]

92. Oczkowski, M.; Rembiszewska, A.; Dziendzikowska, K.; Wolińska-Witort, E.; Kołota, A.; Malik, A.; Stachoń, M.; Lachowicz, K.; Gromadzka-Ostrowska, J. Beer consumption negatively regulates hormonal reproductive status and reduces apoptosis in Leydig cells in peripubertal rats. *Alcohol* **2019**. [CrossRef]

93. Silva, A.F.; Faria-Costa, G.; Sousa-Nunes, F.; Santos, M.F.; Ferreira-Pinto, M.J.; Duarte, D.; Rodrigues, I.; Guimarães, J.T.; Leite-Moreira, A.; Moreira-Gonçalves, D.; et al. Anti-remodeling effects of xanthohumol-fortified beer in pulmonary arterial hypertension mediated by ERK and AKT inhibition. *Nutrients* **2019**, *11*, 583. [CrossRef] [PubMed]

94. Merino, P.; Santos-López, J.A.; Mateos, C.J.; Meseguer, I.; Garcimartín, A.; Bastida, S.; Sánchez-Muniz, F.J.; Benedí, J.; González-Muñoz, M.J. Can nonalcoholic beer, silicon and hops reduce the brain damage and behavioral changes induced by aluminum nitrate in young male Wistar rats? *Food Chem. Toxicol.* **2018**. [CrossRef] [PubMed]

95. Lima-Fontes, M.; Costa, R.; Rodrigues, I.; Soares, R. Xanthohumol Restores Hepatic Glucolipid Metabolism Balance in Type 1 Diabetic Wistar Rats. *J. Agric. Food Chem.* **2017**. [CrossRef] [PubMed]

96. Costa, R.; Negrão, R.; Valente, I.; Castela, A.; Duarte, D.; Guardão, L.; Magalhães, P.J.; Rodrigues, J.A.; Guimarães, J.T.; Gomes, P.; et al. Xanthohumol modulates inflammation, oxidative stress, and angiogenesis in type 1 diabetic rat skin wound healing. *J. Nat. Prod.* **2013**. [CrossRef] [PubMed]

97. Negrão, R.; Costa, R.; Duarte, D.; Gomes, T.T.; Coelho, P.; Guimarães, J.T.; Guardão, L.; Azevedo, I.; Soares, R. Xanthohumol-supplemented beer modulates angiogenesis and inflammation in a skin wound healing model. Involvement of local adipocytes. *J. Cell. Biochem.* **2012**. [CrossRef]

98. Jastrzebski, Z.; Gorinstein, S.; Czyzewska-Szafran, H.; Leontowicz, H.; Leontowicz, M.; Trakhtenberg, S.; Remiszewska, M. The effect of short-term lyophilized beer consumption on established hypertension in rats. *Food Chem. Toxicol.* **2007**. [CrossRef]

99. Gorinstein, S.; Caspi, A.; Pawelzik, E.; Deldago-Licon, E.; Libman, I.; Trakhtenberg, S.; Weisz, M.; Martin-Belloso, O. Proteins of beer affect lipid levels in rats. *Nutr. Res.* **2001**. [CrossRef]

100. Gorinstein, S.; Zemser, M.; Weisz, M.; Halevy, S.; Martin-Belloso, O.; Trakhtenberg, S. The influence of alcohol-containing and alcohol-free beverages on lipid levels and lipid peroxides in serum of rats. *J. Nutr. Biochem.* **1998**. [CrossRef]

101. Negrão, R.; Costa, R.; Duarte, D.; Taveira Gomes, T.; Mendanha, M.; Moura, L.; Vasques, L.; Azevedo, I.; Soares, R. Angiogenesis and inflammation signaling are targets of beer polyphenols on vascular cells. *J. Cell. Biochem.* **2010**. [CrossRef]

102. Bourne, L.; Paganga, G.; Baxter, D.; Hughes, P.; Rice-Evans, C. Absorption of ferulic acid from low-alcohol beer. *Free Radic. Res.* **2000**. [CrossRef]

103. Nardini, M.; Natella, F.; Scaccini, C.; Ghiselli, A. Phenolic acids from beer are absorbed and extensively metabolized in humans. *J. Nutr. Biochem.* **2006**. [CrossRef] [PubMed]

104. Ghiselli, A.; Natella, F.; Guidi, A.; Montanari, L.; Fantozzi, P.; Scaccini, C. Beer increases plasma antioxidant capacity in humans. *J. Nutr. Biochem.* **2000**. [CrossRef]

105. Pérez-Mañá, C.; Farré, M.; Pastor, A.; Fonseca, F.; Torrens, M.; Menoyo, E.; Pujadas, M.; Frias, S.; Langohr, K.; de la Torre, R. Non-linear formation of EtG and FAEEs after controlled administration of low to moderate doses of ethanol. *Alcohol Alcohol.* **2017**. [CrossRef] [PubMed]

106. Krnic, M.; Modun, D.; Budimir, D.; Gunjaca, G.; Jajic, I.; Vukovic, J.; Salamunic, I.; Sutlovic, D.; Kozina, B.; Boban, M. Comparison of acute effects of red wine, beer and vodka against hyperoxia-induced oxidative stress and increase in arterial stiffness in healthy humans. *Atherosclerosis* **2011**. [CrossRef] [PubMed]

107. Codoñer-Franch, P.; Hernández-Aguilar, M.T.; Navarro-Ruiz, A.; López-Jaén, A.B.; Borja-Herrero, C.; Valls-Bellés, V. Diet supplementation during early lactation with non-alcoholic beer increases the antioxidant properties of breastmilk and decreases the oxidative damage in breastfeeding mothers. *Breastfeed. Med.* **2013**. [CrossRef] [PubMed]

108. Martínez Alvarez, J.R.; Bellés, V.V.; López-Jaén, A.B.; Marín, A.V.; Codoñer-Franch, P. Effects of alcohol-free beer on lipid profile and parameters of oxidative stress and inflammation in elderly women. *Nutrition* **2009**. [CrossRef] [PubMed]

109. Croft, K.D.; Puddey, I.B.; Rakic, V.; Abu-Amsha, R.; Dimmitt, S.B.; Beilin, L.J. Oxidative susceptibility of low-density lipoproteins—Influence of regular alcohol use. *Alcohol. Clin. Exp. Res.* **1996**. [CrossRef]

110. Franco, L.; Bravo, R.; Galan, C.; Sanchez, C.; Rodriguez, A.B.; Barrigs, C.; Cubero Juánez, J. Effects of beer, Hops (Humulus lupulus) on total antioxidant capacity in plasma of stressed subjects. *Cell Membr. Free Radic. Res.* **2013**, *5*, 232–235.

111. Hernández-Quiroz, F.; Nirmalkar, K.; Villalobos-Flores, L.E.; Murugesan, S.; Cruz-Narváez, Y.; Rico-Arzate, E.; Hoyo-Vadillo, C.; Chavez-Carbajal, A.; Pizano-Zárate, M.L.; García-Mena, J. Influence of moderate beer consumption on human gut microbiota and its impact on fasting glucose and β-cell function. *Alcohol* **2020**. [CrossRef]

112. Quifer-Rada, P.; Chiva-Blanch, G.; Jáuregui, O.; Estruch, R.; Lamuela-Raventós, R.M. A discovery-driven approach to elucidate urinary metabolome changes after a regular and moderate consumption of beer and nonalcoholic beer in subjects at high cardiovascular risk. *Mol. Nutr. Food Res.* **2017**. [CrossRef]

113. Chiva-Blanch, G.; Magraner, E.; Condines, X.; Valderas-Martínez, P.; Roth, I.; Arranz, S.; Casas, R.; Navarro, M.; Hervas, A.; Sisó, A.; et al. Effects of alcohol and polyphenols from beer on atherosclerotic biomarkers in high cardiovascular risk men: A randomized feeding trial. *Nutr. Metab. Cardiovasc. Dis.* **2015**. [CrossRef] [PubMed]

114. Chiva-Blanch, G.; Condines, X.; Magraner, E.; Roth, I.; Valderas-Martínez, P.; Arranz, S.; Casas, R.; Martínez-Huélamo, M.; Vallverdú-Queralt, A.; Quifer-Rada, P.; et al. The non-alcoholic fraction of beer increases stromal cell derived factor 1 and the number of circulating endothelial progenitor cells in high cardiovascular risk subjects: A randomized clinical trial. *Atherosclerosis* **2014**. [CrossRef] [PubMed]

115. Quifer-Rada, P.; Martínez-Huélamo, M.; Chiva-Blanch, G.; Jáuregui, O.; Estruch, R.; Lamuela-Raventós, R.M. Urinary isoxanthohumol is a specific and accurate biomarker of beer consumption. *J. Nutr.* **2014**. [CrossRef] [PubMed]

116. Scherr, J.; Nieman, D.C.; Schuster, T.; Habermann, J.; Rank, M.; Braun, S.; Pressler, A.; Wolfarth, B.; Halle, M. Nonalcoholic beer reduces inflammation and incidence of respiratory tract illness. *Med. Sci. Sports Exerc.* **2012**. [CrossRef]

117. Gorinstein, S.; Caspi, A.; Libman, I.; Leontowicz, H.; Leontowicz, M.; Tashma, Z.; Katrich, E.; Jastrzebski, Z.; Trakhtenberg, S. Bioactivity of beer and its influence on human metabolism. *Int. J. Food Sci. Nutr.* **2007**. [CrossRef]

118. Zilkens, R.R.; Burke, V.; Hodgson, J.M.; Barden, A.; Beilin, L.J.; Puddey, I.B. Red wine and beer elevate blood pressure in normotensive men. *Hypertension* **2005**. [CrossRef]

119. Di Giuseppe, R.; De Lorgeril, M.; Salen, P.; Laporte, F.; Di Castelnuovo, A.; Krogh, V.; Siani, A.; Arnout, J.; Cappuccio, F.P.; Van Dongen, M.; et al. Alcohol consumption and n-3 polyunsaturated fatty acids in healthy men and women from 3 European populations. *Am. J. Clin. Nutr.* **2009**. [CrossRef]

120. Padro, T.; Muñoz-García, N.; Vilahur, G.; Chagas, P.; Deyà, A.; Antonijoan, R.M.; Badimon, L. Moderate beer intake and cardiovascular health in overweight individuals. *Nutrients* **2018**. [CrossRef]

121. Godos, J.; Sinatra, D.; Blanco, I.; Mulè, S.; La Verde, M.; Marranzano, M. Association between dietary phenolic acids and hypertension in a mediterranean cohort. *Nutrients* **2017**, *9*, 1069. [CrossRef]

122. González-Zancada, N.; Redondo-Useros, N.; Díaz, L.E.; Gómez-Martínez, S.; Marcos, A.; Nova, E. Association of Moderate Beer Consumption with the Gut Microbiota and SCFA of Healthy Adults. *Molecules* **2020**, *25*, 4772. [CrossRef]

123. Le Roy, C.I.; Wells, P.M.; Si, J.; Raes, J.; Bell, J.T.; Spector, T.D. Red Wine Consumption Associated With Increased Gut Microbiota α-Diversity in 3 Independent Cohorts. *Gastroenterology* **2020**. [CrossRef] [PubMed]

124. Chao, C.; Haque, R.; Van Den Eeden, S.K.; Caan, B.J.; Poon, K.Y.T.; Quinn, V.P. Red wine consumption and risk of prostate cancer: The California men's health study. *Int. J. Cancer* **2010**. [CrossRef]

125. Mehlig, K.; Skoog, I.; Guo, X.; Schütze, M.; Gustafson, D.; Waern, M.; Östling, S.; Björkelund, C.; Lissner, L. Alcoholic beverages and incidence of dementia: 34-Year follow-up of the prospective population study of women in Göteborg. *Am. J. Epidemiol.* **2008**. [CrossRef]

126. Petti, S.; Scully, C. Association between different alcoholic beverages and leukoplakia among non- to moderate-drinking adults: A matched case-control study. *Eur. J. Cancer* **2006**. [CrossRef]

127. Nettleton, J.A.; Harnack, L.J.; Scrafford, C.G.; Mink, P.J.; Barraj, L.M.; Jacobs, D.R. Dietary flavonoids and flavonoid-rich foods are not associated with risk of type 2 diabetes in postmenopausal women. *J. Nutr.* **2006**. [CrossRef]

128. Petti, S.; Scully, C. Oral cancer: The association between nation-based alcohol-drinking profiles and oral cancer mortality. *Oral Oncol.* **2005**. [CrossRef]

129. Schoonen, W.M.; Salinas, C.A.; Kiemeney, L.A.L.M.; Stanford, J.L. Alcohol consumption and risk of prostate cancer in middle-aged men. *Int. J. Cancer* **2005**. [CrossRef]

130. Infante-Rivard, C.; Krajinovic, M.; Labuda, D.; Sinnett, D. Childhood acute lymphoblastic leukemia associated with parental alcohol consumption and polymorphisms of carcinogen-metabolizing genes. *Epidemiology* **2002**. [CrossRef]

131. Ussar, S.; Griffin, N.W.; Bezy, O.; Fujisaka, S.; Vienberg, S.; Softic, S.; Deng, L.; Bry, L.; Gordon, J.I.; Kahn, C.R. Interactions between gut microbiota, host genetics and diet modulate the predisposition to obesity and metabolic syndrome. *Cell Metab.* **2015**. [CrossRef]
132. Chávez-Carbajal, A.; Nirmalkar, K.; Pérez-Lizaur, A.; Hernández-Quiroz, F.; Ramírez-Del-Alto, S.; García-Mena, J.; Hernández-Guerrero, C. Gut microbiota and predicted metabolic pathways in a sample of Mexican women affected by obesity and obesity plus metabolic syndrome. *Int. J. Mol. Sci.* **2019**, *20*, 438. [CrossRef]
133. Zhao, H.; Chen, W.; Lu, J.; Zhao, M. Phenolic profiles and antioxidant activities of commercial beers. *Food Chem.* **2010**. [CrossRef]
134. Nardini, M.; Garaguso, I. Characterization of bioactive compounds and antioxidant activity of fruit beers. *Food Chem.* **2020**, *305*, 125437. [CrossRef] [PubMed]
135. Legua, P.; Serrano, M.; Melgarejo, P.; Valero, D.; Martínez, J.J.; Martínez, R.; Hernández, F. Quality parameters, biocompounds and antioxidant activity in fruits of nine quince (Cydonia oblonga Miller) accessions. *Sci. Hortic. (Amsterdam)* **2013**, *154*, 61–65. [CrossRef]
136. Wojdyło, A.; Oszmiański, J.; Bielicki, P. Polyphenolic composition, antioxidant activity, and polyphenol oxidase (PPO) activity of quince (cydonia oblonga miller) varieties. *J. Agric. Food Chem.* **2013**. [CrossRef]
137. Zapata, P.J.; Martínez-Esplá, A.; Gironés-Vilaplana, A.; Santos-Lax, D.; Noguera-Artiaga, L.; Carbonell-Barrachina, Á.A. Phenolic, volatile, and sensory profiles of beer enriched by macerating quince fruits. *LWT* **2019**. [CrossRef]
138. Sun, J.; Li, L.; You, X.; Li, C.; Zhang, E.; Li, Z.; Chen, G.; Peng, H. Phenolics and polysaccharides in major tropical fruits: Chemical compositions, analytical methods and bioactivities. *Anal. Methods* **2011**. [CrossRef]
139. Gasinski, A.; Kawa-Rygielska, J.; Szumny, A.; Czubaszek, A.; Gasior, J.; Pietrzak, W. Volatile compounds content, physicochemical parameters, and antioxidant activity of beers with addition of mango fruit (Mangifera Indica). *Molecules* **2020**, *25*, 3033. [CrossRef]
140. Lako, J.; Trenerry, V.C.; Wahlqvist, M.; Wattanapenpaiboon, N.; Sotheeswaran, S.; Premier, R. Phytochemical flavonols, carotenoids and the antioxidant properties of a wide selection of Fijian fruit, vegetables and other readily available foods. *Food Chem.* **2007**. [CrossRef]
141. Humia, B.V.; Santos, K.S.; Schneider, J.K.; Leal, I.L.; de Abreu Barreto, G.; Batista, T.; Machado, B.A.S.; Druzian, J.I.; Krause, L.C.; da Costa Mendonça, M.; et al. Physicochemical and sensory profile of Beauregard sweet potato beer. *Food Chem.* **2020**. [CrossRef]
142. Talhaoui, N.; Taamalli, A.; Gómez-Caravaca, A.M.; Fernández-Gutiérrez, A.; Segura-Carretero, A. Phenolic compounds in olive leaves: Analytical determination, biotic and abiotic influence, and health benefits. *Food Res. Int.* **2015**, *77*, 92–108. [CrossRef]
143. Guglielmotti, M.; Passaghe, P.; Buiatti, S. Use of olive (Olea europaea L.) leaves as beer ingredient, and their influence on beer chemical composition and antioxidant activity. *J. Food Sci.* **2020**, *85*, 2278–2285. [CrossRef] [PubMed]
144. Deng, Y.; Lim, J.; Nguyen, T.T.H.; Mok, I.-K.; Piao, M.; Kim, D. Composition and biochemical properties of ale beer enriched with lignans from Schisandra chinensis Baillon (omija) fruits. *Food Sci. Biotechnol.* **2020**, *29*, 609–617. [CrossRef] [PubMed]
145. Bhuyan, D.J.; Barooah, M.S.; Bora, S.S.; Singaravadivel, K. Biochemical and nutritional analysis of rice beer of North East India. *Indian J. Tradit. Knowl.* **2014**, *13*, 142–148.
146. Agu, R.C. Comparative study of experimental beers brewed from millet, sorghum and barley malts. *Process. Biochem.* **1995**. [CrossRef]
147. Palmer, G.H.; Etokakpan, O.U.; Igyor, M.A. Sorghum as brewing material. *MIRCEN J. Appl. Microbiol. Biotechnol.* **1989**, *5*, 265–275. [CrossRef]
148. Serna-Saldivar, S.O. Maize: Foods from Maize. In *Encyclopedia of Food Grains*, 2nd ed.; Wrigley, C., Corke, H., Seetharaman, K., Faubion, J., Eds.; Academic Press: Oxford, UK, 2016; pp. 97–109. ISBN 978-0-12-394786-4.
149. Escalante-Aburto, A.; Ramírez-Wong, B.; Torres-Chávez, P.I.; Manuel Barrón-Hoyos, J.; de Dios Figueroa-Cárdenas, J.; López-Cervantes, J. The nixtamalization process and its effect on anthocyanin content of pigmented maize, A review. *Rev. Fitotec. Mex.* **2013**. [CrossRef]
150. Flores-Calderón, A.M.D.; Luna, H.; Escalona-Buendía, H.B.; Verde-Calvo, J.R. Chemical characterization and antioxidant capacity in blue corn (Zea mays L.) malt beers. *J. Inst. Brew.* **2017**. [CrossRef]
151. Romero-Medina, A.; Estarrón-Espinosa, M.; Verde-Calvo, J.R.; Lelièvre-Desmas, M.; Escalona-Buendiá, H.B. Renewing Traditions: A Sensory and Chemical Characterisation of Mexican Pigmented Corn Beers. *Foods* **2020**, *9*, 886. [CrossRef]
152. Fumi, M.D.; Galli, R.; Lambri, M.; Donadini, G.; De Faveri, D.M. Effect of full-scale brewing process on polyphenols in Italian all-malt and maize adjunct lager beers. *J. Food Compos. Anal.* **2011**, *24*, 568–573. [CrossRef]
153. Klose, C.; Mauch, A.; Wunderlich, S.; Thiele, F.; Zarnkow, M.; Jacob, F.; Arendt, E.K. Brewing with 100% oat malt. *J. Inst. Brew.* **2011**. [CrossRef]
154. Di Ghionno, L.; Sileoni, V.; Marconi, O.; De Francesco, G.; Perretti, G. Comparative study on quality attributes of gluten-free beer from malted and unmalted teff [Eragrostis tef (zucc.) trotter]. *LWT Food Sci. Technol.* **2017**. [CrossRef]
155. Liu, Y.; Ma, T.J.; Chen, J. Changes of the flavonoid and phenolic acid content and antioxidant activity of tartary buckwheat beer during the fermentation. *Adv. Mater. Res.* **2013**, *781–784*, 1619–1624. [CrossRef]
156. Kordialik-Bogacka, E.; Bogdan, P.; Pielech-Przybylska, K.; Michałowska, D. Suitability of unmalted quinoa for beer production. *J. Sci. Food Agric.* **2018**. [CrossRef] [PubMed]
157. Zarnkow, M.; Faltermaier, A.; Back, W.; Gastl, M.; Arendt, E.K. Evaluation of different yeast strains on the quality of beer produced from malted proso millet (*Panicum miliaceum* L.). *Eur. Food Res. Technol.* **2010**. [CrossRef]
158. Lyumugabe, F.; Bajyana Songa, E.; Wathelet, J.P.; Thonart, P. Volatile compounds of the traditional sorghum beers "ikigage" brewed with Vernonia amygdalina "umubirizi.". *Cerevisia* **2013**. [CrossRef]

159. Onyenekwe, P.C.; Erhabor, G.O.; Akande, S.A. Characterisation of aroma volatiles of indigenous alcoholic beverages: Burukutu and pito. *Nat. Prod. Res.* **2016**. [CrossRef]
160. Handique, P.; Deka, A.K.; Deka, D.C. Antioxidant Properties and Phenolic Contents of Traditional Rice-Based Alcoholic Beverages of Assam, India. *Natl. Acad. Sci. Lett.* **2020**. [CrossRef]
161. Kordialik-Bogacka, E.; Bogdan, P.; Diowksz, A. Malted and unmalted oats in brewing. *J. Inst. Brew.* **2014**. [CrossRef]
162. Wu, D.; Cai, G.; Li, X.; Li, B.; Lu, J. Cloning and expression of ferulic acid esterase gene and its effect on wort filterability. *Biotechnol. Lett.* **2018**. [CrossRef]
163. Cui, Y.; Wang, A.; Zhang, Z.; Speers, R.A. Enhancing the levels of 4-vinylguaiacol and 4-vinylphenol in pilot-scale top-fermented wheat beers by response surface methodology. *J. Inst. Brew.* **2015**. [CrossRef]
164. Xu, L.; Zhang, H.; Cui, Y.; Zeng, D.; Bao, X. Increasing the level of 4-vinylguaiacol in top-fermented wheat beer by secretory expression of ferulic acid decarboxylase from Bacillus pumilus in brewer's yeast. *Biotechnol. Lett.* **2020**. [CrossRef]
165. Brányik, T.; Silva, D.P.; Baszczyňski, M.; Lehnert, R.; Almeida e Silva, J.B. A review of methods of low alcohol and alcohol-free beer production. *J. Food Eng.* **2012**, *108*, 493–506. [CrossRef]
166. Liguori, L.; De Francesco, G.; Russo, P.; Perretti, G.; Albanese, D.; Di Matteo, M. Quality Attributes of Low-Alcohol Top-Fermented Beers Produced by Membrane Contactor. *Food Bioprocess. Technol.* **2016**. [CrossRef]
167. De Fusco, D.O.; Madaleno, L.L.; Del Bianchi, V.L.; da Silva Bernardo, A.; Assis, R.R.; de Almeida Teixeira, G.H. Development of low-alcohol isotonic beer by interrupted fermentation. *Int. J. Food Sci. Technol.* **2019**. [CrossRef]
168. Desbrow, B.; Murray, D.; Leveritt, M. Beer as a sports drink? Manipulating beer's ingredients to replace lost fluid. *Int. J. Sport Nutr. Exerc. Metab.* **2013**. [CrossRef] [PubMed]
169. Maughan, R.J. The sports drink as a functional food: Formulations for successful performance. *Proc. Nutr. Soc.* **1998**. [CrossRef]
170. Menz, G.; Vriesekoop, F.; Zarei, M.; Zhu, B.; Aldred, P. The growth and survival of food-borne pathogens in sweet and fermenting brewers' wort. *Int. J. Food Microbiol.* **2010**. [CrossRef] [PubMed]
171. Meneses, N.G.T.; Martins, S.; Teixeira, J.A.; Mussatto, S.I. Influence of extraction solvents on the recovery of antioxidant phenolic compounds from brewer's spent grains. *Sep. Purif. Technol.* **2013**. [CrossRef]
172. Cooray, S.T.; Chen, W.N. Valorization of brewer's spent grain using fungi solid-state fermentation to enhance nutritional value. *J. Funct. Foods* **2018**. [CrossRef]
173. Ferk, F.; Mišík, M.; Nersesyan, A.; Pichler, C.; Jäger, W.; Szekeres, T.; Marculescu, R.; Poulsen, H.E.; Henriksen, T.; Bono, R.; et al. Impact of xanthohumol (a prenylated flavonoid from hops) on DNA stability and other health-related biochemical parameters: Results of human intervention trials. *Mol. Nutr. Food Res.* **2016**. [CrossRef]
174. Wunderlich, S.; Zürcher, A.; Back, W. Enrichment of xanthohumol in the brewing process. *Mol. Nutr. Food Res.* **2005**. [CrossRef] [PubMed]
175. Karabín, M.; Jelínek, L.; Kinčl, T.; Hudcová, T.; Kotlíková, B.; Dostálek, P. New approach to the production of xanthohumol-enriched beers. *J. Inst. Brew.* **2013**. [CrossRef]
176. Machado, J.C.; Faria, M.A.; Melo, A.; Martins, Z.E.; Ferreira, I.M.P.L.V.O. Modeling of α-acids and xanthohumol extraction in dry-hopped beers. *Food Chem.* **2019**. [CrossRef] [PubMed]
177. Magalhães, P.J.; Dostalek, P.; Cruz, J.M.; Guido, L.F.; Barros, A.A. The impact of a xanthohumol-enriched hop product on the behavior of xanthohumol and isoxanthohumol in pale and dark beers: A pilot scale approach. *J. Inst. Brew.* **2008**. [CrossRef]
178. Benelli, R.; Ven, R.; Ciarlo, M.; Carlone, S.; Barbieri, O.; Ferrari, N. The AKT/NF-κB inhibitor xanthohumol is a potent anti-lymphocytic leukemia drug overcoming chemoresistance and cell infiltration. *Biochem. Pharmacol.* **2012**. [CrossRef]
179. EFSA Panel on Dietetic Products, Nutrition and Allergies (NDA). Scientific Opinion on the substantiation of health claims related to polyphenols in olive and protection of LDL particles from oxidative damage (ID 1333, 1638, 1639, 1696, 2865), maintenance of normal blood HDL cholesterol concentrations (ID 1639), mainte. *EFSA J.* **2011**, *9*, 2033. [CrossRef]
180. EFSA Panel on Dietetic Products, Nutrition and Allergies (NDA). Scientific Opinion on the modification of the authorisation of a health claim related to cocoa flavanols and maintenance of normal endothelium-dependent vasodilation pursuant to Article 13(5) of Regulation (EC) No 1924/2006 following a request in accordan. *EFSA J.* **2014**, *12*, 3654. [CrossRef]
181. Merinas-Amo, T.; Tasset-Cuevas, I.; Díaz-Carretero, A.M.; Alonso-Moraga, Á.; Calahorro, F. In vivo and in vitro studies of the role of lyophilised blond Lager beer and some bioactive components in the modulation of degenerative processes. *J. Funct. Foods* **2016**. [CrossRef]

Article

Terroir Effect on the Phenolic Composition and Chromatic Characteristics of Mencía/Jaen Monovarietal Wines: Bierzo D.O. (Spain) and Dão D.O. (Portugal)

Fernanda Cosme [1], Alice Vilela [1], Luís Moreira [2], Carla Moura [2], José A. P. Enríquez [2], Luís Filipe-Ribeiro [2] and Fernando M. Nunes [3,*]

[1] CQ-VR, Chemistry Research Centre-Vila Real, Food and Wine Chemistry Lab, Biology and Environment Department, University of Trás-os-Montes and Alto Douro, 5001-801 Vila Real, Portugal; fcosme@utad.pt (F.C.); avimoura@utad.pt (A.V.)

[2] CQ-VR, Chemistry Research Centre-Vila Real, Food and Wine Chemistry Lab, School of Life Sciences and Environment, University of Trás-os-Montes and Alto Douro, 5001-801 Vila Real, Portugal; al57742@utad.eu (L.M.); marina.prazelos@gmail.com (C.M.); apenri@gmail.com (J.A.P.E.); fmota@utad.pt (L.F.-R.)

[3] CQ-VR, Chemistry Research Centre-Vila Real, Food and Wine Chemistry Lab, Chemistry Department, University of Trás-os-Montes and Alto Douro, 5001-801 Vila Real, Portugal

* Correspondence: fnunes@utad.pt; Tel.: +351-259-350-000

Received: 29 November 2020; Accepted: 15 December 2020; Published: 18 December 2020

Abstract: 'Mencía'/'Jaen' it's an important red grape variety, exclusive of the Iberian Peninsula, used in wine production namely in Bierzo D.O. and Dão D.O., respectively. This work evaluates the effect of the two different "terroirs" on the phenolic composition and chromatic characteristics of 'Mencía'/'Jaen' monovarietal wines produced at an industrial scale in the same vintage. Using Principal Component Analysis (PCA), Partial Least Squares-Discrimination Analysis (PLS-DA), and Orthogonal PLS-DA (OPLS-DA) it was found that peonidin-3-coumaroylglucoside, petunidin-3-glucoside, malvidin-3-coumaroylglucoside, peonidin-3-glucoside, malvidin-3-acetylglucoside, malvidin-3-glucoside, and ferulic acid were the phenolic compounds with the highest differences between the two regions. PLS regression allowed to correlate the differences in lightness (L*) and redness (a*) of wines from 'Jaen' and 'Mencía' to differences in colored anthocyanins, polymeric pigments, total pigments, total anthocyanins, cyanidin-3-acetylglucoside, delphinidin-3-acetylglucoside, delphinidin-3-glucoside, peonidin-3-coumaroylglucoside, petunidin-3-glucoside and malvidin-3-glucoside in wines, and the colorless ferulic, caffeic, and coutaric acids, and ethyl caffeate. The wines a* values were more affected by colored anthocyanins, ferulic acid, total anthocyanins, delphinidin-3-acetylglucoside, delphinidin-3-glucoside and petunidin-3-acetylglucoside, and catechin. The positive influence of ferulic acid in the a* values and ferulic, caffeic, coutaric acids, and ethyl caffeate on the L* values can be due to the co-pigmentation phenomena. The higher dryness and lower temperatures during the September nights in this vintage might explain the differences observed in the anthocyanin content and chromatic characteristics of the wines.

Keywords: red wine; Mencía; Jean; terroir; anthocyanins; phenolic acids; flavonols; wine color

---

1. Introduction

'Mencía' in Spain and 'Jaen' in Portugal is a grape variety present almost exclusively in the northwest of the Iberian Peninsula whose economic importance has been growing in the last years. 'Mencía' is extensively cultivated in Galicia, north-western Spain, to produce quality red wines with five Denomination of Origin (D.O.) Rías Baixas in the province of Pontevedra, Ribeiro, Valdeorras and Monterrei in the province of Ourense and Ribeira Sacra between the province of Ourense and Lugo. However, this grape variety also predominates in the Bierzo region (Figure 1) located in the northwest of the province of León (Castile and León, Spain) originating the 'Mencía' red wines with Bierzo D.O. [1,2]. In Portugal, this grape variety is known as 'Jaen' and is almost exclusively cultivated in the Dão Demarcated Region of Portugal (Figure 1) [1,2], whose importance for wine production in this region is growing, and a significant increase in the vineyard area of this grape variety between 2013 to 2017, from 1 731 ha [3] to 3 789 ha [3], respectively, has been observed.

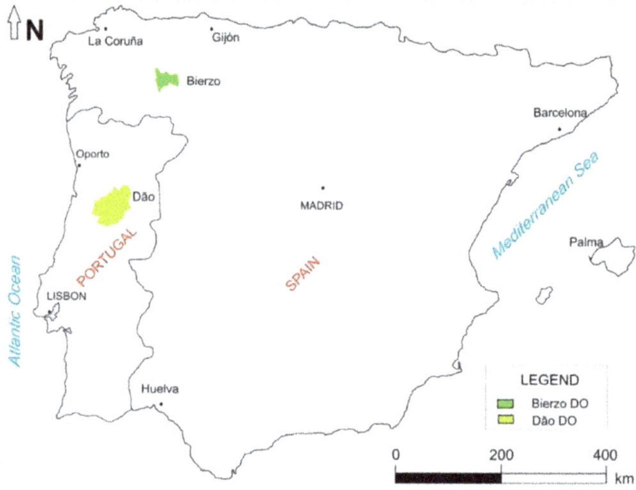

Figure 1. Location of Bierzo D.O. and Dão D.O. where 'Mencía' and 'Jaen' monovarietal wines are produced.

The wine quality is determined by the conditions of the growing area, of the vintage, agricultural practices that will influence the grape composition, and by the winemaking technology used. The same variety, growing in two different viticultural regions with diverse climatic conditions, results in wines with different content of phenolic compounds in the same vintage [4]. Several factors such as soil type, environmental conditions, agricultural practices, climatic conditions, vine phenology, or winemaking processesall contribute to the "terroir" effect that can change the chemical composition of grapes and wines [5–8]. In the resolution OIV/VITI 333/2010 [9] the International Organization of Vine and Wine (OIV), defines vitivinicultural "terroir"-vitivinicultural "terroir" as a concept that refers to an area in which collective knowledge of the interactions between the identifiable physical and biological environment and applied vitivinicultural practices develop, providing distinctive characteristics for the products originating from this area. "Terroir" includes specific soil, topography, climate, landscape characteristics, and biodiversity features.

The phenolic, aroma, and mineral composition, as well as the sensory descriptors of red wines vinified with 'Mencía' grape variety, were studied by several authors. There are various works concerning the characterization of the aroma profile of 'Mencía' red wines [10–16], with an observed change in the 'Mencía' wine aroma profile depending on the grape's geographic origin and vintage [17,18]. Also, pre-fermentative maceration techniques (enzymes, refrigeration, and cryomaceration) were shown to influence the aroma profile of 'Mencía' wines [19,20]. The use of indigenous yeasts has been used as a tool to increase the diversity of 'Mencía' monovarietal wines [21]. Other studies were performed on the content of minerals [22,23], and wine sensory descriptors [24].

Although phenolic compounds, including phenolic acids, anthocyanins, and tannins, contribute to wine color, bitterness, astringency, and antioxidant capacity [25,26], fewer studies were performed on the phenolic composition of 'Mencía' wines [27–31]. These studies did not include wines produced in the Bierzo D.O. and no studies were reported for the wines produced with Jaen grape variety in the Dão D.O.

It is expected that the geographical origin, related to the different edaphoclimatic conditions of the two regions, will impact differently on the composition of the wines and this information is of importance for the understanding of the enological potential and diversity of this grape variety in the two different regions. Therefore, the purpose of this work was to study the phenolic composition and chromatic characteristics of monovarietal 'Mencía' wines of the Bierzo D.O. and monovarietal 'Jaen' wines produced at the Dão D.O. at an industrial scale in the 2015 vintage to evaluate the impact of "terroir" on the wines produced from this important Iberian grape variety.

2. Results and Discussion

2.1. Effect of Bierzo D.O. and Dão D.O. "Terroirs" on the Anthocyanins and Phenolic Acid Composition of 'Mencía'/'Jaen' Wines—Unsupervised Analysis

To study the effect of Bierzo D.O. and Dão D.O. "terroirs" on the phenolic composition of 'Mencía'/'Jaen' monovarietal wines, the concentration of the individual phenolic compounds was determined by RP-HPLC-DAD (Reversed-Phase High-Performance Liquid Chromatography with Diode Array Detector, Tables 1 and 2). In all wines analyzed gallic, *trans*-caftaric, coutaric isomer, coutaric, caffeic, *p*-coumaric, and ferulic acids along with the ethyl esters of caffeic and coumaric acids, and the flavanol catechin was present (Table 1). The concentrations determined for the phenolic acids, phenolic esters, and catechin agreed with those described by García-Falcón et al. [27] and Alén-Ruiz et al. [28] for 'Mencía' wines produced in the Galician Ribeiro D.O. (Ourense).

Eleven and thirteen anthocyanins were identified and quantified in the 'Mencía' red wine produced in the Bierzo D.O. and in the 'Jaen' red wine produced in the Dão D.O., respectively (Table 2). For the 'Mencía' wines, the anthocyanin composition and concentrations were in accordance with those obtained by Rivas-Gonzalo et al. [32], García-Falcón et al. [27], and Soto Vázquez et al. [29], although García-Falcón et al. [27] only detected seven anthocyanins at the end of malolactic fermentation.

A principal component analysis (PCA) was applied to the anthocyanins and phenolic acids determined by HPLC-DAD (after normalization to zero mean and unit standard deviation) to reduce in an unbiased and unsupervised way the dimensionality of this multivariate dataset and identify the differences or similarities among the samples (Figure 2a) and to identify phenolic compounds responsible for the grouping or separation among the samples (Figure 2b). Figure 2a shows the score plot of the first two PC´s showing a clear separation of 'Mencía' and 'Jaen' wines. Additionally, replicate samples of the same label were grouped in the same cluster, although they were not overlapped, indicating that bottle to bottle variation occurred. The first PC (Principal Component), which explained 42% of the total variance of the original data set, correlated positively with malvidin-3-coumaroylglucoside (M3CG), petunidin-3-glucoside (Pet3G), malvidin-3-glucoside (M3G), malvidin-3-acetylglucoside (M3AG), peonidin-3-glucoside (Peo3G), cyanidin-3-glucoside (C3G), peonidin-3-coumaroylglycoside (Peo3CG), delphinidin-3-coumaroylglucoside (D3CG), and peonidin-3-acetylglucoside (Peo3AG). The second PC, which explained 14% of the total variance, correlated negatively with coutaric acid (Cout), ethyl ester of caffeic acid (EtCaf), caftaric acid (Caft), and coumaric acid (Coum) (Figure 2b). Therefore, the results from PCA allowed for the conclusion that 'Jaen' wine samples from Dão D.O. are separated from 'Mencía' wine samples from Bierzo D.O. according to PC1, showing a higher relative amount of M3CG, Pet3G, M3G, M3AG, Peo3G, C3G, Peo3CG, D3CG, and Peo3AG (Figure 2a,b).

Table 1. 'Mencía' and 'Jaen' red wines phenolic profile (phenolic acid and catechin) expressed in mg/L.

Wine Samples	Gallic Acid	Catechins	*trans*-Caftaric Acid	Coutaric Acid Isomer	Coutaric Acid	Caffeic Acid	*p*-Coumaric Acid	Ferulic Acid	Caffeic Acid Ethyl Ester	Coumaric Acid Ethyl Ester
Mencía	21.42 ± 0.02 [d]	8.14 ± 0.27 [ab]	33.77 ± 0.43 [d]	0.83 ± 0.01 [b]	21.00 ± 0.21 [g]	2.95 ± 0.94 [ab]	5.63 ± 0.16 [g]	0.81 ± 0.03 [ab]	2.22 ± 0.10 [e]	0.41 ± 0.00 [a]
Mencía	20.76 ± 0.01 [d]	7.28 ± 0.47 [a]	5.10 ± 0.02 [a]	1.25 ± 0.01 [c]	5.63 ± 0.00 [b]	3.96 ± 0.01 [b]	2.58 ± 0.02 [d]	2.40 ± 0.00 [d]	1.40 ± 0.14 [d]	2.06 ± 0.06 [c]
Mencía	7.93 ± 0.10 [ab]	7.10 ± 0.07 [a]	35.42 ± 0.58 [d]	0.60 ± 0.03 [a]	19.76 ± 0.35 [fg]	2.19 ± 0.30 [a]	3.38 ± 0.04 [e]	3.22 ± 0.02 [e]	2.93 ± 0.07 [f]	0.51 ± 0.33 [ab]
Mencía	19.54 ± 0.34 [cd]	8.79 ± 2.38 [b]	17.28 ± 3.29 [b]	0.56 ± 0.07 [a]	9.83 ± 0.29 [c]	2.38 ± 0.05 [a]	5.78 ± 0.10 [f]	0.92 ± 0.01 [ab]	1.27 ± 0.02 [d]	2.72 ± 0.09 [d]
Mencía	17.46 ± 0.44 [c]	2.37 ± 0.00 [a]	34.37 ± 0.39 [c]	0.72 ± 0.00 [b]	13.29 ± 0.19 [d]	3.78 ± 0.05 [b]	0.86 ± 0.01 [b]	0.66 ± 0.01 [a]	0.40 ± 0.01 [ab]	0.51 ± 0.02 [a]
Mencía	21.88 ± 0.33 [d]	6.11 ± 0.18 [a]	36.50 ± 2.39 [e]	1.30 ± 0.13 [c]	19.91 ± 1.26 [fg]	2.91 ± 0.018 [ab]	4.01 ± 0.26 [f]	1.62 ± 0.13 [c]	1.43 ± 0.19 [d]	2.38 ± 0.14 [cd]
Mencía	21.61 ± 0.03 [d]	7.26 ± 80.71 [a]	29.78 ± 1.16 [d]	0.73 ± 0.01 [b]	15.77 ± 0.63 [e]	2.80 ± 0.15 [a]	1.31 ± 0.04 [c]	2.70 ± 0.10 [d]	1.02 ± 0.15 [c]	0.65 ± 0.03 [b]
Mencía	6.16 ± 0.35 [a]	11.40 ± 0.85 [bc]	4.90 ± 0.24 [a]	0.33 ± 0.04 [a]	1.68 ± 0.42 [a]	2.13 ± 0.52 [a]	1.59 ± 0.01 [c]	0.79 ± 0.00 [ab]	0.11 ± 0.07 [a]	0.47 ± 0.02 [a]
Mencía	22.55 ± 2.25 [d]	10.35 ± 1.27 [b]	30.24 ± 0.12 [d]	0.81 ± 0.00 [b]	16.34 ± 0.08 [e]	2.82 ± 0.01 [a]	4.28 ± 0.14 [f]	0.61 ± 0.05 [a]	0.94 ± 0.09 [c]	0.58 ± 0.04 [b]
Mencía	29.04 ± 2.75 [e]	10.28 ± 3.52 [b]	18.14 ± 0.38 [b]	0.69 ± 0.17 [ab]	6.91 ± 0.23 [b]	3.46 ± 0.02 [b]	3.01 ± 0.04 [de]	2.44 ± 0.14 [d]	0.22 ± 0.02 [a]	0.85 ± 0.10 [b]
Mencía	18.98 ± 0.18 [cd]	10.45 ± 4.14 [b]	23.61 ± 0.21 [c]	1.11 ± 0.02 [bc]	10.15 ± 0.05 [c]	2.80 ± 0.13 [a]	1.59 ± 0.01 [c]	0.84 ± 0.08 [ab]	0.49 ± 0.01 [b]	0.92 ± 0.00 [b]
Mencía	20.55 ± 0.06 [d]	8.52 ± 0.64 [b]	33.47 ± 2.62 [d]	1.00 ± 0.25 [bc]	18.81 ± 1.18 [f]	2.68 ± 0.19 [a]	3.17 ± 0.18 [e]	0.84 ± 0.02 [ab]	0.70 ± 0.05 [b]	0.85 ± 0.17 [b]
Mencía	22.01 ± 0.10 [d]	6.22 ± 0.01 [a]	39.24 ± 0.77 [ef]	0.23 ± 0.09 [a]	14.93 ± 0.29 [de]	1.96 ± 0.03 [a]	0.93 ± 0.02 [b]	0.99.1 ± 0.00 [b]	0.71 ± 0.02 [bc]	0.62 ± 0.01 [b]
Jaen	17.52 ± 0.37 [c]	13.91 ± 0.54 [bc]	29.64 ± 0.24 [d]	0.64 ± 0.11 [a]	9.60 ± 0.24 [c]	3.76 ± 0.05 [b]	1.01 ± 0.15 [b]	4.28 ± 0.17 [f]	0.64 ± 0.02 [b]	0.67 ± 0.08 [b]
Jaen	19.52 ± 0.46 [cd]	14.11 ± 0.28 [bc]	42.18 ± 1.21 [f]	0.46 ± 0.09 [a]	16.61 ± 0.29 [e]	3.44 ± 0.08 [b]	0.59 ± 0.09 [ab]	2.51 ± 0.04 [d]	0.16 ± 0.02 [a]	0.17 ± 0.05 [a]
Jaen	20.87 ± 1.08 [d]	17.03 ± 0.80 [c]	28.18 ± 0.69 [c]	0.77 ± 0.02 [b]	15.81 ± 0.21 [e]	3.56 ± 0.09 [b]	0.36 ± 0.00 [a]	2.67 ± 0.03 [d]	0.24 ± 0.08 [a]	0.26 ± 0.01 [a]
Jaen	16.32 ± 0.50 [c]	7.36 ± 0.22 [a]	17.14 ± 0.33 [b]	0.55 ± 0.26 [a]	15.81 ± 0.21 [e]	3.69 ± 0.12 [b]	0.43 ± 0.33 [a]	3.74 ± 0.16 [c]	0.14 ± 0.01 [a]	0.35 ± 0.01 [a]
Jaen	11.18 ± 0.51 [b]	10.26 ± 1.37 [b]	26.06 ± 0.36 [c]	1.93 ± 0.16 [d]	10.15 ± 0.12 [c]	2.18 ± 0.17 [a]	0.96 ± 0.08 [b]	1.92 ± 0.07 [c]	0.56 ± 0.03 [b]	0.51 ± 0.03 [a]

Values are presented as mean ± standard deviation ($n = 2$); Means within a column followed by the same superscript letter are not significantly different (Tukey $p < 0.05$).

Table 2. 'Mencía' and 'Jaen' red wines monomeric anthocyanins profile expressed in mg/L.

Wine Samples	D3G	C3G	Pet3G	Peo3G	M3G	D3AG	C3AG	Pet3AG	Peo3AG	M3AG	D3CG	Peo3CG	M3CG
Mencía	1.98 ± 0.01 [c]	6.31 ± 0.19 [c]	9.84 ± 0.19 [c]	10.02 ± 0.10 [b]	43.44 ± 0.69 [b]	12.82 ± 0.48 [f]	1.96 ± 0.15 [f]	1.00 ± 0.19 [b]	3.57 ± 0.04 [e]	5.08 ± 0.21 [ab]	nd	nd	3.01 ± 0.08 [b]
Mencía	2.76 ± 0.02 [d]	5.52 ± 0.07 [b]	8.19 ± 0.03 [bc]	13.19 ± 0.06 [bc]	44.31 ± 0.10 [b]	3.34 ± 0.01 [a]	0.52 ± 0.00 [b]	1.05 ± 0.04 [b]	2.23 ± 0.01 [c]	5.84 ± 0.05 [b]	nd	nd	3.45 ± 0.03 [bc]
Mencía	4.52 ± 0.14 [e]	11.04 ± 1.30 [e]	16.10 ± 0.26 [d]	16.80 ± 0.18 [c]	83.43 ± 1.30 [c]	3.06 ± 0.03 [a]	nd	1.95 ± 0.01 [c]	1.99 ± 0.01 [c]	11.96 ± 0.16 [c]	nd	nd	5.75 ± 0.06 [d]
Mencía	1.69 ± 0.02 [b]	4.58 ± 0.05 [b]	8.29 ± 0.04 [bc]	9.28 ± 0.04 [b]	48.57 ± 0.38 [b]	2.96 ± 0.02 [a]	0.49 ± 0.02 [b]	1.09 ± 0.00 [b]	1.58 ± 0.05 [b]	7.31 ± 0.03 [b]	nd	nd	4.17 ± 0.03 [c]
Mencía	0.81 ± 0.02 [a]	1.72 ± 0.01 [a]	4.12 ± 0.10 [a]	4.07 ± 0.10 [a]	36.05 ± 0.54 [ab]	3.67 ± 0.83 [b]	0.84 ± 0.01 [c]	0.89 ± 0.01 [b]	nd	3.83 ± 0.03 [a]	nd	nd	1.72 ± 0.02 [a]
Mencía	2.99 ± 0.29 [d]	8.00 ± 0.62 [c]	14.89 ± 1.15 [d]	14.30 ± 0.14 [c]	79.00 ± 5.97 [c]	7.37 ± 0.49 [d]	1.27 ± 0.06 [d]	0.56 ± 0.02 [a]	2.80 ± 0.16 [d]	10.15 ± 0.75 [c]	nd	nd	7.52 ± 0.57 [e]
Mencía	2.09 ± 0.15 [c]	4.00 ± 0.36 [b]	6.30 ± 0.56 [ab]	7.06 ± 0.66 [a]	28.64 ± 2.40 [a]	5.01 ± 0.24 [bc]	0.91 ± 0.10 [c]	0.79 ± 0.0 [b]	0.84 ± 0.09 [a]	2.88 ± 0.30 [a]	nd	nd	1.89 ± 0.18 [a]
Mencía	2.04 ± 0.09 [c]	3.51 ± 0.14 [ab]	4.98 ± 0.17 [a]	4.36 ± 0.08 [a]	30.27 ± 0.37 [a]	1.97 ± 0.64 [a]	0.28 ± 0.05 [a]	0.16 ± 0.08 [a]	0.76 ± 0.01 [a]	3.18 ± 0.01 [a]	nd	nd	2.30 ± 0.07 [a]
Mencía	1.36 ± 0.00 [b]	4.63 ± 0.01 [b]	7.24 ± 0.04 [bc]	9.86 ± 0.03 [b]	36.78 ± 0.17 [ab]	9.75 ± 0.00 [e]	1.46 ± 0.00 [de]	0.70 ± 0.01 [b]	1.48 ± 0.02 [b]	3.77 ± 0.00 [a]	nd	nd	2.34 ± 0.02 [a]
Mencía	1.78 ± 0.09 [b]	4.71 ± 0.22 [b]	7.09 ± 0.18 [bc]	9.24 ± 0.40 [b]	31.43 ± 1.35 [a]	8.30 ± 0.46 [de]	1.14 ± 0.01 [c]	0.61 ± 0.03 [a]	1.18 ± 0.09 [ab]	3.51 ± 0.19 [a]	nd	nd	2.87 ± 0.14 [b]
Mencía	2.22 ± 0.01 [c]	4.31 ± 0.07 [b]	7.29 ± 0.08 [bc]	13.14 ± 0.07 [bc]	35.46 ± 0.42 [ab]	2.58 ± 0.05 [a]	0.51 ± 0.02 [b]	0.89 ± 0.02 [b]	1.06 ± 0.01 [a]	2.83 ± 0.03 [a]	nd	nd	2.94 ± 0.0 3 [b]
Mencía	1.77 ± 0.26 [bc]	5.51 ± 0.63 [b]	7.84 ± 1.94 [bc]	9.56 ± 3.12 [b]	34.73 ± 5.59 [ab]	4.08 ± 0.66 [b]	0.75 ± 0.15 [b]	0.70 ± 0.19 [b]	1.10 ± 0.17 [a]	3.34 ± 0.45 [a]	nd	nd	2.63 ± 0.42 [ab]
Mencía	0.65 ± 0.01 [a]	8.79 ± 0.03 [d]	14.70 ± 0.26 [d]	27.01 ± 0.10 [e]	133.27 ± 0.59 [de]	1.83 ± 0.04 [c]	0.45 ± 0.01 [b]	0.95 ± 0.02 [b]	2.64 ± 0.03 [d]	6.64 ± 0.23 [b]	nd	nd	2.07 ± 0.06 [a]
Jaen	1.55 ± 0.01 [b]	19.70 ± 0.19 [f]	40.10 ± 0.41 [g]	29.89 ± 0.06 [f]	202.51 ± 0.24 [g]	7.34 ± 0.01 [d]	0.69 ± 0.01 [bc]	0.82 ± 0.01 [b]	8.10 ± 0.03 [g]	35.56 ± 0.12 [f]	0.45 ± 0.01 [a]	3.59 ± 0.17 [c]	24.72 ± 0.15 [h]
Jaen	2.02 ± 0.08 [c]	27.54 ± 0.05 [g]	41.78 ± 0.03 [g]	32.22 ± 0.11 [f]	198.12 ± 0.00 [g]	13.33 ± 0.13 [f]	1.68 ± 0.20 [ef]	0.99 ± 0.03 [b]	3.07 ± 0.15 [d]	26.88 ± 0.05 [e]	0.54 ± 0.15 [a]	3.91 ± 0.16 [d]	28.02 ± 0.01 [i]
Jaen	2.36 ± 0.03 [c]	10.84 ± 0.02 [e]	21.52 ± 0.20 [e]	21.48 ± 0.28 [d]	123.52 ± 1.17 [d]	8.34 ± 0.10 [de]	1.62 ± 0.03 [e]	0.96 ± 0.01 [b]	3.65 ± 0.05 [e]	16.85 ± 0.13 [d]	nd	1.53 ± 0.14 [a]	13.68 ± 0.01 [f]
Jaen	2.07 ± 0.04 [c]	10.27 ± 1.08 [de]	27.41 ± 1.63 [f]	19.98 ± 1.97 [d]	162.69 ± 4.04 [f]	6.16 ± 0.10 [c]	0.74 ± 0.03 [b]	0.76 ± 0.39 [b]	7.11 ± 0.21 [f]	27.40 ± 2.26 [e]	n.d.	2.12 ± 0.12 [b]	18.21 ± 0.05 [g]
Jaen	3.11 ± 0.13 [a]	8.13 ± 0.23 [c]	13.66 ± 0.98 [d]	24.57 ± 2.62 [e]	141.99 ± 12.8 [e]	5.96 ± 0.50 [c]	1.00 ± 0.07 [c]	0.79 ± 0.12 [b]	2.67 ± 0.28 [d]	19.39 ± 0.95 [d]	n.d.	3.74 ± 0.03 [cd]	13.64 ± 0.76 [f]

Values are presented as mean ± standard deviation (n = 2); Means within a column followed by the same superscript letter are not significantly different (Tukey $p < 0.05$). Delphinidin-3-glucoside (D3G), cyanidin-3-glucoside (C3G), peonidin-3-glucoside (Peo3G), petunidin-3-glucoside (Pet3G), malvidin-3-glucoside (M3G), delphinidin-3-acetylglucoside (D3AG), cyanidin-3-acetylglucoside (C3AG), peonidin-3-acetylglucoside (Peo3AG), petunidin-3-acetylglucoside (Pet3AG), malvidin-3-acetylglucoside (M3AG), delphinidin-3-coumaroylglucoside (D3CG), peonidin-3-coumaroylglucoside (Peo3CG), malvidin-3-coumaroylglucoside (M3CG).

(**a**)

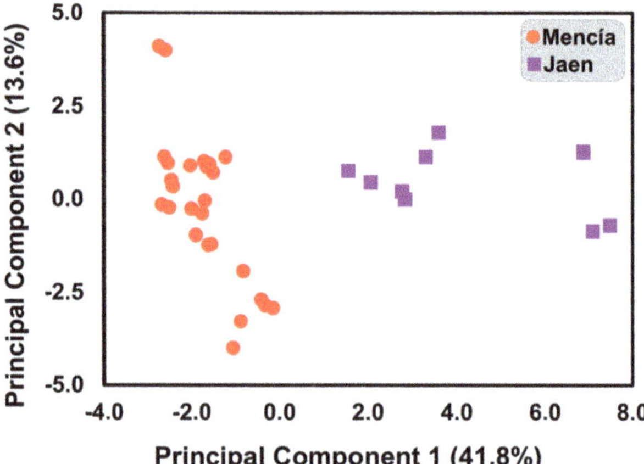

(**b**)

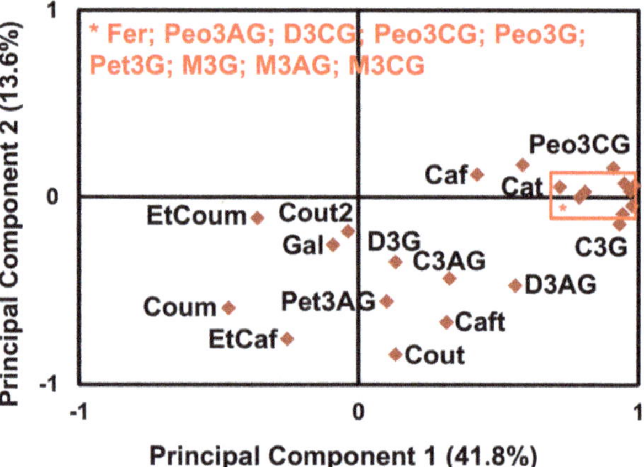

Figure 2. Sample scores projection on the first and second principal component (**a**) and variable loading on the first and second principal component (**b**) Delphinidin-3-glucoside (D3G), cyanidin-3-glucoside (C3G), peonidin-3-glucoside (Peo3G), petunidin-3-glucoside (Pet3G), malvidin-3-glucoside (M3G), delphinidin-3-acetylglucoside (D3AG), cyanidin-3-acetylglucoside (C3AG), peonidin-3-acetylglucoside (Peo3AG), petunidin-3-acetylglucoside (Pet3AG), malvidin-3-acetylglucoside (M3AG), delphinidin-3-coumaroylglucoside (D3CG), peonidin-3-coumaroylglucoside (Peo3CG), malvidin-3-coumaroylglucoside (M3CG), gallic acid (Gal), caftaric acid (Caft), coutaric acid (Cout), coutaric acid isomer (Cout2), caffeic acid (Caf), coumaric acid (Coum), catechin (Cat), ethyl ester of caffeic acid (EtCaf), ethyl ester of coumaric acid (EtCoum).

To identify what phenolic compounds were more affected by the "terroir" of Bierzo D.O. and Dão D.O., a volcano plot was used (Figure 3), by plotting the negative logarithm of statistical significance (*p* values) of Mann–Whitney non-parametric test on the y-axis and the normal logarithm of the fold change on the x-axis. Bierzo D.O. wines contained significantly less M3CG, Pet3G, M3G, M3AG, Peo3G, C3G, Peo3CG, D3CG, and Peo3AG and significantly higher levels of Coum and EtCoum. These results supported the results obtained by PCA.

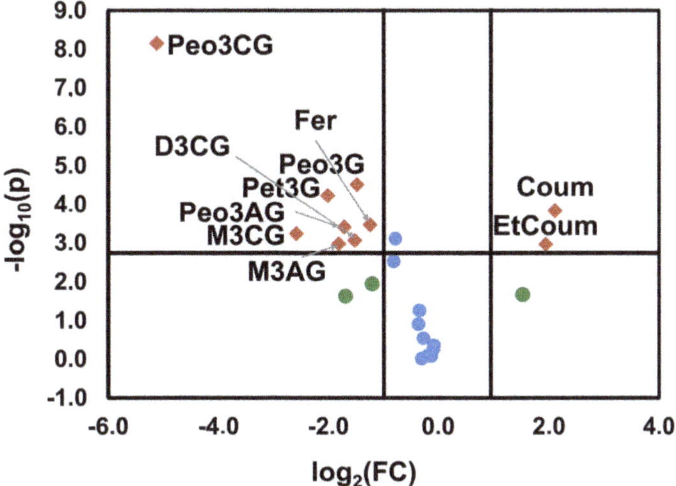

Figure 3. Volcano plot representing the statistical significance (*p*-values) on the Mann–Whitney non-parametric test and the fold change (FC) for the anthocyanins, phenolic acids, and catechin of wines 'Mencía'/'Jaen' wines from Bierzo D.O. and Dão D.O., respectively. The horizontal line represents the threshold of significance corrected for multiple comparisons by Bonferroni's method ($p = 0.00217$). Delphinidin-3-glucoside (D3G), cyanidin-3-glucoside (C3G), peonidin-3-glucoside (Peo3G), petunidin-3-glucoside (Pet3G), malvidin-3-glucoside (M3G), delphinidin-3-acetylglucoside (D3AG), cyanidin-3-acetylglucoside (C3AG), peonidin-3-acetylglucoside (Peo3AG), petunidin-3-acetylglucoside (Pet3AG), malvidin-3-acetylglucoside (M3AG), delphinidin-3-coumaroylglucoside (D3CG), peonidin-3-coumaroylglucoside (Peo3CG), malvidin-3-coumaroylglucoside (M3CG), gallic acid (Gal), caftaric acid (Caft), coutaric acid (Cout), coutaric acid isomer (Cout2), caffeic acid (Caf), coumaric acid (Coum), catechin (Cat), ethyl ester of caffeic acid (EtCaf), ethyl ester of coumaric acid (EtCoum).

2.2. Effect of Bierzo D.O. and Dão D.O. Terroirs on the Anthocyanins and Phenolic Acid Composition of 'Mencía'/'Jaen' Wines—Discrimination and Variable Importance

Partial least squares-discriminant analysis (PLS-DA) and orthogonal PLS-DA (OPLS-DA) were used to access their prediction efficiency and most importantly determine the main characteristics that distinguish the wines from the two terroirs concerning their phenolic composition. The PLS-DA model was developed using the standardized phenolic composition of 'Mencía' and 'Jaen' wines (Figure 4). Figure 4a shows the score plot of the first two components of the PLS-DA model obtained. It was similar to the PCA score plot (Figure 2a), however, the separation of wines according to geographical origin appears more obvious. This might be explained by the fact that the PLS-DA algorithm maximizes the variance between groups rather than within the group. The PLS-DA loadings for the calibration model (Figure 4b) were similar to those observed in the PCA analysis (Figure 2b). The accuracy, R^2, and Q^2 for the PLS-DA calibration model were 1, 0.9364, and 0.83821, respectively (two PLS latent variables). The calibration statistics indicated that the model developed was acceptable to classify new samples.

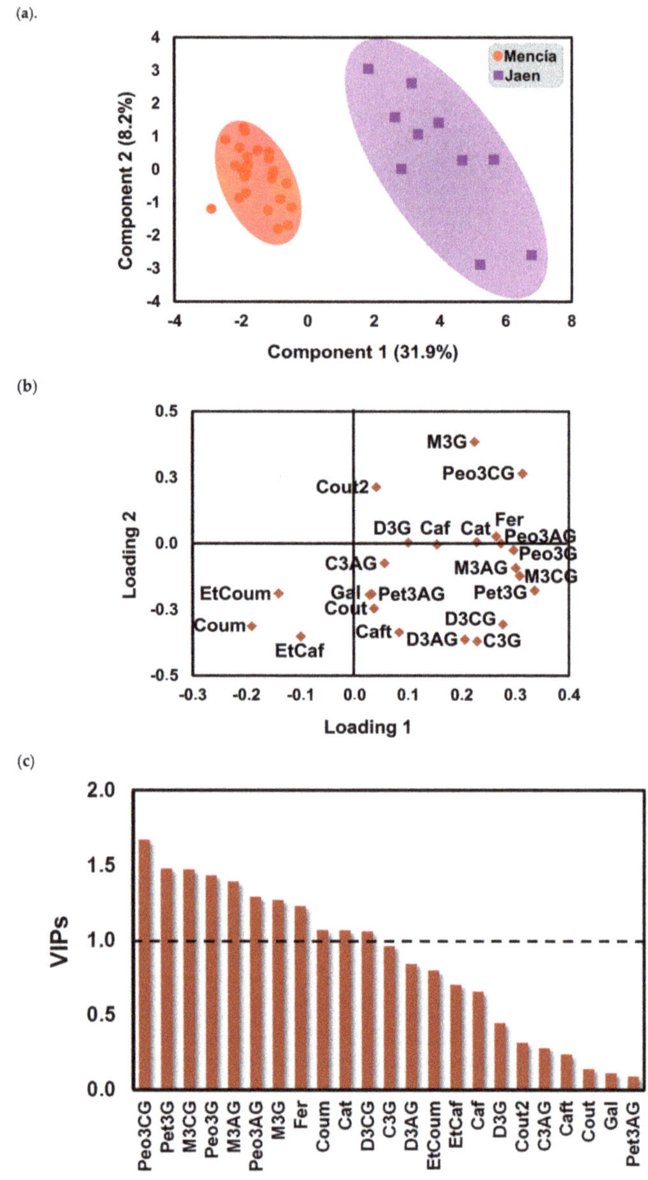

Figure 4. (**a**) Scores plot and (**b**) loadings plot of the first two factors of the Partial Least Squares-Discrimination Analysis (PLS-DA) model built with the phenolic compounds profile of the 'Mencía'/'Jaen' monovarietal wines from the Bierzo D.O. and Dão D.O., respectively. Phenolic compounds ranked by variable in projection (VIP) scores (**c**). Delphinidin-3-glucoside (D3G), cyanidin-3-glucoside (C3G), peonidin-3-glucoside (Peo3G), petunidin-3-glucoside (Pet3G), malvidin-3-glucoside (M3G), delphinidin-3-acetylglucoside (D3AG), cyanidin-3-acetylglucoside (C3AG), peonidin-3-acetylglucoside (Peo3AG), petunidin-3-acetylglucoside (Pet3AG), malvidin-3-acetylglucoside (M3AG), delphinidin-3-coumaroylglucoside (D3CG), peonidin-3-coumaroylglucoside (Peo3CG), malvidin-3-coumaroylglucoside (M3CG), gallic acid (Gal), caftaric acid (Caft), coutaric acid (Cout), coutaric acid isomer (Cout2), caffeic acid (Caf), coumaric acid (Coum), catechin (Cat), ethyl ester of caffeic acid (EtCaf), ethyl ester of coumaric acid (EtCoum).

To access the variable importance in the discrimination of 'Mencía'/'Jaen' wines, the variable in projection (VIP) was used (Figure 4c). According to the VIP results, Peo3CG, Pet3G, M3CGlc, Peo3G, M3AG, M3G, and Fer were the most influential variables in the discrimination between 'Mencía' and 'Jaen' wines from Bierzo D.O. and Dão D.O., respectively.

Generally, PCA and PLS-DA are used to distinguish samples and to explain the differences. When the number of variables were higher than the number of observations, PCA and PLS-DA models were suited to handle these data sets for discriminant analysis [33,34]. Nevertheless, for large data sets with a number of observations higher than the number of variables, PCA and PLS-DA scores and loadings can become rotated due to the presence in the data of strong systematic variations unrelated to the response, making more difficult the interpretation of the models. The OPLS-DA (Orthogonal partial least squares discriminant analysis) integrated an orthogonal signal correction filter to separate the variations in the data that are related to the prediction of a quantitative response from the variations not related or orthogonal to the prediction [35]. The advantage of OPLS-DA compared to PLS-DA is that the model is rotated so that class separation is found in the 1st predictive component, t_p, also referred to as the correlated variation, and variation not related to class separation is seen in orthogonal components, t_o, also referred to as the uncorrelated variation, facilitating model interpretation. The quality of the OPLS-DA model was validated by the values of the parameters $R^2X = 0.259$, $R^2Y = 0.877$, $Q^2 = 0.852$, which demonstrated the potential usefulness of the OPLS-DA model (Figure 5).

(a)

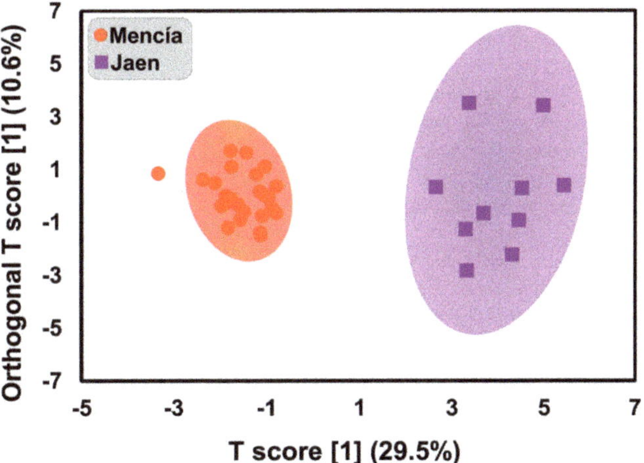

Figure 5. *Cont.*

(b)

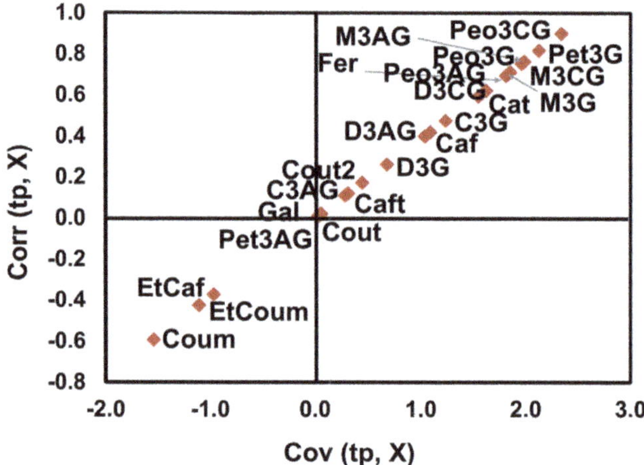

Figure 5. OPLS-DA (Orthogonal Partial Least Squares Discriminant Analysis) score plot for 'Mencía'/'Jaen' monovarietal wines from Bierzo D.O. and Dão D.O. (**a**) and S-Plot for visualization of the variable influence on the OPLS-DA model by representing the covariance (Cov) and correlation (Corr) loading profiles resulting from the projection on the predictive component, tp (**b**). Delphinidin-3-glucoside (D3G), cyanidin-3-glucoside (C3G), peonidin-3-glucoside (Peo3G), petunidin-3-glucoside (Pet3G), malvidin-3-glucoside (M3G), delphinidin-3-acetylglucoside (D3AG), cyanidin-3-acetylglucoside (C3AG), peonidin-3-acetylglucoside (Peo3AG), petunidin-3-acetylglucoside (Pet3AG), malvidin-3-acetylglucoside (M3AG), delphinidin-3-coumaroylglucoside (D3CG), peonidin-3-coumaroylglucoside (Peo3CG), malvidin-3-coumaroylglucoside (M3CG), gallic acid (Gal), caftaric acid (Caft), coutaric acid (Cout), coutaric acid isomer (Cout2), caffeic acid (Caf), coumaric acid (Coum), catechin (Cat), ethyl ester of caffeic acid (EtCaf), ethyl ester of coumaric acid (EtCoum).

Besides, in the OPLS-DA model, the S-plot was proposed as a tool that provides both the covariance and correlation loadings between the analytes and the t_p, thus helping to identify both statistically and biochemically significant analytes [36]. In Figure 5b, the S-plot shows that peonidin-3-coumaroylglycoside (Peo3CG), petunidin-3-glucoside (Pet3G), peonidin-3-glucoside (Peo3G), malvidin-3-acetylglucoside (M3AG), and malvidin-3-coumaroylglucoside (M3CG) had relatively higher covariance and correlation loading values, and they were likely to be the key phenolic compounds for 'terroir' differentiation.

The composition of wines, especially the phenolic composition, is dependent on the chemical composition of grapes and on the winemaking technology used for its production. As wines analyzed in this work were from different wineries in both regions, and the differences observed between the two regions was significantly higher than the differences observed between the wines from each region, the winemaking technology used for wine production was not responsible for the differences observed in the phenolic composition of wines between the two regions. It has been shown that the main factors involved in terroir expression are soil, climate, and grape variety, and these factors interact [37]. Temperature and precipitation are the main factors influencing grapevine phenology, grape yield, as well as wine quality [38]. The wine composition produced in a particular region is influenced by the baseline climate, while the vintage effect is the result of climatic variability between vintages [39]. The monthly mean, minimum, and maximum temperature for this particular vintage (2015) in both regions are presented in Figure 6, along with the precipitation, and the bioclimatic indices commonly used for accessing the suitability of a particular region for wine production (Table 3). Winkler index (WI) measures the heat accumulation during the growing season [40], both regions are included in the Winkler's Region II and classified as moderately cold. When considering the Huglin Heliothermic Index (HI) Dão D.O. corresponds to temperate warm and Bierzo to warm viticultural regions. The Dryness Index (DI) allows to access the availability of soil water content for the vine in the growing season, considering precipitation and reference evapotranspiration [41]. According to the values between April and October, Dao D.O. was a sub-humid region and Bierzo DO was a moderately dry region. The cool night index (CI) [42] considers the minimum temperatures during the grape maturation period, providing complementary information about the thermal regime in this period. The ripening stage is determining for the titratable acidity, pH, phenolic compounds, and anthocyanins as well for the flavor/aroma potential, and therefore CI is a reliable index for accessing the potential quality associated with viticultural climates [43]. Bierzo D.O. had CI values that correspond to very cool nights, whereas Dão D.O. had cool nights. The hydrothermic index of Branas, Bernon, and Levadoux (BI) [44] considers the influence of both temperature and precipitation on grape yield and wine quality. This index estimates the risk of downy mildew disease, which is a common limiting factor for grapevine yield [45]. Both regions presented a low risk of mildew disease. The bioclimatic indexes obtained for 2015 are in accordance with those described for Dão Region in the 1950–2000 period [46] and for the Bierzo region from 2007 up to 1967 [47], showing that 2015 was a typical climatic year. The major differences between the two regions in this particular vintage were the lower minimum temperature at night between April and September, being found a total number of days with a temperature below 10 °C during the maturation months (August and September) much higher in the Bierzo region (32 days) when compared to the Dão region (12 days). Also, the number of days with a maximum temperature above 30 °C was much higher in the Bierzo region (14 days) when compared to the Dão region (7 days). This different temperature profile between the two regions might explain the differences observed in the anthocyanins content of the wines from the two regions. In warmer climates, higher temperatures may result in negative changes in fruit composition. A significantly lower anthocyanins concentration at maturity was observed in grapevines exposed to 30 rather than 20 °C temperature treatments [48]. Another study found that high temperatures (35 °C) both inhibited anthocyanin production and degraded the anthocyanins that were produced [49]. High and low temperatures during ripening, likely affect the production and/or degradation of abscisic acid (ABA) in berry skins that affect the expression of VvmybA1 that controls the expression of the anthocyanin's biosynthetic enzyme genes [50]. A day temperature of 35 °C completely inhibited anthocyanins synthesis in 'Tokay' berries, regardless of night temperature [51]. The effect of the weather variables in the phenolic composition of 'Mencía'/'Jaen' wines were in accordance with the previous study of Vilanova et al. [18], who observed that the composition of 'Mencía' grapes, including the phenolic composition, was more affected by the vintage (from 2009 to 2012) than the geographic origin when studying the 'Mencía' cultivar located in different geographic areas from NW Spain (Amandi, Chantada, Quiroga-Bibei, Ribeiras do Sil, and Ribeiras do Miño).

Table 3. Bioclimatic index of Bierzo D.O. and Dão D.O. for 2015 and classes of the viticultural climate of the grape-growing regions.

Bioclimatic Index	Bierzo DO		Dão DO	
Wrinkler index	1549	Moderately cold	1638	Moderately cold
Huglin Heliothermic index	2473	Warm	2217	Temperate warm
Drying Index	−16	Moderately dry	130	Sub-humid
Cool nigh index	7.6	Very cool nights	12.3	Cool nights
Branas Hydrothermic Index	1818	Low risk	2279	Low risk
N. of hot days (>30 °C) [a]	14		7	
N. of cold nights (<10 °C) [a]	32		12	

[a] in August and September months.

(a)

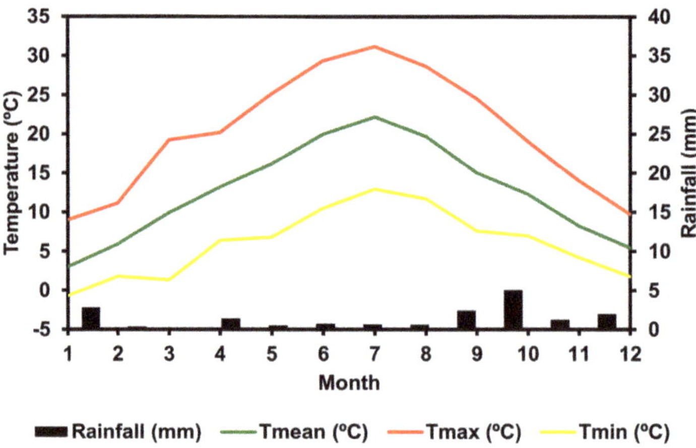

(b)

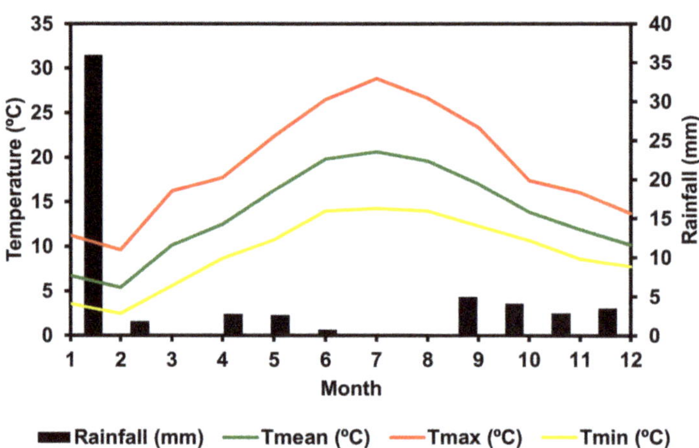

Figure 6. Distribution of rainfall and maximum (Tmax), minimum (Tmin), and mean temperatures (Tmean) in the (**a**) Bierzo D.O. and (**b**) Dão D.O. in 2015.

2.3. Impact of the Phenolic Composition on the Color of Red Wines from 'Mencía'/'Jaen'

As expected from the phenolic composition of 'Mencía'/'Jaen' wines, 'Jaen' wines from Dão D.O. presented a significantly higher color intensity when compared to 'Mencía' wines from Bierzo D.O. (14.1 vs. 11.3 a.u., $p < 0.00687$, respectively, Table 4). The color intensity for 'Mencía' wines from Bierzo D.O. obtained in this work was in accordance with the values described by Revilla et al. [52] for young red wines produced with 'Mencía' grapes in AOC Valdeorras (Galicia) (range from 6.43 to 17.21 a.u.), by García-Falcón et al. [27] also for 'Mencía' wines (range from 11.2 to 12.6 a.u.), by Bouzas-Cid et al. [16] (range from 9.1 to 10.8 a.u.), and by Soto Vázquez et al. [29] (range from 7.37 to 11.57 a.u.). However, Blanco et al. [21] obtained lower color intensity values for 'Mencía' monovarietal wines produced in Galicia (range from 6.87 to 7.67 a.u.).

Table 4. 'Mencía' and 'Jaen' red wines chromatic characteristics.

Wine Samples	Color Intensity	Hue	Chromatic Characteristics				
	(a.u.)		L*	a*	b*	C*	°h
Mencía	15.36 ± 0.36 [h]	0.712 ± 0.041 [b]	68.9 ± 0.0 [b]	30.63 ± 0.71 [f]	7.51 ± 0.08 [fg]	31.54 ± 0.71 [g]	13.79 ± 0.17 [f]
Mencía	13.91 ± 0.05 [g]	0.700 ± 0.002 [ab]	73.6 ± 0.1 [e]	27.37 ± 0.54 [e]	7.82 ± 0.24 [g]	28.46 ± 0.58 [f]	15.96 ± 0.17 [h]
Mencía	13.22 ± 0.01 [f]	0.750 ± 0.004 [bc]	73.8 ± 0.1 [e]	18.43 ± 0.19 [a]	6.43 ± 0.16 [e]	19.52 ± 0.24 [ab]	19.25 ± 0.26 [i]
Mencía	8.18 ± 0.06 [b]	0.791 ± 0.003 [c]	81.3 ± 0.5 [i]	20.45 ± 0.04 [b]	7.12 ± 0.00 [f]	21.66 ± 0.04 [c]	19.20 ± 0.04 [i]
Mencía	6.76 ± 0.06 [a]	0.795 ± 0.011 [c]	84.3 ± 0.5 [j]	19.37 ± 0.07 [ab]	4.95 ± 0.06 [d]	19.99 ± 0.08 [b]	14.33 ± 0.11 [fg]
Mencía	16.23 ± 0.56 [i]	0.685 ± 0.012 [a]	66.7 ± 0.3 [a]	36.74 ± 0.32 [i]	6.19 ± 0.10 [e]	37.26 ± 0.33 [i]	9.57 ± 0.07 [d]
Mencía	13.06 ± 0.04 [f]	0.782 ± 0.004 [c]	72.2 ± 0.2 [d]	31.70 ± 0.23 [fg]	6.22 ± 0.15 [e]	32.30 ± 0.26 [g]	11.10 ± 0.19 [e]
Mencía	10.43 ± 0.04 [cd]	0.720 ± 0.005 [b]	83.5 ± 0.7 [j]	24.12 ± 0.01 [c]	10.70 ± 0.12 [h]	26.39 ± 0.04 [d]	23.94 ± 0.25 [k]
Mencía	11.95 ± 0.01 [e]	0.784 ± 0.003 [c]	76.8 ± 0.3 [f]	18.37 ± 0.05 [a]	3.47 ± 0.15 [b]	18.69 ± 0.08 [a]	10.71 ± 0.43 [e]
Mencía	10.57 ± 0.03 [d]	0.758 ± 0.006 [bc]	77.5 ± 0.5 [fg]	26.77 ± 0.08 [e]	7.94 ± 0.07 [g]	27.92 ± 0.05 [ef]	16.53 ± 0.18 [h]
Mencía	11.05 ± 0.09 [d]	0.775 ± 0.001 [c]	78.6 ± 0.3 [gh]	18.32 ± 0.12 [a]	4.79 ± 0.08 [d]	18.93 ± 0.09 [ab]	14.67 ± 0.33 [g]
Mencía	9.67 ± 0.02 [c]	0.731 ± 0.009 [b]	79.3 ± 0.3 [h]	25.66 ± 0.07 [d]	4.46 ± 0.08 [cd]	26.05 ± 0.06 [d]	9.86 ± 0.20 [d]
Mencía	6.12 ± 0.02 [a]	0.688 ± 0.017 [a]	89.1 ± 0.1 [k]	18.46 ± 0.07 [a]	7.29 ± 0.01 [f]	18.95 ± 0.06 [b]	21.57 ± 0.08 [j]
Jaen	14.06 ± 0.05 [g]	0.710 ± 0.002 [b]	70.6 ± 0.2 [c]	32.60 ± 0.37 [g]	1.82 ± 0.02 [a]	32.65 ± 0.37 [g]	3.20 ± 0.01 [a]
Jaen	13.88 ± 0.45 [g]	0.643 ± 0.032 [a]	72.1 ± 0.1 [d]	35.30 ± 0.12 [h]	3.10 ± 0.11 [b]	35.45 ± 0.12 [h]	5.01 ± 0.17 [b]
Jaen	14.89 ± 0.16 [h]	0.799 ± 0.037 [c]	71.2 ± 0.4 [cd]	25.92 ± 0.09 [d]	18.14 ± 0.16 [j]	31.57 ± 0.07 [g]	35.00 ± 0.15 [l]
Jaen	15.26 ± 0.07 [h]	0.714 ± 0.043 [b]	68.4 ± 0.3 [b]	34.53 ± 0.40 [h]	4.27 ± 0.20 [c]	34.81 ± 0.39 [h]	7.05 ± 0.25 [c]
Jaen	12.53 ± 0.14 [ef]	0.684 ± 0.011 [a]	79.8 ± 0.0 [h]	24.82 ± 0.09 [c]	11.19 ± 0.04 [i]	27.22 ± 0.06 [e]	24.29 ± 0.16 [k]

The values are presented as mean ± standard deviation; Means within a column followed by the same superscript letter are not significantly different (Tukey $p < 0.05$). L*—lightness, a*—redness, b*—yellowness, C*—chroma; °h—hue angle, a.u.—absorbance units.

As can be observed in Table 4, the 'Jaen' monovarietal wines from Dão D.O. presented a significantly lower lightness, L* value, when compared to the 'Mencía' wines from Bierzo D.O. (72.40 vs. 77.35, $p < 0.02684$, respectively). On the other hand, 'Jaen' monovarietal wines presented a significantly higher a* value (redness) than 'Mencía' wines (30.64 vs. 24.34, $p < 0.00509$, respectively). For the b* values (yellowness) there was no significant difference between 'Jaen' and 'Mencía' wines (7.70 vs. 6.53, $p < 0.399$, respectively).

To understand the relative contribution of the phenolic composition of 'Mencía' and 'Jaen' wines on the significantly different chromatic characteristics (L* and a* values), a Partial Least Squares (PLS1) regression of the standardized chromatic characteristics (Y) on the standardized phenolic and monomeric anthocyanins composition of wines (X) was performed. The number of factors for each dependent variable analyzed was estimated by 7-fold cross-validation and the prediction ability of the model obtained was determined using an independent validation set randomly selected. The regression curves obtained for L* and a* are presented in Figure 7a,b, respectively. For the L* value, 63.2% of the variance in the three PLS components related to phenolic and monomeric anthocyanin composition of wines explained 86.5% of the variation in L* values of 'Mencía'/'Jaen' monovarietal wines with a Q^2 cumulative of 0.6870. For the validation set, a R^2 value of 0.7936 was obtained, showing a medium predictive ability. For the a* value, the phenolic composition and monomeric anthocyanin composition of 'Mencía'/'Jaen' wines allowed to obtain a good model, with 69.6% of the variance in four PLS components related to the phenolic and monomeric anthocyanin composition of wines explaining 92.6% of the variation in a* values (Q^2 cumulative = 0.7321). The predictive ability for the validation set was bad, showing that this model cannot accurately predict the a* value from the phenolic and monomeric anthocyanin composition of wines (Figure 7b).

(a)

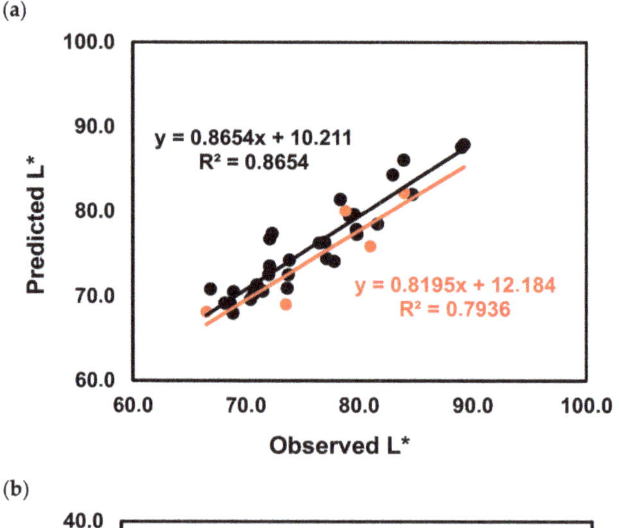

(b)

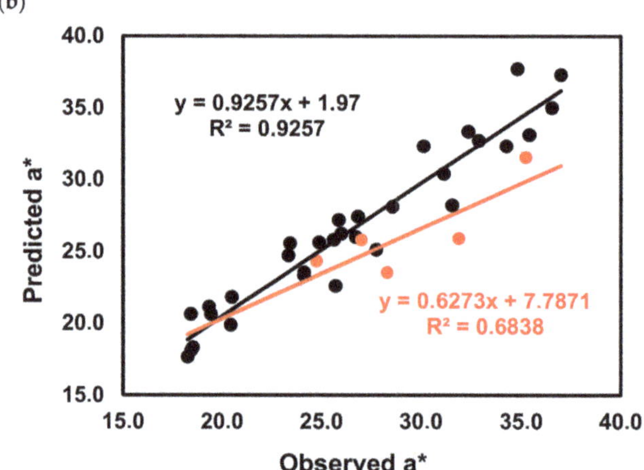

Figure 7. Calibration curves were obtained after Partial Least Squares (PLS) regression of the L* values (**a**) and of the a* values (**b**) of 'Mencía'/'Jaen' wines using the phenolic composition as independent variables. Black circles calibration samples, red circles test samples.

This medium to the bad predictive ability of L* and a* values using only the individual phenolic composition of 'Mencía'/'Jaen' red wines can be explained by the fact that in wine monomeric anthocyanins are involved in multiple equilibriums depending on the wine pH [53], co-pigmentation phenomenon [54], sulfur dioxide bleaching [55], and besides the monomeric anthocyanins, polymeric anthocyanins can also be present [56]. Therefore, although these are young wines and the monomeric anthocyanins are the most abundant anthocyanins, probably the wine's chromatic characteristics cannot be described solely by the levels of the individual monomeric anthocyanins present. Therefore, the classic colorimetric methods of polymeric pigments, colored anthocyanins, total pigments, and total anthocyanins were added to the X matrix (Table 5) to access if these variables could increase the explanation of the dependent variables, the wines L* and a* values.

Table 5. 'Mencía' and 'Jaen' red wines total anthocyanins, colored anthocyanins, polymeric, and total pigments.

Wine Samples	Total Anthocyanins (mg/L)	Colored Anthocyanins (a.u.)	Total Pigments (a.u.)	Polymeric Pigments (a.u.)
Mencía	277 ± 2 [e]	2.83 ± 0.05 [a]	20.00 ± 2.29 [cd]	4.78 ± 0.11 [g]
Mencía	257 ± 8 [d]	4.23 ± 0.08 [b]	17.88 ± 0.43 [cd]	2.83 ± 0.07 [fg]
Mencía	243 ± 4 [cd]	4.80 ± 0.05 [c]	19.04 ± 0.93 [cd]	1.75 ± 0.06 [c]
Mencía	208 ± 2 [b]	2.28 ± 0.01 [a]	18.53 ± 0.21 [cd]	1.72 ± 0.03 [c]
Mencía	190 ± 6 [b]	2.15 ± 0.14 [a]	9.80 ± 0.71 [ab]	1.19 ± 0.04 [b]
Mencía	333 ± 1 [f]	5.20 ± 0.17 [cd]	20.65 ± 0.007 [cd]	3.07 ± 0.02 [g]
Mencía	293 ± 6 [e]	3.87 ± 0.02 [b]	16.77 ± 2.57 [c]	2.42 ± 0.04 [e]
Mencía	203 ± 1 [b]	4.34 ± 0.00 [bc]	12.42 ± 0.00 [b]	0.90 ± 0.01 [a]
Mencía	254 ± 9 [cd]	2.98 ± 0.14 [a]	11.62 ± 0.14 [b]	2.52 ± 0.014 [e]
Mencía	232 ± 9 [c]	3.03 ± 0.03 [a]	10.61 ± 0.86 [ab]	2.20 ± 0.04 [d]
Mencía	231 ± 3 [c]	3.05 ± 0.11 [a]	10.66 ± 0.07 [ab]	2.24 ± 0.09 [d]
Mencía	236 ± 2 [cd]	2.88 ± 0.01 [a]	10.40 ± 0.14 [ab]	2.05 ± 0.001 [d]
Mencía	147 ± 8 [a]	2.36 ± 0.00 [a]	7.37 ± 0.00 [a]	0.93 ± 0.025 [a]
Jaen	515 ± 9 [i]	5.04 ± 0.15 [cd]	31.41 ± 0.71 [e]	2.02 ± 0.06 [d]
Jaen	525 ± 8 [i]	5.32 ± 0.86 [cd]	21.56 ± 1.07 [d]	2.38 ± 0.01 [e]
Jaen	382 ± 5 [g]	3.69 ± 0.22 [b]	27.52 ± 0.21 [e]	2.74 ± 0.05 [f]
Jaen	434 ± 1 [h]	5.89 ± 0.18 [d]	38.08 ± 1.00 [f]	1.71 ± 0.10 [c]
Jaen	381 ± 7 [g]	2.96 ± 0.01 [a]	16.97 ± 0.14 [c]	1.77 ± 0.01 [c]

The values are presented as mean ± standard deviation; Means within a column followed by the same superscript letter are not significantly different (Tukey $p < 0.05$). a.u.—absorbance units.

As can be observed in Figure 8a, the model quality for predicting the wine L* values and especially the wine a* values increased both in the calibration, and more importantly on the validation test sets. From the analysis of the standardized coefficients (Figure 8b), it can be observed that the variables with the highest influence in predicting the wines L* values were colored anthocyanins, polymeric pigments, total pigments, total anthocyanins C3AG, D3AG, D3G, PeoCG, Pet3G and M3G, but also the colorless Fer, Caf, Cout, EtCaf.

(a)

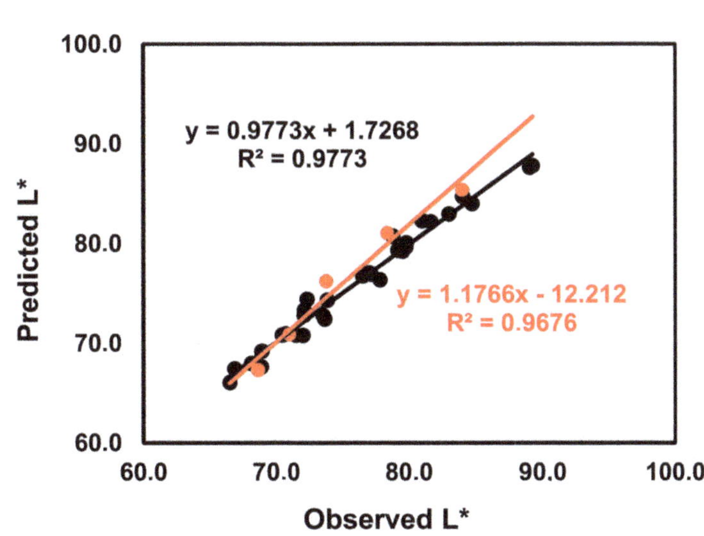

Figure 8. *Cont.*

(b)

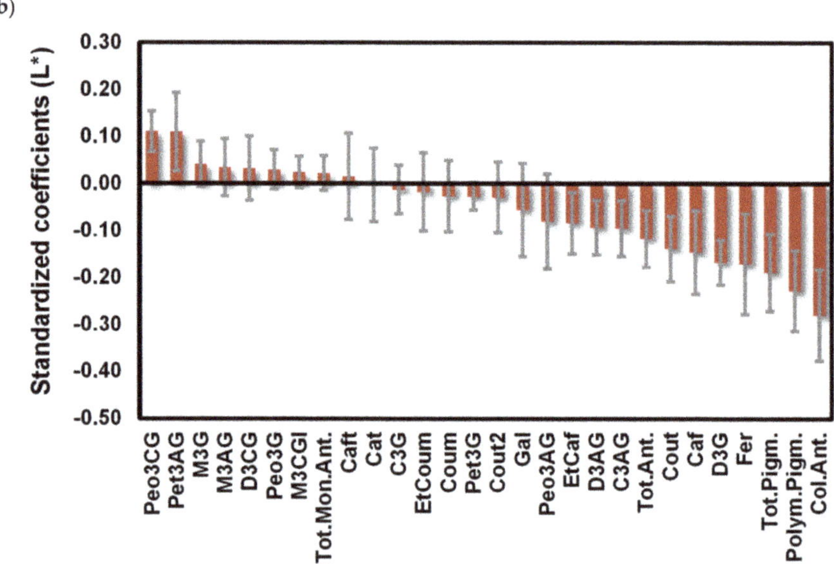

Figure 8. Calibration curves obtained after PLS regression of the L* values of 'Mencía'/'Jaen' wines using the phenolic composition and colorimetric determination of colored anthocyanins, polymeric pigments, total pigments, total anthocyanins as independent variables (black circles calibration samples, red circles test samples) (**a**) and B coefficients plots (**b**). Delphinidin-3-glucoside (D3G), cyanidin-3-glucoside (C3G), peonidin-3-glucoside (Peo3G), petunidin-3-glucoside (Pet3G), malvidin-3-glucoside (M3G), delphinidin-3-acetylglucoside (D3AG), cyanidin-3-acetylglucoside (C3AG), peonidin-3-acetylglucoside (Peo3AG), petunidin-3-acetylglucoside (Pet3AG), malvidin-3-acetylglucoside (M3AG), delphinidin-3-coumaroylglucoside (D3CG), peonidin-3-coumaroylglucoside (Peo3CG), malvidin-3-coumaroylglucoside (M3CG), gallic acid (Gal), caftaric acid (Caft), coutaric acid (Cout), coutaric acid isomer (Cout2), caffeic acid (Caf), coumaric acid (Coum), catechin (Cat), ethyl ester of caffeic acid (EtCaf), ethyl ester of coumaric acid (EtCoum), total monomeric anthocyanins (Tot.Mon.Ant.), total pigment (Tot.Pigm.), polymeric pigments (Polym.Pigm.), colored anthocyanins (Col.Ant.), total anthocyanins measured colorimetrically (Tot.Ant.).

For the a* values of wines, also the most important variables in its prediction were the colored anthocyanins, Fer, total anthocyanins, D3AG, D3G and Pet3AG, Cat (Figure 9b).

The positive influence of Fer in the a* values and Fer, Caf, Cout, EtCaf on the L* values can be due to the co-pigmentation phenomena of anthocyanins. Recent research assigned more relevance concerning co-pigmentation to hydroxycinnamic acid derivatives than flavan-3-ols. [57,58]. The latter result suggests that the levels of co-pigments in red wine are at least as important as the levels of anthocyanins in determining the differences in the color of red wines from 'Mencía'/'Jaen' from Bierzo D.O and Dão D.O. in the 2015 vintage.

(**a**)

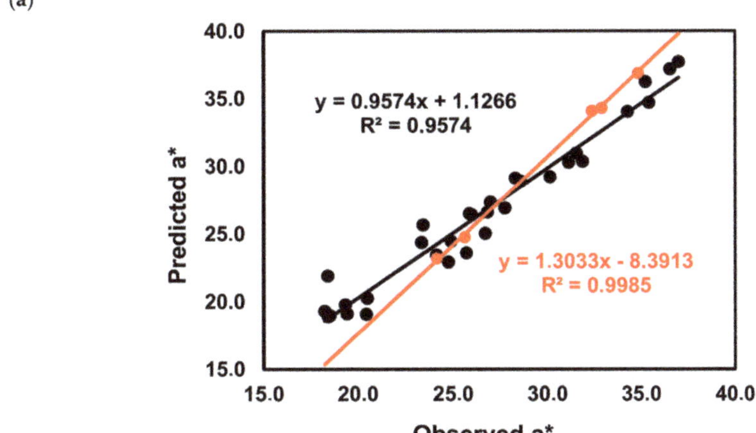

(**b**)

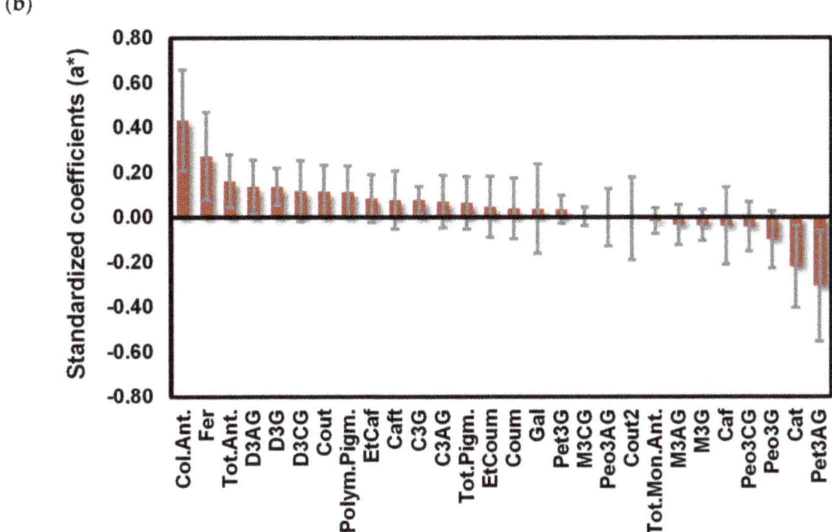

Figure 9. Calibration curves obtained after PLS regression of the a* values of 'Mencía'/'Jaen' wines using the phenolic composition and colorimetric determination of colored anthocyanins (Col.Ant.), polymeric pigments (Polym.Pigm.), total pigments (Tot.Pigm.), total anthocyanins (Tot.Ant.) as independent variables (black circles calibration samples, red circles test samples) (**a**) and B coefficients plots (**b**). Delphinidin-3-glucoside (D3G), cyanidin-3-glucoside (C3G), peonidin-3-glucoside (Peo3G), petunidin-3-glucoside (Pet3G), malvidin-3-glucoside (M3G), delphinidin-3-acetylglucoside (D3AG), cyanidin-3-acetylglucoside (C3AG), peonidin-3-acetylglucoside (Peo3AG), petunidin-3-acetylglucoside (Pet3AG), malvidin-3-acetylglucoside (M3AG), delphinidin-3-coumaroylglucoside (D3CG), peonidin-3-coumaroylglucoside (Peo3CG), malvidin-3-coumaroylglucoside (M3CG), gallic acid (Gal), caftaric acid (Caft), coutaric acid (Cout), coutaric acid isomer (Cout2), caffeic acid (Caf), coumaric acid (Coum), catechin (Cat), ethyl ester of caffeic acid (EtCaf), ethyl ester of coumaric acid (EtCoum), total monomeric anthocyanins (Tot.Mon.Ant.), total pigment (Tot.Pigm.), polymeric pigments (Polym.Pigm.), colored anthocyanins (Col.Ant.), total anthocyanins measured colorimetrically (Tot.Ant.).

3. Materials and Methods

3.1. Wine Samples

Thirteen 'Mencía' red wine samples were from the Bierzo D.O located in the northwest of the province of León (Castile and León, Spain), and five 'Jaen' wines samples from the Dão D.O. All of them produced in Vintage 2015 in different wineries at an industrial scale and analyzed in 2016. The number of wine samples from each region was the one that the producers certified that the wines were exclusively produced from 'Mencía'/'Jaen' grape varieties. Each producer supplied two bottles of wine.

3.2. High-Performance Liquid Chromatography (HPLC) Analysis of Anthocyanins, Catechin, and Phenolic Acids

The phenolic profile was performed by HPLC-DAD using an Ultimate 3000 HPLC system (Dionex Corporation, Sunnyvale, CA, USA) equipped with a photodiode array detector (PDA-100, Dionex Corporation, Sunnyvale, CA, USA). An ACE 5 C-18 column (250 × 4.6 mm) was used. The eluent was constituted by 5% aqueous formic acid (solvent A) and methanol (solvent B). The elution program was as follows: 5% of B from zero to 5 min followed by a linear gradient up to 65% of B until 65 min and from 65 to 67 min down to 5% of B. The photodiode detector assembly was operated between 200–600 nm and the chromatographic profile was recorded at 280, 325, and 525 nm. Then, 50 µL of the sample was injected at a flow rate of 1 mL/min and then the column was maintained at a temperature of 35 °C [59]. Quantification was performed with calibration curves with standard caffeic acid, coumaric acid, ferulic acid, gallic acid, and catechin. *Trans*-caftaric acid, 2-*S*-glutathionylcaftaric acid (GRP), and caffeic acid ethyl ester were expressed as caffeic acid equivalents, coutaric acid and coumaric acid ethyl ester were expressed as coumaric acid equivalents. A calibration curve of malvidin-3-glucoside, peonidin-3-glucoside, and cyanidin-3-glucoside was used for quantification of these anthocyanins. Using the coefficient of molar absorptivity (ε) and by extrapolation, it was possible to obtain the slopes for delphinidin-3-glucoside, petunidin-3-glucoside, and malvidin-3-coumaroylglucoside and perform the quantification. The results of delphinidin-3-acetylglucoside, petunidin-3-acetylglucoside, peonidin-3-acetylglucoside, cyanidin-3-acetylglucoside, and cyanidin-3-coumaroylglucoside were expressed as respective glucoside equivalents [60,61]. Analysis of each bottle were performed in duplicate.

3.3. Color Intensity, Total Anthocyanins, Colored Anthocyanins, Pigments, and Chromatic Characteristics

Red wine color intensity ($A_{420\,nm} + A_{520\,nm} + A_{620\,nm}$) and hue ($A_{420\,nm}/A_{520\,nm}$) were quantified as described in the OIV methods [62]. The concentration of total anthocyanins from red wine was determined by the SO_2 bleaching procedure using the method described by Ribéreau-Gayon and Stronestreet [63], and the colored anthocyanin's ($A_{520} \times 10 - A_{520}^{SO2} \times 10$), total pigments ($A_{520}^{HCl} \times 101$) and polymeric pigments (A_{520} bis × 10) from red wine were determined according to Somers and Evans [64]. For the chromatic characteristics of red wine, the absorption spectra of wine samples were scanned from 380 to 780 nm using a 1-cm path length quartz cell, and the wines' chromatic characteristics L* (lightness), a* (redness), and b* (yellowness) coordinates were calculated using the CIELab method according to OIV [62]. The Chroma ($C^* = [(a^*)^2 + (b^*)^2]^{1/2}$]) and hue-angle ($^{o}h = tang^{-1}(b^*/a^*)$) values were also determined. Analysis of each bottle were performed in duplicate.

3.4. Climacteric Data, Bioclimatic Indexes, and Soil Characteristics

The climacteric data of Bierzo (Spain) and Dão (Portugal) for the 2015 harvesting year were obtained from Ponferrada-Carracedelo (Spain, latitude 42°36′13″, longitude 6°30′02″, elevation above sea level of 800 m) and Viseu (Portugal, latitude 40°39′39″, longitude 7°54′34″, elevation above sea level 469 m), respectively, and included the maximum (Tmax), minimum (Tmin), and mean (Tmean) daily temperature along with the daily precipitation (P) [65,66]. To simplify the global description of weather conditions in both locations during the growing season in the two regions, the Winkler Index (WI) [40], Huglin Heliothermal Index (HI) [67], Cool Night Index (CI) [42], Dryness Index [41], and Branas hydrothermic index [44] were calculated, which quantify the impact of weather by single aggregate values.

The Winkler Index (WI) is based on heat summation or growing degree-days exceeding the threshold of 10 °C during the growing season (April 1 through October 31), calculated according to the following Equation (1):

$$WI = \sum_{1/04}^{30/10} \frac{(T_{max} - T_{min})}{2} - 10 \tag{1}$$

Negative values are calculated as zero when the addition is performed. According to the WI, geographical areas are divided into five climate regions: Region I (cold) WI ≤ 1390; Region II (moderately cold) 1391 ≤ WI ≤ 1670; Region III (warm) 1671 ≤ WI ≤ 1940: Region IV (moderately warm) 1941 ≤ WI ≤ 2220; Region V (hot) WI > 2200.

The Huglin Heliothermal Index (HI) is a heat summation index that takes daily mean temperature and daily maximum temperature into account as well as an adjustment for day length. It is calculated by summing the values from April 1 to October 31 using the following Equation (2):

$$HI = \sum_{1/04}^{31/10} \frac{(T_{mean} - 10) + (T_{max} - 10)}{2} d \tag{2}$$

where d is the length of day coefficient ranging from 1.02 to 1.06 between 40° and 50° of latitude. Viticultural zones can be classified as [42]: Very cool (HI − 3; HI ≤ 1500); Cool (HI − 2; 1500 < HI ≤ 1800); Temperate (HI − 1; 1800 < HI ≤ 2100); Temperate warm (HI + 1; 2100 < HI ≤ 2400); Warm (HI + 2; 2400 < HI ≤ 3000); Very warm (HI + 3; HI > 3000).

Dryness Index (DI) indicates the potential water availability in the soil, related to the level of dryness in a region. DI is calculated according to the following Equation (3):

$$W = \sum_{1/04}^{31/10} (W_0 + P - Tv - Es) \tag{3}$$

where W_0 is soil water reserve, P is precipitation, Tv is potential transpiration, and Es is soil evaporation (all variables in mm). Daily variables are used for this calculation, being distinguished four climate classes: very dry, where viticulture is limited by severe dryness (DI + 2: DI ≤ −100 mm); moderately dry (−100 < DI ≤ 50 mm); sub-humid (50 < DI ≤ 150 mm); and humid (DI > 150 mm) (defined in Tonietto and Carbonneau [42]). W is the estimate of soil water reserve at the end of the 1 April–31 October at modelled growing season period. To compute Tv and Es it is also necessary to compute the monthly total potential evapotranspiration. This was approximated by the Hargreaves method, which produces comparable results in arid and semiarid environments and requires temperature data only [68]. The result is mm of water in the soil. The initial W_0 is usually taken as 200 mm [42,69].

The Cool Night Index (CI) is a night coolness variable determined using Equation (4):

$$CI = \sum_{1/09}^{30/9} \frac{T_{min}}{30} \tag{4}$$

According to the CI values, viticultural zones can be classified as [42]: very cool nights (CI + 2; CI ≤ 12); cool nights (CI + 1: 12 < CI ≤ 14); temperate nights (CI − 1: 14 < CI ≤ 18); warm nights (CI − 2: CI > 18).

The hydrothermic index of Branas, Bernon, and Levadoux (*BI*) [44] is the sum of the products of monthly mean temperature and monthly accumulated precipitation amount (in mm) during the April to August season (Equation 5):

$$BI = \sum_{1/04}^{30/08} T_{mean}P \tag{5}$$

This index estimates the risk of downy mildew disease, which is a common limiting factor for grapevine yield [43]. This risk is usually considered low when BI values are below 2500 °C mm, high for values higher than 5100 °C mm, and very high for values higher than 7500 °C mm [70].

In Bierzo DO region, the predominant composition of the vine soils is a mixture of quartzites, sandstones, limestones, clays, and shales. The texture of the soil is predominantly clay-loam. The soil acidity ranged from 4–8.5 with levels above 6 in the valleys. The calcium oxide content is low and with organic material reaching 1% [71]. In the DO Dão, 97.4% of the soils are granitic, and the vineyards to produce DO Dão wines must be installed predominantly on granitic soils with brown non-humic litholic soils and in schist outcrops with brown non-humic Mediterranean soils. The soils presented an acid pH, poor in organic material and extractable mineral, with poor water retention capacity and, therefore, with low fertility [72,73].

3.5. Statistical Analysis

For statistical analysis of the chemical data, a one-way analysis of variance ($\alpha = 0.05$) was applied. When this test was significant, means were compared using the Tukey test, using the STATISTICA 2010 software 10 (StatSoft, Tulsa, OK, USA) program.

PCA is a chemometric method for data reduction and exploratory analysis of high-dimensional data sets. PCA decomposes the original matrix into the multiplication of loading (the phenolic composition of wines) and score (wine samples) matrices. The principal components are linear combinations of the original variables. The principal components are uncorrelated and account for the total variance of the original variables. PCA is an unsupervised method of pattern recognition in the sense that no grouping of the data must be known before the analysis. The new sub-space defined by the principal components leads to a model that is easier to interpret than the original data set. From these results, it should be possible to highlight several characteristics and correlate them to the chemical composition of the different wine samples analyzed. Partial least squares-discriminant analysis (PLS-DA) is a regression method commonly used in multivariate statistics, to establish the relationship between 2 data information sets, referred to as X being the phenolic compounds of the wines, and Y, a binary vector value. Orthogonal PLS-DA (OPLS-DA) is a modification of PLS-DA, which separates the systematic variation in X into 2 parts, one is linearly related to Y (t_p) and the other is orthogonal to Y (to). The quality of the OPLS-DA model is evaluated by the goodness-of-fit parameter (R^2X), the proportion of the variance of the response variable that is explained by the model (R^2Y), and the predictive ability parameter (Q^2), which is calculated by a 7-round internal cross-validation of the data, using a default option of MetaboAnalyst 3.0. The parameters R^2X and R^2Y represent the fraction of the variance of matrix X and matrix Y, respectively, and Q^2 represents the predictive accuracy of the model. R^2X, R^2Y, and Q^2 values close to 1 indicate an excellent OPLS-DA model, and values higher than 0.5 indicate an OPLS-DA model of good quality [74]. In addition to the evaluation of the OPLS-DA models by calculating their R^2X, R^2Y, and Q^2 values, we also analyzed more wine samples for further model validation. These samples were different from the wine samples used for model establishment. By calculating the recognition degree between wine samples used for model establishment and validation, we could validate the practical authenticating ability of these OPLS-DA models. The S-plot reflects the variable influence in an OPLS-DA model, which combines the covariance (magnitude) and correlation (reliability) loading profiles correlated with the predictive component in X (t_p). In our research, analytes (phenolic compounds) that have higher covariance and correlation values also have higher concentrations and more repetitions in different samples during the

OPLS-DA modelling, respectively. Using S-plot analysis, we can find out which analytes play a role both statistically significant (with high covariance values) and potentially biochemically significant (with high correlation values) in differentiating wine samples [36]. The selection of key phenolic compounds in sample differentiation needs to consider both covariance and correlation values.

4. Conclusions

Although the results obtained in this study correspond only to the 2015 vintage, they clearly show that the 'Mencía' monovarietal wines from Bierzo D.O. and the 'Jean' monovarietal wines from Dão D.O. produced at an industrial scale presented a significantly different phenolic composition. The most significant differences between the wines from the two regions were related to the levels of peonidin-3-coumaroylglucoside, petunidin-3-glucoside, malvidin-3-coumaroylglucoside, peonidin-3-glucoside, malvidin-3-acetylglucoside, malvidin-3-glucoside, and ferulic acid. These differences resulted in wines with significantly different chromatic characteristics, with 'Mencía' wines presenting a significantly higher lightness and lower red color. For the L* value, the differences observed were mainly related to the colored anthocyanins, polymeric pigments, total pigments, total anthocyanins, cyanidin-3-acetylglucoside, delphinidin-3-acetylglucoside, delphinidin-3-glucoside, peonidin-3-coumaroylglucoside, petunidin-3-glucoside, and malvidin-3-glucoside, but also to the colorless ferulic acid, caffeic acid, coutaric acid, and ethyl ester of caffeic acid. For the wines, a* values, also the most important variables were the colored anthocyanins, ferulic acid, total anthocyanins, delphinidin-3-acetylglucoside, delphinidin-3-glucoside, and petunidin-3-acetylglucoside and catechin. The positive influence of ferulic acid in the a* values and ferulic acid, caffeic acid, coutaric acid, and ethyl ester of caffeic acid in the L* values can be due to the co-pigmentation phenomena of anthocyanins. These differences in the phenolic composition, namely in the anthocyanin content wines produced with 'Mencía'/'Jean' grape variety in Bierzo D.O. and Dão D.O., respectively, can be explained by the different climacteric conditions of these two "terroirs" in 2015, with Bierzo D.O. presenting the characteristics of a warm, moderately dry viticultural region with very cold nights and Dão D.O. presenting the character of a temperate warm, sub-humid viticultural region with cold nights.

Author Contributions: Conceptualization, F.C., and F.M.N.; formal analysis, F.C., A.V., L.F.-R., F.M.N.; investigation, L.M., C.M., J.A.P.E.; methodology, F.C., F.M.N., L.F.-R.; supervision, F.C., A.V.; writing—original draft, F.C., L.F.-R., F.M.N.; writing—review and editing, F.C., A.V., L.M., C.M., J.A.P.E., F.M.N. All authors have read and agreed to the published version of the manuscript.

Funding: This research was funded by European Investment Funds FEDER/COMPETE/POCI under Project POCI-01-0145-FEDER-007728 (CQ-VR) and Funds from the Fundação para a Ciência Tecnologia (Portugal) to CQ-VR (UIDB/00616/2020 and UIDP/00616/2020).

Conflicts of Interest: The authors declare no conflict of interest.

References

1. Martín, J.P.; Santiago, J.L.; Pinto-Carnide, O.; Leal-Martínez, F.M.C.; Ortiz, J.M. Determination of relationships among autochthonous grapevine varieties (*Vitis vinifera* L.) in the Northwest of the Iberian Peninsula by using microsatellite markers. *Genet. Resour. Crop Evol.* **2006**, *53*, 1255–1261. [CrossRef]

2. Cunha, J.; Zinelabidine, L.H.; Teixeira-Santos, M.; Brazão, J.; Fevereiro, P.; Martínez-Zapater, J.M.; Ibáñez, J.; Eiras-Dias, J.E. Grapevine cultivar 'Alfrocheiro' or 'Bruñal' plays a primary role in the relationship among Iberian grapevines. *Vitis* **2015**, *54*, 59–65.

3. Instituto da Vinha e do Vinho (IVV). Available online: https://www.ivv.gov.pt/np4/35/ (accessed on 30 September 2020).

4. Jiang, B.A.O.; Zhang, Z.W.E.N.; Zhang, X.Z. Influence of terrain on phenolic compounds and antioxidant activities of Cabernet Sauvignon wines in loess plateau region of China. *J. Chem. Soc. Pak.* **2011**, *33*, 900.

5. Cortell, J.M.; Halbleib, M.; Gallagher, A.V.; Righetti, T.L.; Kennedy, J.A. Influence of Vine Vigor on Grape (*Vitis vinifera* L. Cv. Pinot Noir) and Wine Proanthocyanidins. *J. Agric. Food Chem.* **2005**, *53*, 5798–5808. [CrossRef]

6.   Russell, K.; Zivanovic, S.; Morris, W.C.; Penfield, M.; Weiss, J. The Effect of Glass Shape on the Concentration of Polyphenolic Compounds and Perception of Merlot Wine. *J. Food Qual.* **2005**, *28*, 377–385. [CrossRef]

7.   Kumsta, M.; Pavlousek, P.; Kupsa, J. Influence of terroir on the concentration of selected stilbenes in wines of the cv. Riesling in the Czech Republic. *Hortic. Sci.* **2012**, *39*, 38–46. [CrossRef]

8.   Roullier-Gall, C.; Boutegrabet, L.; Gougeon, R.D.; Schmitt-Kopplin, P. A grape and wine chemodiversity comparison of different appellations in Chemical Signatures of Two "Climats de Bourgogne" Burgundy: Vintage vs terroir effects. *Food Chem.* **2014**, *152*, 100–107. [CrossRef]

9.   *Definition of Vitivinicultural "Terroir"*; Resolution OIV/Viti 333/2010; OIV: Paris, France, 2010. Available online: http://www.oiv.int/public/medias/379/viti-2010-1-en.pdf (accessed on 30 September 2020).

10. Rebolo, S.; Peña, R.M.; Latorre, M.J.; García, S.; Botana, A.M.; Herrero, C. Characterization of Galician (NW Spain) Ribeira Sacra wines using pattern recognition analysis. *Anal. Chim. Acta* **2000**, *417*, 211–220. [CrossRef]

11. Calleja, A.; Falque, E. Volatile composition of Mencia wines. *Food Chem.* **2005**, *90*, 357–363. [CrossRef]

12. Noguerol-Pato, R.; Gonzalez-Barreiro, C.; Cancho-Grande, B.; Simal-Gandara, J. Quantitative determination and characterisation of the main odorants of Mencia monovarietal red wines. *Food Chem.* **2009**, *117*, 473–484.

13. Noguerol-Pato, R.; González-Rodríguez, R.M.; González-Barreiro, C.; Cancho-Grande, B.; Simal-Gándara, J. Influence of tebuconazole residues on the aroma composition of Mencía red wines. *Food Chem.* **2011**, *124*, 1525–1532. [CrossRef]

14. González-Rodríguez, R.M.; Cancho-Grande, B.; Torrado-Agrasar, A.; Simal-Gándara, J.; Mazaira-Pérez, J. Evolution of tebuconazole residues through the winemaking process of Mencía grapes. *Food Chem.* **2009**, *117*, 529–537. [CrossRef]

15. Canosa, P.; Oliveira, J.M.; Masa, A.; Vilanova1, M. Study of the Volatile and Glycosidically Bound Compounds of Minority Vitis vinifera Red Cultivars from NW Spain. *J. Inst. Brew.* **2011**, *117*, 462–471. [CrossRef]

16. Bouzas-Cid, Y.; Trigo-Córdoba, E.; Orriols, I.; Falqué, E.; Mirás-Avalos, J.M. Influence of Soil Management on the Red Grapevine (*Vitis vinifera* L.) Mencía Must Amino Acid Composition and Wine Volatile and Sensory Profiles in a Humid Region. *Beverages* **2018**, *4*, 76. [CrossRef]

17. Vilanova, M.; Campo, E.; Escudero, A.; Graña, M.; Masa, A.; Cacho, J. Volatile composition and sensory properties of Vitis vinifera red cultivars from North West Spain: Correlation between sensory and instrumental analysis. *Anal. Chim. Acta* **2012**, *720*, 104–111. [CrossRef]

18. Vilanova, M.; Rodríguez, I.; Canosa, P.; Otero, I.; Gamero, E.; Moreno, D.; Talaverano, I.; Valdés, E. Variability in chemical composition of *Vitis vinifera* cv Mencía from different geographic areas and vintages in Ribeira Sacra (NW Spain). *Food Chem.* **2015**, *169*, 187–196. [CrossRef]

19. Añón, A.; López, J.F.; Hernando, D.; Orriols, I.; Revilla, E.; Losada, M.M. Effect of five enological practices and of the general phenolic composition on fermentation-related aroma compounds in Mencia young red wines. *Food Chem.* **2014**, *148*, 268–275. [CrossRef]

20. Mihnea, M.; González-SanJosé, M.L.; Ortega-Heras, M.; Pérez-Magariño, S. A comparative study of the volatile content of Mencía wines obtained using different pre-fermentative maceration techniques. *LWT-Food Sci. Technol.* **2015**, *64*, 32–41. [CrossRef]

21. Blanco, P.; Mirás-Avalos, J.M.; Pereira, E.; Fornos, D.; Orriols, I. Modulation of chemical and sensory characteristics of red wine from Mencía by using indigenous Saccharomyces cerevisiae yeast strains. *J. Int. Sci. Vigne. Vin.* **2014**, *48*, 63–74. [CrossRef]

22. Nuñez, M.; Peña, R.M.; Herrero, C.; García-Martín, S. Analysis of some metals in wine by means of capillary electrophoresis. Application to the differentiation of Ribeira Sacra Spanish red wines. *Analusis* **2000**, *28*, 432–437.

23. Peña, R.M.; Latorre, M.J.; García, S.; Botana, A.M.; Herrero, C. Pattern recognition analysis applied to classification of Galician (NW Spain) wines with certified brand of origin Ribeira Sacra. *J. Food Sci.* **1999**, *79*, 2052. [CrossRef]

24. Vilanova, M.; Soto, B. The impact of geographic origin on sensory properties of Vitis vinifera cv. Mencía. *J. Sens. Stud.* **2005**, *20*, 503–511. [CrossRef]

25. Harbertson, J.F.; Spayd, S. Measuring phenolics in the winery. *Am. J. Enol. Vitic.* **2006**, *57*, 280–288.

26. Downey, M.O.; Dokoozlian, N.K.; Krstic, M.P. Cultural practice and environmental impacts on the flavonoid composition of grapes and wine: A review of recent research. *Am. J. Enol. Vitic.* **2006**, *57*, 257–268.

27. Garcia-Falcón, M.S.; Perez-Lamela, C.; Martinez-Carballo, E.; Simal-Gandara, J. Determination of phenolic compounds in wines: Influence of bottle storage of young red wines on their evolution. *Food Chem.* **2007**, *105*, 248–259. [CrossRef]

28. Alén-Ruiz, F.; García-Falcón, M.S.; Pérez-Lamela, M.C.; Martínez-Carballo, E.; Simal-Gándara, J. Influence of major polyphenols on antioxidant activity in Mencía and Brancellao red wines. *Food Chem.* **2009**, *113*, 53–60. [CrossRef]

29. Soto Vázquez, E.; Río Segade, S.; Orriols Fernández, I. Effect of the winemaking technique on phenolic composition and chromatic characteristics in young red wines. *Eur. Food Res. Technol.* **2010**, *231*, 789–802. [CrossRef]

30. Ortega-Heras, M.; Perez-Magariño, S.; Gonzalez-Sanjose, M.L. Comparative study of the use of maceration enzymes and cold prefermentative maceration on phenolic and anthocyanic composition and color of a Mencía red wine. *LWT-Food Sci. Technol.* **2012**, *48*, 1–8. [CrossRef]

31. Moreno, A.; Castro, M.; Falqué, E. Evolution of trans- and cis-resveratrol content in red grapes (*Vitis vinifera* L. cv Mencía, Albarello and Merenzao) during ripening. *Eur. Food Res. Technol.* **2008**, *227*, 667–674.

32. Rivas-Gonzalo, J.C.; Gutierrez, Y.; Hebrero, E.; Santos-Buelga, C. Comparisons of methods for the determination of anthocyanins in red wines. *Am. J. Enol. Vitic.* **1992**, *43*, 210–214.

33. Perez-Enciso, M.; Tenenhaus, M. Prediction of clinical outcome with microarray data: A partial least squares discriminant analysis (PLS-DA) approach. *Hum. Genet.* **2003**, *112*, 581–592. [CrossRef] [PubMed]

34. Jonsson, P.; Bruce, S.J.; Moritz, T.; Trygg, J.; Sjostrom, M.; Plumb, R.; Granger, J.; Maibaum, E.; Nicholson, J.K.; Holmes, E.; et al. Extraction, interpretation and validation of information for comparing samples in metabolic LC/MS data sets. *Analyst* **2005**, *130*, 701–707. [CrossRef] [PubMed]

35. Wold, S.; Antti, H.; Lindgren, F.; Ohman, J. Orthogonal signal correction of near-infrared spectra. *Chemom. Intell. Lab. Syst.* **1998**, *44*, 175–185. [CrossRef]

36. Wiklund, S.; Johansson, E.; Sjoestroem, L.; Mellerowicz, E.J.; Edlund, U.; Shockcor, J.P.; Gottfries, J.; Moritz, T.; Trygg, J. Visualization of GC/TOF-MS-based metabolomics data for identification of biochemically interesting compounds using OPLS class models. *Anal. Chem.* **2008**, *80*, 115–122. [CrossRef]

37. Van Leeuwen, C.; Roby, J.-P.; de Rességuier, L. Soil-related terroir factors: A review. *OENO One* **2018**, *52*, 2. [CrossRef]

38. Costa, R.; Fraga, H.; Fonseca, A.; Cortázar-Atauri, I.G.; Val, M.C.; Carlos, C.; Reis, S.; Santos, J.A. Grapevine Phenology of cv. Touriga Franca and Touriga Nacional in the Douro Wine Region: Modelling and Climate Change Projections. *Agronomy* **2019**, *9*, 210. [CrossRef]

39. Lorenzo, M.N.; Taboada, J.J.; Lorenzo, J.F.; Ramos, A.M. Influence of climate on grape production and wine quality in the Rías Baixas, north-western Spain. *Reg. Environ. Chang.* **2013**, *13*, 887–896. [CrossRef]

40. Amerine, M.A.; Winkler, A.J. Composition and quality of musts and wines of California grapes. *Hilgardia* **1944**, *15*, 493–675. [CrossRef]

41. Riou, C.; Becker, N.; Sotes Ruiz, V.; Gomez-Miguel, V.; Carbonneau, A.; Panagiotou, M.; Calo, A.; Costacurta, A.; De Castro, R.; Pinto, A.; et al. *Le déterminisme Climatique de la Maturation du Raisin: Application au Zonage de la Teneur en Sucre Dans la Communauté Européenne*; Office des Publications Officielles des Communautés Européennes: Luxembourg, 1994; 322p.

42. Tonietto, J.; Carbonneau, A. A multicriteria climatic classification system for grape-growing regions worldwide. *Agric. For. Meteorol.* **2004**, *124*, 81–97. [CrossRef]

43. Jackson, D.I.; Lombard, P.B. Environmental and management practices affecting grape composition and wine quality-a review. *Am. J. Enol. Vitic.* **1993**, *44*, 409–430. Available online: http://www.ajevonline.org/content/44/4/409 (accessed on 30 September 2020).

44. Branas, J.; Bernon, G.; Levadoux, L. *Élements de Viticulture Générale*; Imp. De Dehan: Montpellier/Bordeaux, France, 1974.

45. Carbonneau, A. Ecophysiologie de la vigne et terroir. In *Terroir, Zonazione, Viticoltura*; Schuster, D., Paoletti, A., Eds.; Trattato Internazionale, Phytoline: Piacenza, Italy, 2003; pp. 61–102.

46. Fraga, H.; Malheiro, A.C.; Moutinho-Pereira, J.; Jones, G.V.; Alves, F.; Pinto, J.G.; Santos, J.A. Very high resolution bioclimatic zoning of Portuguese wine regions: Present and future scenarios. *Reg. Environ. Chang.* **2014**, *14*, 295–306. [CrossRef]

47. Blanco-Ward, D.; García Queijeiro, J.M.; Jones, G.V. Spatial climate variability and viticulture in the Miño River Valley of Spain. *Vitis* **2007**, *46*, 63–70.

48. Yamane, T.; Jeong, S.T.; Goto-Yamamoto, N.; Koshita, Y.; Kobayashi, S. Effects of Temperature on Anthocyanin Biosynthesis in Grape Berry Skins. *Am. J. Enol. Vitic.* **2006**, *57*, 54–59.

49. Mori, K.; Goto-Yamamoto, N.; Kitayama, M.; Hashizume, K. Loss of anthocyanins in red-wine grape under high temperature. *J. Exp. Bot.* **2007**, *58*, 1935–1945. [CrossRef] [PubMed]

50. Mori, K.; Sugaya, S.; Gemma, H. Decreased anthocyanin biosynthesis in grape berries grown under elevated night temperature condition. *Sci. Hort.* **2005**, *105*, 319–330. [CrossRef]

51. Kliewer, W.M.; Torres, R.E. Effect of Controlled Day and Night Temperatures on Grape Coloration. *Am. J. Enol. Vitic.* **1972**, *23*, 71–77.

52. Revilla, E.; Losada, M.M.; Gutiérrez, E. Phenolic Composition and Color of Single Cultivar Young Red Wines Made with Mencia and Alicante-Bouschet Grapes in AOC Valdeorras (Galicia, NW Spain). *Beverages* **2016**, *2*, 18. [CrossRef]

53. Cheminat, A.; Brouillard, R. PMR investigation of 3-*O*-(β-Dglucosyl) malvidin structural transformations in aqueous solutions. *Tetrahedron Lett.* **1986**, *27*, 4457–4460. [CrossRef]

54. Boulton, R. The copigmentation of anthocyanins and its role in the color of red wine: A critical review. *Am. J. Enol. Vitic.* **2001**, *52*, 67–87.

55. Berke, B.; Chèze, C.; Vercauteren, J.; Deffieux, G. Bisulfite addition to anthocyanins: Revisited structures of colourless adducts. *Tetrahedron Lett.* **1998**, *39*, 5771–5774. [CrossRef]

56. Somers, T. The polymeric nature of wine pigments. *Phytochemistry* **1971**, *10*, 2175–2186. [CrossRef]

57. Darias-Martín, J.; Carrillo, M.; Díaz, E.; Boulton, R.B. Enhancement of red wine color by pre-fermentation addition of copigments. *Food Chem.* **2001**, *73*, 217–220. [CrossRef]

58. Darias-Martín, J.; Martín-Luis, B.; Carrillo-López, M.; LamuelaRaventós, R.; Díaz-Romero, C.; Boulton, R. Effect of caffeic acid on the color of red wine. *J. Agric. Food Chem.* **2002**, *50*, 2062–2067. [CrossRef] [PubMed]

59. Guise, R.; Filipe-Ribeiro, L.; Nascimento, D.; Bessa, O.; Nunes, F.M.; Cosme, F. Comparison between different types of carboxylmethylcellulose and other oenological additives used for white wine tartaric stabilization. *Food Chem.* **2014**, *156*, 250–257. [CrossRef] [PubMed]

60. Filipe-Ribeiro, L.; Milheiro, J.; Matos, C.C.; Cosme, F.; Nunes, F.M. Reduction of 4-ethylphenol and 4-ethylguaiacol in red wine by activated carbons with different physicochemical characteristics: Impact on wine quality. *Food Chem.* **2017**, *229*, 242–251. [CrossRef]

61. Filipe-Ribeiro, L.; Milheiro, J.; Matos, C.C.; Cosme, F.; Nunes, F.M. Data on changes in red wine phenolic compounds, headspace aroma compounds and sensory profile after treatment of red wines with activated carbons with different physicochemical characteristics. *Data Brief* **2017**, *12*, 188–202. [CrossRef]

62. OIV Organisation International de la Vigne et du Vin Recueil de Méthodes Internationales d'Analyse des Vins et des Moûts. Edition Officielle. Paris. 2015. Available online: http://www.oiv.int/fr/normes-et-documents-techniques (accessed on 30 September 2020).

63. Ribéreau-Gayon, P.; Stonestreet, E. Le dosage des anthocyanes dans le vin rouge. *Bull. Soc. Chim. Fr.* **1965**, *9*, 2649–2652.

64. Somers, T.C.; Evans, M.E. Spectral evaluation of young red wines: Anthocyanin equilibria, total phenolics, free and molecular O2, "Chemical age". *J. Sci. Food Agric.* **1977**, *28*, 279–287. [CrossRef]

65. Available online: http://www.inforiego.org/opencms/opencms/info_meteo/index.html (accessed on 1 December 2020).

66. Available online: https://en.tutiempo.net/climate/2015/ws-85600.html (accessed on 1 December 2020).

67. Huglin, P. *New Method of Evaluating the Solar Thermal Possibilities of a Wine-Growing Environment*; Reports from the French Academy of Agriculture; Académie d'Agriculture de France: Paris, France, 1978; Volume 64, pp. 1117–1126.

68. Hargreaves, G.H.; Asce, F.; Allen, R.G. History and evaluation of Hargreaves evapotranspiration equation. *J. Irrig. Drain. Eng. ASCE* **2003**, *129*, 53–63. [CrossRef]

69. Allen, R.G.; Pruitt, W.O.; Wright, J.L.; Howell, T.A.; Ventura, F.; Snyder, R.; Itenfisu, D.; Steduto, P.; Berengena, J.; Baselga, J.; et al. A recommendation on standardized surface resistance for hourly calculation of reference ETo by the FAO56 Penman–Monteith method. *Agric. Water Manag.* **2006**, *81*, 1–22. [CrossRef]

70. Santos, J.A.; Malheiro, A.C.; Pinto, J.G.; Jones, G.V. Macroclimate and viticultural zoning in Europe: Observed trends and atmospheric forcing. *Clim. Res.* **2012**, *51*, 89–103. [CrossRef]

71. Specifications of the PDO <<Bierzo>> PDO-ES-A0615 Reglamento_1_1601369319.pdf. Available online: http://www.crdobierzo.es/en/regulator-board/legislation/ (accessed on 1 December 2020).

72. Instituto da Vinha e do Vinho (IVV). Available online: https://www.ivv.gov.pt/np4/%7BclientServletPath%7D/ ?newsId=8617&fileName=DOP_D_O___Caderno_de_especifica__es_fina.pdf (accessed on 1 December 2020).

73. Gouveia, J.; Lopes, C.M.; Pedroso, V.; Martins, S.; Rodrigues, P.; Alves, I. Effect of irrigation on soil water depletion, vegetative growth, yield and berry composition of the grapevine variety Touriga Nacional. *Ciência Téc. Vitiv.* **2012**, *27*, 115–122.

74. Rubert, J.; Lacina, O.; Fauhl-Hassek, C.; Hajslova, J. Metabolic fingerprinting based on high-resolution tandem mass spectrometry: A reliable tool for wine authentication? *Anal. Bioanal. Chem.* **2014**, *406*, 6791–6803. [CrossRef] [PubMed]

Sample Availability: Samples of the compounds used are not available from the authors.

Publisher's Note: MDPI stays neutral with regard to jurisdictional claims in published maps and institutional affiliations.

Review

Wine or Beer? Comparison, Changes and Improvement of Polyphenolic Compounds during Technological Phases

Sanja Radonjić [1,*], **Vesna Maraš** [1], **Jovana Raičević** [1] **and Tatjana Košmerl** [2]

[1] "13. Jul Plantaže" a.d., Research and Development Sector, Put Radomira Ivanovića 2, 81000 Podgorica, Montenegro; vesnam@t-com.me (V.M.); jovana_raicevic@yahoo.com (J.R.)

[2] Biotechnical Faculty, University of Ljubljana, Jamnikarjeva 101, 1000 Ljubljana, Slovenia; tatjana.kosmerl@bf.uni-lj.si

* Correspondence: sanja.radonjic511@gmail.com

Academic Editor: Mirella Nardini

Received: 25 September 2020; Accepted: 21 October 2020; Published: 27 October 2020

Abstract: Wine and beer are nowadays the most popular alcoholic beverages, and the benefits of their moderate consumption have been extensively supported by the scientific community. The main source of wine and beer's antioxidant behavior are the phenolic substances. Phenolic compounds in wine and beer also influence final product quality, in terms of color, flavor, fragrance, stability, and clarity. Change in the quantity and quality of phenolic compounds in wine and beer depends on many parameters, beginning with the used raw material, its place of origin, environmental growing conditions, and on all the applied technological processes and the storage of the final product. This review represents current knowledge of phenolic compounds, comparing qualitative and quantitative profiles in wine and beer, changes of these compounds through all phases of wine and beer production are discussed, as well as the possibilities for increasing their content. Analytical methods and their importance for phenolic compound determination have also been pointed out. The observed data showed wine as the beverage with a more potent biological activity, due to a higher content of phenolic compounds. However, both of them contain, partly similar and different, phenolic compounds, and recommendations have to consider the drinking pattern, consumed quantity, and individual preferences. Furthermore, novel technologies have been developing rapidly in order to improve the polyphenolic content and antioxidant activity of these two beverages, particularly in the brewing industry.

Keywords: wine; beer; polyphenols; antioxidant activity; winemaking; brewing

---

1. Introduction

Wine has existed on Earth for more than 6000 years [1], while beer has existed for over 5000 years [2]. Throughout history, both drinks were produced in Ancient Egypt and regions of Mesopotamia. Wine was used in various therapies and treatments, while beer was an essential part of diet, first to appear when people began agriculture. The brewing industry is more linked to northern Europe, where due to cold conditions viticulture development was inhibited. Both of these beverages are very complex in terms of their ingredients, and besides their long traditions, there are so many characteristics and parameters that determine their final quality, from the quality of raw material (malt and hop for beer and grape for wine), yeast, regimes of alcoholic fermentation, conditions of aging etc. However, the parameters of all the phases of production and composition of these two beverages have been very well studied by many researchers, since the early 20th century. Besides their flavor, which determines their use, wine and beer are known as rich with bioactive compounds, i.e., antioxidants that increase the interest in their nutritional profile. A great number of studies and comprehensive reviews have dealt with the bioactive compounds responsible for the possible health benefits due to moderate wine and beer consumption, and with the different methods of improvement of the antioxidant compounds in these two beverages [3–14]. Much of this research supports the thesis that moderate consumption of alcoholic beverages, such are red wine and beer, positively influences the decrease of cardiovascular disease [3]. Key roles as antioxidants in wine and beer belong to the phenolic compounds, and many of them, such as flavonoids, have an effect on cardiovascular and chronic degenerative diseases [15,16], non-flavonoids (stilbenes, hydroxycinnamic, and hydroxybenzoic acids) also positively affect the cardiovascular system [17]. In addition, it has been recently shown that there is a relation between beer consumption and higher protection against coronary diseases, compared to other spirits, and beer is also associated with bone density increase, and with immunological and cardiovascular benefits [18–20]. However, there are huge differences between the phenolic profile and content among red wine and beer, primarily due to the different raw material used in their production. The importance of phenolic compounds for wine and beer is very significant, as their presence influences the final quality of these products. Some polyphenol classes can only be found in beer (chalcones and flavanones) and others are mainly found in wine (stilbenes, proanthocyanidins), while flavanols and flavan-3-ols are found in similar concentrations in both beverages. In beer quality they play a key role, as they influence the time of transport and storage, flavor stability, clarity, and color of beer. Additionally, phenolic compounds are essential in wine, because they determine the sensorial wine characteristics (taste and fragrance), color, microbiological and oxidative stability, and chemical properties of wine, as they interact with other compounds including other polyphenols, proteins, and polysaccharides.

Production of wine and beer consists of many technological phases, which are influenced by many parameters, and the huge numbers of occurring variables; the changes in biochemistry are very complex. In both beverages, the composition of phenolic compounds is very diverse and depends on many similar parameters, first of all on the genetic factors of the raw material and the environmental conditions during their growth, as well as technological and aging factors [21]. In regards to beer, malt and hops represent the two main ingredients on which antioxidant compounds depend; actually 70–80% are derived from malt, and the remaining from hops [22,23], and this ratio also depends on the type of the beer [24]. Furthermore, during beer making, important technological phases, in which the change of polyphenolic compounds occurs, begins with the malting process (steeping and germination), kilning, mashing, wort separation and boiling, whirlpool rest, through to the fermentation, maturation, and at the end, to the stabilization/filtration and bottling. Primarily classification of beer is made based on the fermentation process [25], and in these terms there are lagers, ales, and lambic types of beer. The most consumed are lagers, produced by low fermentation at lower temperatures (6–15 °C), while in contrast ale-type beer is made by high fermentation at higher temperatures between 16–24 °C, and as a result of spontaneous fermentation there is lambic beer. Exclusively, grape is used as the raw material for wine production, and based on the color of the used grape varieties, wine is classified into red and white. The main difference, and at the same time the most important, between the making of red and white is that during the making of red, along with alcoholic fermentation, maceration i.e., extraction of color and other substances from grape skin and seed occurs, while within the process of the alcoholic fermentation of whites only colorless and clarified grape juice is used in the process of alcoholic fermentation. As for making rose wine, winemakers use limited skin contact in order to extract color and some compounds, depending on the desired degree of complexity. Due to this maceration, occurring along with alcoholic fermentation during red winemaking, in which the phenolic compounds are extracted from grapes, this step represents the key one in determining the content of polyphenolic compounds in red wine. Furthermore, because of this step, it is commonly known that red wine contains more antioxidant compounds, and has been more studied and reviewed by researchers in the last decades [14,26–34]. Another important step during winemaking in which it is possible to increase polyphenolic compounds is ageing in wood barrels, or with addition of oak alternatives.

Considering the fact that the beer and wine markets are becoming more competitive and saturated, and considering that consumers nowadays are more interested in beverages influencing in positive way their health, novel technologies for beer and wine are developing in order to produce products with higher antioxidant potential, as well with special personality. The aim of this review is to present the current state of the knowledge of phenolic compounds in beer and wine as well as the possibilities of their increase during different technological phases in beer and wine production, in moving from the raw material to the final product.

2. Bioactive Compounds in Beer and Wine

The largest group within natural antioxidant compounds is the group of polyphenols, consisting of very diverse chemical compounds that can be classified in many ways, but that generally are divided into two main classes: flavonoids and non-flavonoids. Within the class of non-flavonoids, natural polyphenolic compounds can be present in chemical structures, such as: phenolic acids, stilbenes, lignans, chalcones, and tannins (hydrolysable and condensed) [34–36], Figure 1. Phenolic acids in wine act as copigments, and they do not impact odor and flavor. Stilbenes are the most well known as antioxidants, and within the chalcones group there is xanthohumol; present in beer, and of huge importance, as this compound possess high biological activity. Flavan-3-ols influence bitterness, astringency, and wine structure, and participate in the stabilization of color during aging. Tannins also contribute to the sensory characteristics, particularly of red wine, as they are related to the astringency, they also interact with other macromolecules (proteins and polysaccharides) influencing the colloidal behavior of wine. Condensed tannins (proanthocyanidins) in the brewing industry are interesting as they influence haze formation in beer. Anthocyanins are responsible for the color of red grapes and wines.

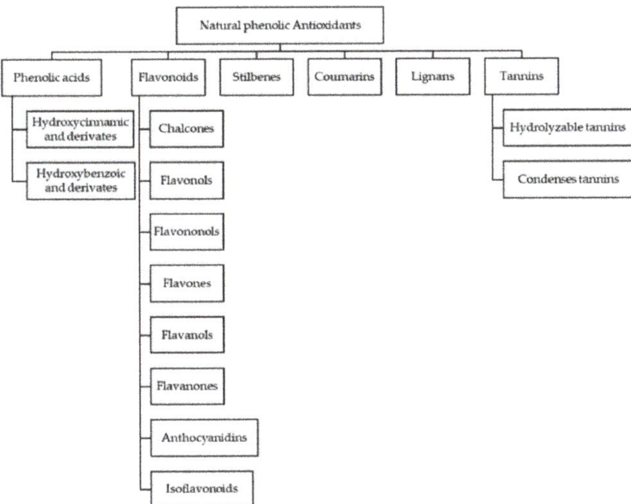

Figure 1. Natural phenolic compound classification [8].

As is expected, due to different used raw material and technological processes, there are differences between wine and beer, in the presence, as well in the concentrations, of phenolic substances. Moreover, the antioxidant compounds in beer belong to different groups of chemical substances such are: thiols [37], SO_2 [38] (product of the Maillard reaction [39,40]), α-acids derived from hops [41–43], and phenolic compounds [44–46]. Thiols have been suggested to correlate with sulfites in the antioxidative mechanism, and are important for beer's oxidative stability. Sulfites were found to be the only compound that was able to delay the formation of radicals [38], and actually give antioxidant and antimicrobe properties in wine too. The main product of the Maillard reaction is melanoidin, which affects the color, flavor, and body of beer. Hop α-acids (also called humulones) represent the main bittering compound in beer, and have shown a high ability to quench radicals, while *iso*-α-acids possess this activity to a lower extent. In addition, *iso*-α-acids can influence beer staling, but not to a high degree. However, within this review, focus will be on the content of phenolic compounds, as they have been recognized as mainly responsible for antioxidant activity in wine and beer.

The polyphenol complexes of beer and red wine, additionally, represent a source of dietary antioxidants. As both beverages are very popular and widely consumed, benefits of the light-moderate consumption of wine and beer are supported by scientific literature data. Polyphenols from red wine and beer could act as antioxidants, and also as anti-inflammatory agents contributing to the defense against atherosclerotic pathologies [19,47]. The beneficial moderate consumption of beer is also based on antioxidant compounds present in beer, i.e., on their redox properties [48]. Antioxidants present in beer improve several diseases, and are associated with benefits to the cardiovascular and immunological system [19,20]. It was shown that after consumption of non-alcoholic beer, the decrease of several inflammation biomarkers, homocysteine, and systolic blood pressure occurred. These influences were mainly attributed to the polyphenolic compounds in beer. Furthermore, several studies have shown that the light-moderate intake of alcohol is associated with lower incidence of diabetes type-2, a higher level of high-density lipoprotein cholesterol, as well as with lipid oxidative stress reduction [49–52]. Torres et al. [53] reported that moderate wine intake, compared to other alcoholic beverages like vodka, rum, and brandy increased total antioxidant capacity, and decreased pro-inflammatory factors along with a fat-enriched diet that was consumed by young healthy volunteers. This is also supported by the phenomenon known as the "French paradox", which indicates that moderate daily drinking of red wine contributes to lower coronary heart diseases incidence, despite their diets possessing a higher amount of saturated fatty acids and total fat [54]. However, excessive intake of alcohol beverages is associated with chronic disease development and other very serious problems, representing the leading risk factor for mortality [55]. Roercke et al. [3] reported there is an important influence of drinking patterns, such are episodic heavy drinking within average moderate drinkers, and some other important influencing parameters in term of health issues like smoking status, age, body mass index, and physical activity, and all of them have to be considered in order to estimate dose of alcohol as well the risks. After all, chronic heavy and episodic drinking should be avoided. In the Dietary Guidelines for Americans (2015–2020), moderate intake of alcohol proposes up to one unit of alcohol per day for women and two for men [56].

However, the positive influence of single polyphenols on human health occurs at higher concentrations than those found in beer and wine, indicating the synergistic action of different polyphenolic mixtures [57]. Ranges of some of the most important phenolic compounds, found in red wine and beer are presented in Table 1. Phenolic acids also possess antioxidant and anti-inflammatory properties [58]. Based on literature data, beer has shown higher upper values for content of *p*-coumaric acid, and all hydroxybenzoic acids, while for other polyphenolic compounds it was mainly the opposite, and higher concentrations dominated in red wine. Flavonols are considered very important bioactive compounds, and have shown positive effects against certain cancers and cardiovascular diseases in some epidemiological studies [59,60]. Concentrations of all three presented flavonols (quercetin, myricetin, and kaempferol) were much higher in red wine than in beer. Stilbenes, particularly resveratrol, are the most associated with wine's beneficial properties. Resveratrol is recognized as an antioxidant, anticancer, cardioprotective, and anti-inflammatory agent. Due to its bioactivity, trans-resveratrol was proposed for many diseases as a therapeutic agent [61]. The content of stilbenes is not comparable for wine and beer, as based on literature data these compounds are rarely, or never, found in beer. It was also indicated that flavan-3-ols may show cardioprotective activity, and their antioxidant activity was shown in some studies [59]. Flavones are also recognized as molecules with important biological activity (anti-tumor, antioxidant, and anti-inflammatory), and were used as treatment for some neurodegenerative disorders and coronary heart diseases [62]. Flavanones also belong to antioxidant agents and the found concentrations in wine and beer were very low, while a lower content of naringerin was found in red wine compared to beer. Tannins also showed potent radical scavenging, anti-inflammatory, and antioxidant activity [63], and much higher levels of condensed tannins were found in red wine. Besides the presented polyphenolic compounds there are also some compounds found in wine and not in beer and the opposite. Within the compounds found in beer, two very important ones are xanthohumol and melanoidin. Both, xanthohumol and melanoidin have shown antimicrobial properties, melanoidin also possess antihypertensive, prebiotic, and antiallergenic properties [64], while xanthohumol showed anti-cancer, anti-inflammatory, anti-obesity, etc. properties [65,66]. Depending on the raw material and brewing process, the content of melanoidin ranges from 0.58 mg/L in alcohol free beer to 1.49 mg/L for dark beer, while in blond beer 0.61 mg/L was determined [67,68]. In wine, among the phenolic compounds with biological activity, there are also anthocyanins. The most common anthocyanins found in red wine, malvidin-3-glucoside and malvidin-3-galactoside, have shown anti-inflammatory effects, and their synergistic activity was observed [69].

Table 1. Ranges of some phenolic compounds in red wine and beer.

	Red Wine	References	Beer	References
	Cinnamic Acids (mg/L)			
ferulic acid	0.05–10.43	[30,70–72]	0.01–5.04	[12,45,46,73–82]
p-coumaric acid	0.02–8.00	[27,30,70–72]	0.003–55.80	[9,12,45,73,78,79]
caffeic acid	0.02–644.50	[27,30,70,71,83]	0.00–23.50	[9,45,46,73–79,81,82,84]
	Hydroxybenzoic Acids (mg/L)			
gallic acid	27.10–66.10	[28,71]	0.00–142.20	[9,46,73–78,80–82]
protocatechuic acid	0.91–1.78	[28,30]	0.01–5.10	[12,75–78,80–82,84]
p-hydroxybenzoic acid	2.75–6.20	[28]	0.00–16.84	[12,45,73,75,76,78]
	Stilbenes (mg/L)			
resveratrol	0,51–11.70	[28,85]	0.002–0.081	[86]
trans-resveratrol	0.21–23.00	[27,70,71,87–89]	-	-
cis-resveratrol	0.01–7.00	[71,87,88]	-	-
total stilbenes	1.00–5.50	[71]	-	-
	Tannins (mg/L)			
hydrolysable tannins	0.4–50.0	[90–94]	1.5	[81]
	Flavan-3-ols (mg/L)			
catechin	6.98–91.99	[27,30,36]	0.03–6.54	[12,73–75,77,80–82]
epicatechin	8.07–52.85	[27,30,36]	0–4.55	[9,74,75,80–82]
	Flavones (mg/L)			
luteolin	0.20–1.00	[95–98]	0.10–0.19	[82]
apigenin	0.00–4.70	[99]	0.80–0.81	[82]
	Flavonols (mg/L)			
myricetin	0.70–30.40	[27,30]	0.15–0.16	[82]
quercetin	1.27–65.90	[27,30,36]	0.06–1.79	[74,80,82]
kaempferol	0.61–26.80	[27]	0.10–1.64	[80,100]
	Flavanones (mg/L)			
naringenin	0.90 to 4.20	[89]	0.06–2.34	[80]

2.1. Analytical Methods for Determination of Antioxidant Activity in Beer and Wine

It is of huge value to mention the importance of the analytical methods applied for determination of antioxidant activity in these beverages. Antioxidant power in different functional food, as well the isolated antioxidant substances, were dependent on applied assays used for their determination, which is also confirmed by Di Pietro and Bamworth [7]. Antioxidant activity in beer has been mainly determined using electron spin resonance spectroscopy, based on spin trapping of radicals that have been forming at 60 °C [38,41]. Differences in the antioxidant activity of examined wines and beers have been shown, and the superior behavior of beer was only demonstrated using the β-carotene bleaching method, while hydroxyl scavenging assay is not reliable for beer assessment, as compounds in beer react with thiobarbituric acid [7]. However, in research, comparing different beverages on an "as is" basis, and equalized according to the alcohol concentrations and the total polyphenol content, red wine performed the best for most used assays, when used samples were not equalized. Recently, Wannenmacher et al. [12] examined in their study two assays, one based on a mechanism consisted of electron transfer, such are ferric reducing antioxidant power (FRAP), and the other an oxygen radical absorbance capacity (ORAC) assay based on the transfer of hydrogen atoms, which scavenge peroxy radicals, in order to determine the antioxidant capacity of beer. Antioxidant value obtained using ORAC correlated positively with free amino nitrogen, total nitrogen, and *p*-coumaric acid, while values obtained using FRAP correlated positively with total anthocyanogens, total polyphenols, and catechin content [12]. Furthermore, both assays correlated significantly with the sum of phenolic compounds in the examined beers.

In winemaking, the most commonly used methods for determination of antioxidant activity and total polyphenol content in wine are spectrophotometric methods. Spectrophotometry is used for monitoring the decrease in absorbance that occurs when a present antioxidant scavenges the added radical in the sample [101]. For these purposes different radicals have been used in order to determine antioxidant capacity in beverages, and among them the most frequently used are DPPH (2,2-diphenyl-1-picrylhydrazyl) [102] and ABTS (2,20-azino-bis(3-ethylbenzothiazoline-6-sulphonic acid) [103]. Moreover, with spectrophotometry, it is possible to monitor the reducing activity of phenolic compounds in wine using FRAP, a method based on regeneration of ferric iron to ferric(II) by phenolic compounds, and using the CUPRAC method based on regeneration of copper(II) to copper (I) ion [104]. Even though these methods remain a classic tool for the evaluation of antioxidant activity and phenolic content, there are also some electrochemical techniques used for determination of antioxidant capacity of wines, and the most used are differential pulse voltammetry on glassy carbon working electrode [105] and cyclic voltammetry [106]. Recently, Ricci et al. [107] examined analytical approaches, using a flow injection system with a sequential diode array and electrochemical amperometric detectors. They concluded that flow injection, coupled with a diode array and electrochemical amperometric detectors, is useful and can be successfully applied for measurement of antioxidant capacity and the total phenolic content in wines. Minkova et al. [108] compared the antiradical capacity of Bulgarian red wines, using two analytical methods (spectrophotometric and chemiluminescent) and showed that chemiluminescent assays were more efficient in the elimination of hypochlorite, compared to the superoxide anion, for all wine samples. García-Guzmán et al. [109] evaluated the polyphenol index in wine and beer samples using a tyrosinase-based amperometric biosensor, obtained via a novel sinusoidal current method, and good analytical performance of this biosensor was achieved in terms of stability, reproducibility, limits of quantification and detection, linear response range, and accuracy, using caffeic acid as a polyphenol reference. However, electroanalytical techniques proved to be an appropriate alternative, considering that they showed quick response and good sensitivity, without sample treatment, simple instrumentation, low cost, etc. In addition, combined use of electroanalytical techniques and enzymes provided good selectivity in determination of polyphenol index [109].

Overall, there are contradictory and insufficient data on the correlation between antioxidant activity and the concentration of individual and total polyphenols in wine [32]. These contrasting data are due to differences in the used raw material, and also due to the proportions and contents of particular phenolic compounds. Nevertheless, it is not easy to compare the observed literature data, and sometimes it is even not possible, due to differences in applied methods, based on different methodological approaches. In Table 2 are presented ranges of total content of polyphenolic compounds and some phenolic classes, as well antioxidant activity measured with different analytical tools. Comparing obtained data, when the same methods were used, it was observed that wine dominated in the content of total polyphenols (FC method) and antioxidant activity (FRAP method), as was expected. However, it should not be forgotten that there are many influencing parameters, due to the different proportions of certain phenolic compounds in wine and beer, as well as due to differences in the used grape varieties in wines, and the type of beer and brewing technology.

Table 2. Ranges of total content of polyphenolic compounds and some phenolic classes, as well as antioxidant activity measured with different analytical tools.

Beer			Wine		
Parameter	Range	Reference	Parameter	Range	Reference
Total polyphenols (FC method); mgGAE/L	127–855	[7,9,12,77]	Total polyphenols (FC method); mgGAE/L	860.2–2912.0	[27,30,31,71]
Total anthocyanogens; mg/L	19.0–84.5	[12,110]	Total anthocyanins; mg/L	21–1011	[71,111,112]
Antioxidant activity; DPPH; mmol TE/L	0.55–6.67	[9,78,113]	Antioxidant activity; ABTS; mmol TE/L	7.5–96.4	[27,30,71]
Antioxidant activity; FRAP; mmol TE/L	0.862–1.271	[12]	Antioxidant activity; FRAP; mmol TE/L	6.9–15.2	[31]

2.2. Non-Flavonoid Polyphenols in Wine and Beer

Non-flavonoid phenolic compounds of wine consist of three main groups: two types of phenolic acids (cinnamic and benzoic), and stilbenes [95]. Beer also contains the non-flavonoid polyphenols, and within this group there are present the monophenolic compounds, chalcone (xanthohumol) and resveratrol [11]. Phenolic acids in both beverages can be found in the form of (hydroxy-) benzoic and (hydroxy-) cinnamic acids derivates, and after are classified according to the nature and the type of their ring substituent [36], Figure 2. The greater part of the determined phenolic acids in beer were found in bound form as esters, glycosides, and bound complexes. In wine, these phenolic acids can also be found as esters with tartaric acid, as well in the free form, or esterified with anthocyanins and ethanol. Additionally, in wine hydroxybenzoic and hydroxycinnamic acids act as copigments.

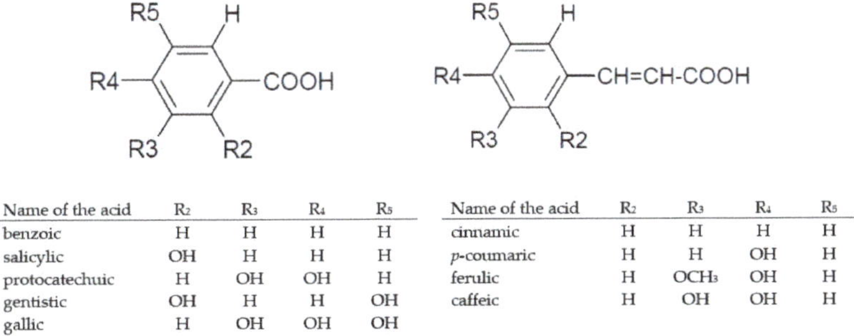

Name of the acid	R_2	R_3	R_4	R_5	Name of the acid	R_2	R_3	R_4	R_5
benzoic	H	H	H	H	cinnamic	H	H	H	H
salicylic	OH	H	H	H	p-coumaric	H	H	OH	H
protocatechuic	H	OH	OH	H	ferulic	H	OCH_3	OH	H
gentistic	OH	H	H	OH	caffeic	H	OH	OH	H
gallic	H	OH	OH	OH					

Figure 2. Hydroxybenzoic and hydroxycinnamic acids derivates [74].

2.2.1. Hydroxycinnamic Acids

Hydroxycinnamic acids in both beverages do not impact the odor and flavor, but they are very important compounds, as they act as precursors for volatile phenolic compounds [9]. In regards to beer the majority of studies reported that ferulic and *p*-coumaric acids are the most abundant acids in beer [45,46,73–77,84]. Hydroxycinnamic acids originate from the raw material used for beer production, i.e., barley malt and hops. Moreover, ferulic and *p*-coumaric predominate in barley malt, and their concentrations are influenced by the malting process, and growing environmental conditions, as well as the post-harvest treatment of barley. It was shown that ferulic acid is one of the most abundant acids in Chinese beers [77,114], European beers [74], and in Chilean beers [115]. However, published values for these two acids vary widely due to the differences in used extraction and analytical methods, as well as due to the form of the analyzed acids (free or bound) [116]. The content of phenolic acids was quantified by Floridi et al. [78], and in 23 Italian lager wines, *p*-coumaric acid was found at 1.364 ± 0.709 mg/L, while ferulic content was 2.41 ± 0.875 mg/L. Recently, Habschied et al. [9] quantified the content of caffeic and *p*-coumaric acids in different types of industrially produced beers (black, dark, lager, and pilsner) and noted that the content of caffeic acid was the lowest among all determined phenolic compounds. They also found that light and dark beers had the lowest share of *p*-coumaric acid, while black beers contained higher concentrations of *p*-coumaric acid. It was also shown that in the ale type of beer, the highest share among these acids belongs to caffeic acid [79], while ferulic acid was dominant in black, non-alcoholic, wheat, abbey, bock, and pilsen beers [46]. It is also important to note that hydroxycinnamates in beer can also be found in conjugation with the polyamides hydroxycinnamic acids amides or phenolamides, and after these compounds can be found glycosylated, and as derivatives of hydroxylated agmatine. These compounds, hordatines, and hydroxycinnamoyl agmatines have been detected in final beers, contributing to the beer's astringency. The content of hordatine in final beers (determined as *p*-coumaric acid equivalent), in large number of samples ranged 5.6 ± 3.1 mg/L, showing that this substance group is the most abundant phenolic substance in beer [117].

Hydroxycinnamic acids can be found in their free form, or in bound form as tartaric esters. These acids represent the majority of the nonflavonoid class in red wines, and the majority of phenolics in white wines. These esters can be partially hydrolyzed during the process of alcoholic fermentation, resulting in their free forms. In grape and wine, caffeic, coumaric, and ferulic acids are also the most important in this sub-class of polyphenols [118]. Content of these acids in grapes, grape juice, and wine depends on the grape variety, as well on the environmental growth conditions. These compounds depend on variety; the influence of the vintage is not negligible, as well as the used winemaking technology. However, the content of caffeic acid was found to be the most abundant [70,71,119]. These data are not in accordance with results obtained by Lima et al. [72], in whose study the predominant acid was *p*-coumaric (2.30–6.70 mg/L), followed by caffeic acid (0.52–1.49 mg/L), while the content of ferulic acid was up to 0.16 mg/L, in wines from the most used grape varieties in Portugal.

2.2.2. Hydroxybenzoic Acids

Hydroxybenzoic acids possess a general C6-C1 structure and belong to the phenolic compounds, Figure 2. In beer the most abundant hydroxybenzoic acids are salicylic, *p*-hydroxybenzoic, vanillic, and gallic acids, while in wine there are *p*-hydroxybenzoic acid, syringic acid, vanillic acid, and gallic acid. In regard to the contents of these acids in beer, Floridi et al. [78] found 2.866 ± 1.553 mg/L of salycilic acid, and 16.84 ± 10.988 mg/L of *p*-hydroxybenzoic acid in research on 23 Italian lager beers. Vanillic acid in beer varied from 0.08 to 2.98 mg/L [45,73,75,77,84], and in research from McMurrough et al. [76] varied between 2.5 to 12.7 mg/L, while in the same research gallic acid ranged from 1.1 to 3.5 mg/L, and in study by Zhao et al. [77] this acid valued from 1.81 to 10.39 mg/L. Recently, Wannenmacher et al. [12] determined the content (0.15–0.33 mg/L) of *p*-hydroxybenzoic and (0.07–0.22 mg/L) of protocatechuic acid in beers by varying the type of raw material and brewing technology. Special attention was put on research of gallic acid content in beer, as it was shown that this acid can be an indicator of oxidation during production of beer, due to its high susceptibility to degradation and oxidation [83]. It was shown that gallic acid predominates in Serbian and Brazilian beers [120,121], and in regard to the type of beer, this acid was found to be highest in lager beers [77]. However, recently, Habschied et al. [9] determined the highest concentration of this acid in black beer (14.22 mg/100 mL), and the lowest concentration in light style lager beer (4.12 mg/100 mL). These results aligned with results reported by Zhao et al. [78], while Mitić et al. [120] reported lower concentrations of gallic acid in bar beers, but still it was the major polyphenolic compounds in all samples. As in beer, gallic acid is also considered as the most important acid in wine, considering that this compound is the precursor of all hydrolysable tannins. The origin of gallic acid could therefore be from the hydrolysis of condensed tannin and gallate esters of hydrolysable tannins [122]. Furthermore, because of its three free hydroxyl groups this acid is considered as a very potent antioxidant, and higher concentrations of this acid using longer maceration times in winemaking of reds were obtained. The total amount of hydroxybenzoic acids in wine was found to be up to 218 mg/L [6].

2.2.3. Stilbenes

One of the most important compounds from the non-flavonoid class, which has received attention thanks to its link to beneficial effects on human health, is resveratrol [123,124]. Many researchers reported the health benefits of this compound, and its ability to prevent a number of human diseases [17,125–129]. In regard to its structure, resveratrol possesses two phenol rings connected by the styrene double bond [130], and it can exist in *cis-* and *trans-*configurations, Figure 3. Resveratrol can also be found in the form of 3-β-glucoside, *trans-*, and *cis-*piceid. Reported values of this compound were much lower in beer in comparison to wine, particularly red wine. Both isomers, *trans-* (0.7–6.5 mg/L) and *cis-* (0.1–7 mg/L) were detected in wine [87,88], and their concentrations depended on the used grape variety, terroir, applied viticulture practices, and the type of wine [131]. The content of *trans-*resveratrol was determined in wines from the Montenegrin terroir, and it was shown that contents varied from 0.62 ± 0.02 mg/L in Cabernet Sauvignon, to 1.27 ± 0.11 mg/L in Vranac wine [70]. Additionally, in this research, the content of *cis-*resveratrol in wines was 0.47 ± 0.03 mg/L in Vranac, and 0.57 ± 0.03 mg/L in Kratošija wine, while lower concentrations were observed in Cabernet Sauvignon wines (0.25 ± 0.01 mg/L). Similar results were obtained by Zoechling et al. [89] and Pajović et al. [71]. The concentrations of resveratrol ranged from 1.99 and 81.22 μg/L in 110 commercial beers [86], while these values for red wine were reported from 2.03 to 11.7 mg/L [86]. However, resveratrol has also been found to a lesser extent in alcohol-free and lager beers [132], and in the ale type [133]. This can be explained by the fact that resveratrol is found in hops, and a low amount of hop is in generally added during beer production, particularly when using the classical hoping regime [134].

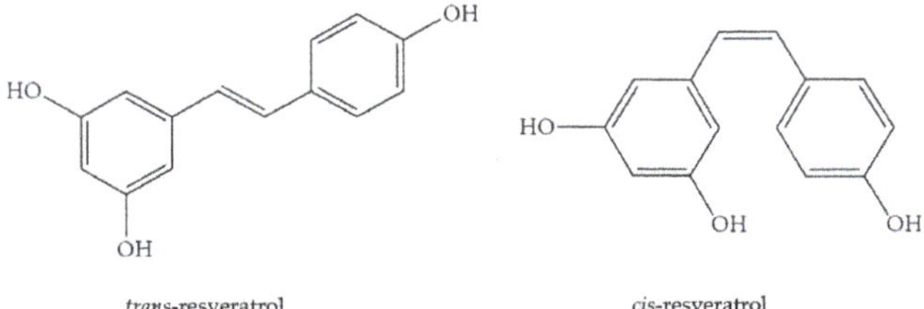

Figure 3. Structure of *trans-* and *cis-*resveratrol [135].

2.2.4. Hydrolysable Tannins

Tannins belong to a very important subgroup of polyphenol compounds, especially in red wine, as they contribute to the sensory characteristics of wine related to the perception of astringency and are also involved in reactions that lead to wine browning, particularly in white wine [6]. According to chemical structure they are divided into to two main classes, i.e., condensed tannins (proanthocyanidins) and hydrolysable tannins. Condensed tannins are polymers of flavan-3-ol, which classifies them into the flavonoid type of phenolic compounds, which will be described more later; they are also present in grapes, and after in wine. Hydrolysable tannins are a natural part of oak barrels, and can be found in wine matured in oak barrels. In red wine, the total concentrations of tannins determined in red wine vary from 1.1 to 3.4 g/L [91,136].

The precursor and basic unit of hydrolysable tannins is gallic and its derivatives, i.e., ellagic acid, and these acids are mainly esterified with sugars, such as glucose, or less commonly quinic or shikimic acid, Figure 4. Due to this esterification they can achieve from 500 to 2800 Da. They are very influenced by pH changes, through which they can be degraded, if some enzymatic or non-enzymatic processes occur. Hydrolysable tannins are usually extracted from oak barrels during wine maturation, therefore aging in oak mainly promotes extraction of ellagitannins into the wine. Depending on the type of the wood used for wine maturation, concentrations of hydrolysable tannins range from 0.40–50 mg/L [90–94]. Hydrolysable tannins in beer originate from hop and malt [11]. Marova et al. [80] determined the concentrations of hydrolysable tannins in 22 samples of commercial beer in amounts of 1.5 mg/L of (+)-catechin gallate and (−)-epicatechin gallate.

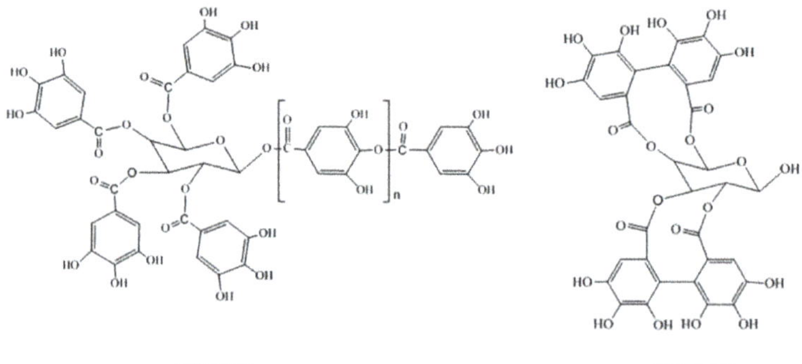

gallotannin   ellagitannin

Figure 4. Structure of hydrolysable tannins, gallotanin, and ellagitannin.

2.3. Flavonoid Compounds

Flavonoids represent the largest group belonging to the polyphenolic compounds, and in wine 85% of the phenolic compounds are accounted for by the flavonoids [137]. The basic structure of flavonoids consists of a system with three-rings: two aromatic and one oxygen-containing central ring [138]. Based on the substitution of the pyran ring and on its oxidation degree, flavonoids are classified into a wide range of subgroups, such as flavones, flavonols, flavanes, flavanols, flavanones, flavononols, anthocyanins, and anthocyanidins, as well as the chalcones and dihydrochalcones [139]. Both beer and wine contain the flavonols, flavanols, anthocyanins etc., but there is one important compound present in beer not found in wine, which is found only in hops, and that is xanthohumol and its cyclization product-flavanone, isoxanthohumol, both of which have shown anti-cancer properties [140,141]. Xanthohumol belongs to the group of prenylated chalcones, and this compound has been widely researched due to its biological activity [142,143]. Its biological and technological aspects, as well the chemistry of this compound, have been deeply reviewed [142]. Concentrations of xanthohumol depend on brewing technology and hopping regime, and the levels of this compound in commercial beer were around 0.2 mg/L [144]. These prenylflavonoids also turned out to be important for beer aromas and flavor, particularly in dark beers [80,83,145], and they are chemically related to the bitter hop acids (which are also biologically active, particularly α-acids) and polyphenols. Beer is considered as the main source of this molecule, with concentrations varied, from 0.002 and 0.628 mg/L [80].

2.3.1. Flavan-3-ols and Condensed Tannins

Flavanols are basically benzopyrans, and these compounds can be found in the form of simple monomers as well in polymers, Figure 5. The best known and important molecules within this group are catechin and its enantiomer, epicatechin. They represent the precursors for the formation of proanthocyanidins that give the structure and astringency to beer and wine [9,137]. Moreover, wine is considered as the beverage with higher concentrations of these compounds (50–120 mg/L) [96,97,146], while in beer the concentrations varied 1–20 mg/L [147]. In some special, and very old, red wines, the content of catechin was noticed even up to 1000 mg/L [148]. Catechin and epicatechin can be found in the grape stems, seeds, and skin, and after in wine [95], while beer catechins are also derived from raw material used for beer production, i.e., from barley/malt and from hop [149]. The flavan-3-ols and their contents were evaluated by many authors [73–75,78,80–82,100,150], and among these compounds, catechin was the one most described and abundant in beer [74,81,82,150]. Besides monomers, in beer also can be found their esters with gallic acids (catechin gallate), and catechin derivatives (epigallocatechin, gallocatechin, epigallocatechin gallate, and epicatechin gallate) were also determined in grape and wines. Recently, Wannenmacher et al. [12] determined the content of (+)-catechin (2.74–6.54 mg/L) in beers made with different technological variations.

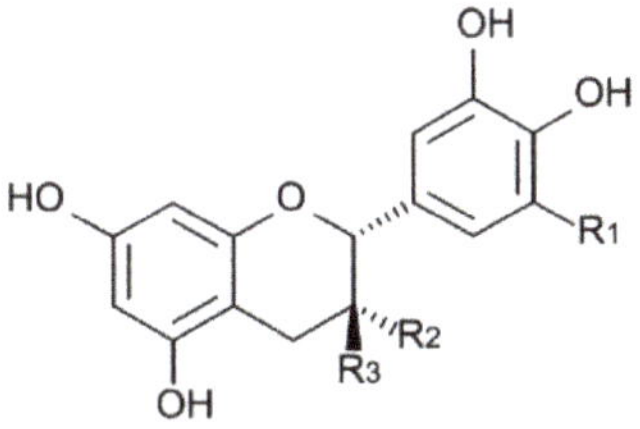

Flavan-3-ols	R_1	R_2	R_3
(+)-catechin	H	H	OH
(-)-epicatechin	H	OH	H
(+)-gallocatechin	OH	H	OH
(-)-epigallocatechin	OH	OH	H

Figure 5. Structure of flavan-3-ols.

Proanthocyanidins are phenolic compounds, and structurally represent oligomers flavan-3-ols, also known as condensed tannins, or in the brewing industry as "anthocyanogens" [12]. These polymeric compounds, transformed to anthocyanidins, can be found in all parts of the grape berry (pulp, skin, and seeds) and are transferred into wine during grape processing (crushing) and during alcoholic fermentation and maceration [151]. The structure of proanthocyanidin depends on the degree of polymerization, hydroxylation pattern and the stereochemistry, the type of the connection between monomers, and the 3-hydroxyl group esterification. Dixon et al. [152] deeply reviewed this type of phenolic compound, and according to the nature of their monomers proanthocyanidins were classified into propelargonidins, prodelphinidins, and procyanidins, Figure 6. Monomers of epicatechin and catechin make up the procyanidins, while epigallocatechin and gallocatechin subunits make up the prodelphinidins, while the propelargonidins consist of mixed oligomers, containing at least one monomer with a 4′-monohydroxyl group. In the brewing industry, proanthocyanidins became interesting due to their relation with the haze formation in beer, and it was shown that the haze in beer increases with higher molecular weight. In addition, proanthocyanidins also became the focus of researchers due to their highly potent antioxidant capacity and possible positive effects on human health, particularly in cardiovascular diseases and cancers [153,154]. Prodelphinidins and procyanidins have been detected in beer and wine as well [95,149]. These two groups in wine lead to delphinidin and cyanidin, and represent the most abundant condensed tannins in grape and wine. Gu et al. [155] determined the proanthocyanidins concentration in beer as 23 mg/L, that is 23-fold lower compared to grape juice, and around 13-fold lower compared to red wine. The main part of proanthocyanidins in beer consist of dimers (11 mg/L), monomers (4 g/L), tetra- to hexamers (4 mg/L), and trimers (3 mg/L), while the main part of proanthocyanidins consist of 37% of tetra- to decamers, 35% of >10-mers, and 28% of mono- to trimers [155]. Proanthocyanidins are very important for wine sensory characteristics as they influence the astringency and bitterness of wine, and also play an important role in the process of wine maturation and aging [95]. The level of astringency and bitterness is affected by the molecular size of the proanthocyanidins, and it was proposed that bitterness comes more from the monomers, while the astringency from the larger molecules [156].

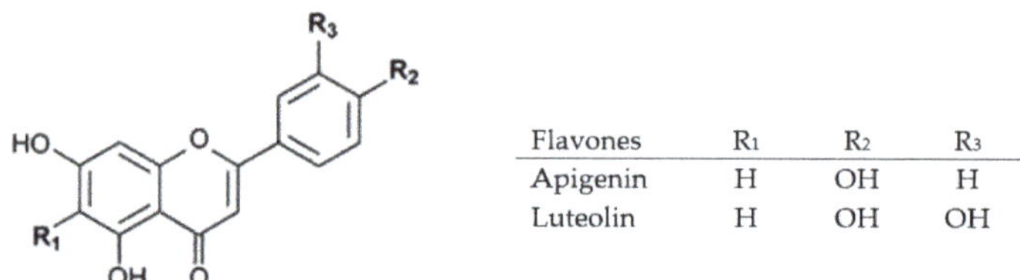

Proanthocyanidins	R
procyanidins	H
prodelphinidins	OH

Figure 6. Structure of proanthocyanidins.

2.3.2. Flavones, Flavonols, and Flavanones

The 4-keto group is the shared characteristic of this group of phenolic compounds. Flavones contain three functional groups, the carbonyl and hydroxyl groups, and a double bond within the flavonoid skeleton, Figure 7 [6]. Flavones were determined in grapes, i.e., in skin and in wine in two forms: aglycones (nonsugars) and glycosides. They were not found in significant levels in grapes and wine, except for luteolin, where concentrations in grape ranged from 0.2 to 1 mg/L [95–98]. It is known that flavones possess important biological characteristics, which are beneficial for human health, including anti-tumor, anti-inflammatory, and antioxidant features [62]. Gerhäuser et al. [157] determined flavones that were isolated in aglycon form from un-stabilized beer, such are apigenin, tricin, and chrysoeriol, and in glucoside form, i.e., apigenin derivatives. It has been shown that the most likely origin of these derivatives is from barley. Kellner et al. [82] determined in commercial beers the content of apigenin, from 0.80 to 0.81 mg/L, while Marova et al. [80] found 0.10 to 0.19 mg/L of luteolin in commercial beers. Apart from flavones, beer also contains the isomer of this phenolic group, i.e., isoflavones, and within them belong daidzein, genistein, formononetin, and biochanin A [82].

Flavones	R_1	R_2	R_3
Apigenin	H	OH	H
Luteolin	H	OH	OH

Figure 7. Structure of flavones [158].

The characteristic functional group for flavonols is a hydroxyl group attached to C3, and because of that they are often named as 3-hydroxyflavones, Figure 8. In red wine are found aglycon forms of flavonols, such are kaempferol, quercetin, rutin, and myricetin, as well as their glycosides, which can be found as galactosides, glucuronides, glucosides, and diglycosides. Concentration of these flavonols was found in wine at levels from 12.7 to 130 mg/L [96–98]. The biological activity of these compounds has been described, particularly as improving cardiovascular health [159]. Speaking about beer raw material, it has been shown that flavonols occur less in cereals, and two of them in aglycon form and their glycosides (quercetin and kaempferol) have been found in hops. Moreover, kaempferol and quercetin glycosides, as well as their malonyl esters, have been determined in hop [160]. Recently, Gangopadhyay et al. [161] detected quercetin in barley flour in the content of 15.1 µg/g dry weight. Relatively high concentrations of quercetin were measured in lager beers (1.72–1.79 mg/L) [74], and a high concentration of kaempferol (1.64 mg/L) was measured in beer [100]. Comparing this literature data, it is obvious that wine is richer in content of this phenolic group.

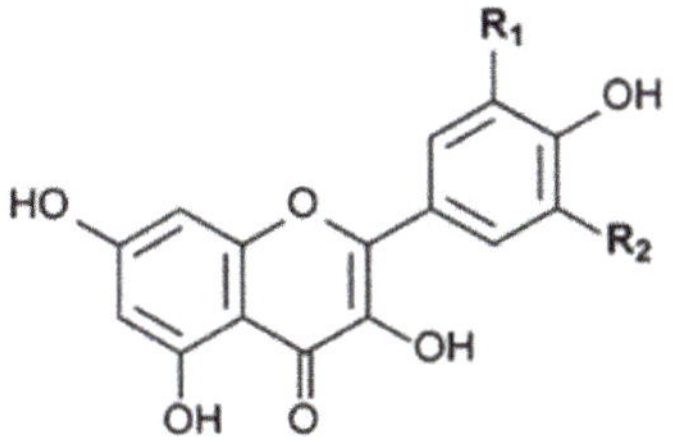

Flavonols	R_1	R_2
kaempferol	H	H
myrcietin	OH	OH
quercetin	OH	H

Figure 8. Structure of flavonols [158]. Reproduced with permission from [Drake V.J.], [Linus Pauling Institute]; published by [Oregon State University], 2005.

When it comes to the flavanones, beer is much more interesting, as it contains four prenylflavanones, isoxanthohumol (the most abundant one) in concentrations from 0.04 to 3.44 mg/L [162], and then there are 6- and 8-prenylnaringenin and 6-geranylnaringenin [141], Figure 9. Isoxanthohumol is formed during the brewing process by isomerization of xanthohumol. Marova et al. [80] determined the content of naringenin in 22 commercial beers, and it varied from 0.06 to 2.34 mg/L, while the total naringenin content in four German red wines varied from 0.9 to 4.2 mg/L [89].

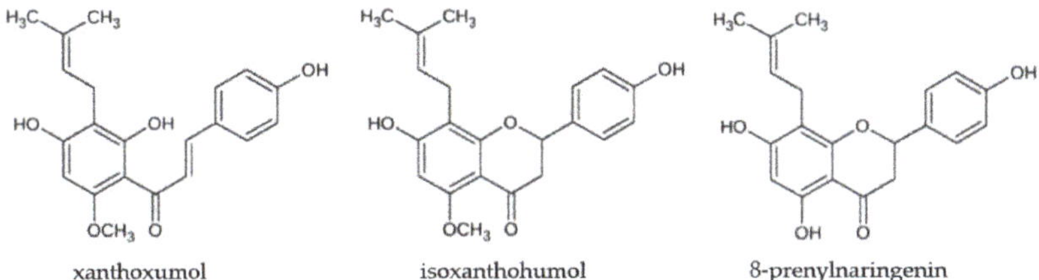

Figure 9. Structure of xanthohumol, isoxanthohumol and 8-prenylnaringenin [163].

2.3.3. Anthocyanins

When it comes to the anthocyanins, according to the literature survey this phenolic group of compounds were extensively reported in wine, while there were no reports of anthocyanin content in beer. Anthocyanins are proved to be promising agents against some diseases [6], and there are many reports on their protective role against coronary heart disease [164–167]. Different anthocyanin derivatives were determined in grapes and wine, as well as in the medium similar to the wine [95]. In regard to the chemical structure of anthocyanins, anthocyanins represent the glycosylated form of the anthocyanidins, and both contain as a nucleus the flavylium (the 2-phenylbenzopyrylium) cation, with methoxyl and hydroxyl groups attached to the different positions [6], Figure 10. Anthocyanins are principally responsible for the grape and wine color, and there are six anthocyanins detected, namely cyanidin, malvidin, peonidin, petunidin, delphinidin, and pelargonidin in red grapes and wines [6], as well as their 3-*O*-monoglucosides [151,168]. Anthocyanins are mainly found in the grape skin, and malvidin is one of the highest representatives in *Vitis vinifera*. The content of anthocyanins is influenced by the grape variety, i.e., it is relatively stable for each grape variety, and concentrations can vary between the vintages due to environmental conditions, i.e., clime conditions and terroir, also their concentration depends on the winemaking process, particularly during maceration, i.e., extraction that occurs during alcoholic fermentation. In addition, anthocyanins are involved in important reactions, such are polymerization, oxidation, and formation of new pigments during the process of winemaking and wine maturation. The total content of anthocyanins in wine can widely range depending on grape variety, from 32.5 to 1011 mg/L [71,111,112,119].

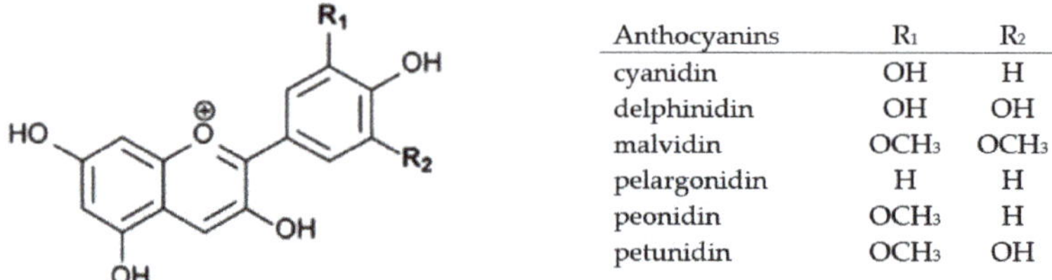

Anthocyanins	R₁	R₂
cyanidin	OH	H
delphinidin	OH	OH
malvidin	OCH₃	OCH₃
pelargonidin	H	H
peonidin	OCH₃	H
petunidin	OCH₃	OH

Figure 10. Structure of anthocyanins [158].

3. Impact of Technologies in Order to Increase Phenolics in Wine and Beer

The only material for wine production is grape, while for beer production there is malt (sometimes along with some adjuncts such as rice, sugar, corn, and wheat), water, and hop. The process within both the production of beer and wine is alcoholic fermentation and, for that purpose, usually commercial dry yeasts are used. In this section, through the processes of making wine and beer, will be highlighted possible methods used for improving the phenolic content in these two beverages. Basic brewing and winemaking technology are presented at Schemes 1 and 2, respectively.

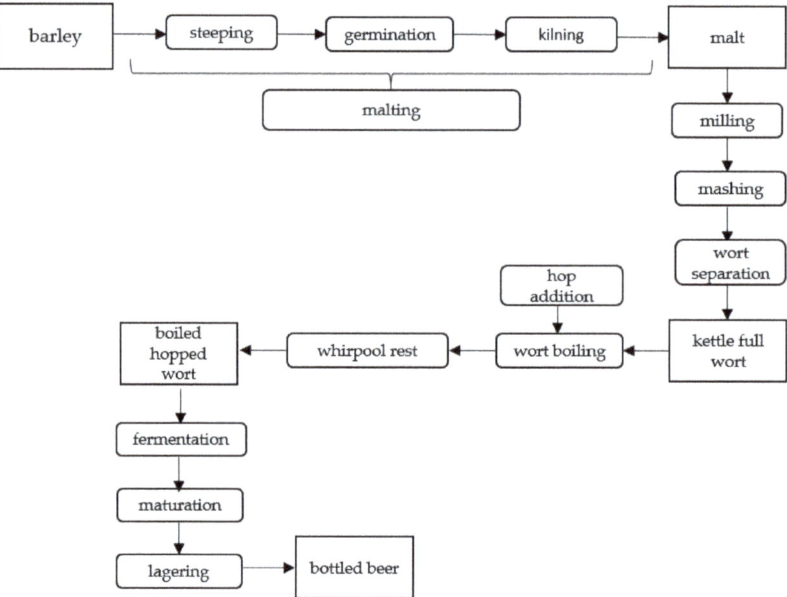

Scheme 1. Basic steps in brewing technology.

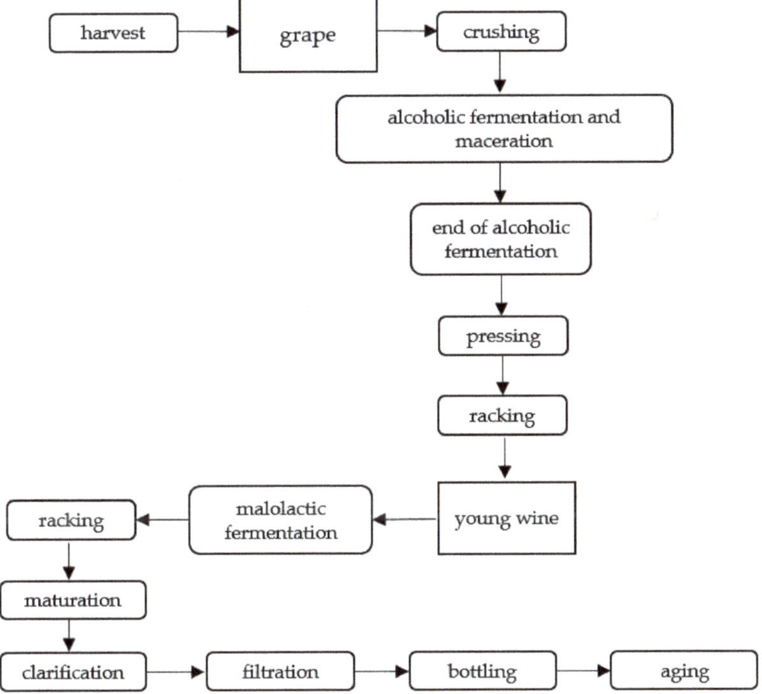

Scheme 2. Basic steps in winemaking technology.

3.1. Raw Material

As was mentioned, phenolic content of grapes depends on grape variety, terroir (clime and vineyard location, soil type), applied vinicultural agro-techniques, harvest date, applied oenological practices, and wine maturation. Nowadays, in regard to grape growing, it is very demanding and challenging to produce healthy grapes, with the optimal level of maturity (phenological and technological), and not over-ripened, due to climate changes with which the world today is faced. Climate changes significantly influence grape ripening, for example in hot and dry vintages the process of grape ripening is very fast, giving the grapes a non-balanced maturity with high sugar content, but with the lack of phenolic compounds, which results in wines with astringent and green tannins [169,170]. In order to deal with this issue, some novel agricultural practices have been investigated. Some of them are cluster thinning and defoliation [171–174], which have been shown to have a beneficial influence on the synthesis particularly of anthocyanins and flavonoids. Additionally, the influence of reduced yield on total phenolic content was observed [175], showing that there is no strong correlation among yield and the content of total polyphenols and anthocyanins, similarly to the antioxidant potential. Cluster thinning and early leaf removal showed an increase of proanthocyanidin and anthocyanin levels in wines of Cabernet Sauvignon and Vranac [171]. Besides these examined agro-techniques, there are also some reports that used elicitors in order to enhance the resveratrol content [176,177]. Recently, Giacosa et al. [178] investigated foliar application of *Saccharomyces cerevisiae* inactive dry yeasts, and concluded that the effect of vintage was very important, as in one year there were significant differences between treated and control grapes, while in another vintage treated wines obtained a lower phenolic compounds content. Therefore, this foliar application could be efficient in conditions that are critical for synthesis and thermal degradation of some phenolic groups, such are anthocyanins [178].

Two main beer ingredients from which phenolic compounds originate are cereal (mainly barley) and hop. Barley (*Hordeum vulgare* L.) is a member of the poaceae family, and for the purposes of malting and brewing, usually two-rowed spring barley varieties are used. Barley contributes from 70–80% of total polyphenolic compounds in beer, even if it was shown that barley malt possesses a lower content of total polyphenolic compounds than hops [179,180]. Barley contains various groups of phenolic compounds, which mainly consist of phenolic acids (free and bound form) Their total content varies from 604 to 1346 µg total phenolic acids/g barley flour [181], lignans (3.7 µg/g of total content) [182]. Hordatines in concentrations from 72–178 µg/g dry weight were determined by Kohyama and Ono, [183] as well as flavanols in concentrations from 325 to 527 µg/g barley flour [181]. However, like in grape varieties, in barley the composition and concentration of the polyphenolic content is influenced by the variety of barley and place of origin. There are many reports on individual and total polyphenol content in regard to the different barley varieties [179,181,184–188]. Besides the variety of barley and growing conditions, polyphenolic content is also influenced by the type of barley; Holtekjølen et al. [181] showed that in hulled varieties the total phenolic content is significantly higher. As there is no research regarding increasing the phenolic content of the barley in the field, there are some reports on the phenolic changes due to variation of some technological brewing parameters [12,110]. Changes in regards to the phenolic content that occur during preparing malt and boiled hopped wort are deeply summarized in review by Wannemacher et al. [11]. The antioxidant activity of pale and dark beer was contributed to by malt [24], while antioxidant activity of beer was not influenced significantly by hopping [189]. Malt made of barley represents the most important starch used in the process of beer making. Production of malt consists of the following phases: steeping, germination, and kilning. During steeping there is an increase in water content, rootlets and sprouts develop during germination, and kilning represents drying of the mass. A decrease in phenolic content occurs during the steeping phase, due to the leaching of phenolic substances, but during germination and kilning the content of phenolic compounds increases [11]. Narziss [179] has shown that an increase of polyphenolic content occurs during malting, and highlighted the kilning step as the most important for solubilization of polyphenols. There are five phases during the kilning process: heating up (start up with establishing air flow), removal of free water (i.e., drying (temperature goes from 50 to 60 °C)), increasing air temperature (intermediate drying), bound water removal, and at the end curing, during which the moisture content of grain increases to 5 and 8% [190]. With variations of parameters during the process of malting, such as degree of steeping, and time and temperature of germination, it is possible to adjust the quality of malt. Muñoz-Insa et al. [191] have shown that there is a positive correlation between the germination temperature and the total phenolic content, and a negative one between higher degree of steeping and the total phenolic content. Recently, Wannemacher et al. [12] investigated the impact of malt modification and hopping regime on the antioxidant potential of beer, and concluded that malt with higher raw protein content gives a beer with significantly higher antioxidant activity, determined using ORAC assay.

Hops (*Humulus lupulus* L.) belong to the cannabaceae family and are added in small amounts to beer in order to provide the final aroma and bitterness, as well to impart antibiotic and antifungal properties [192]. Hop is rich with plenty of antioxidant compounds that can be resinous, like the prenylated chalcones and α-acids, or non-resin phenolic compounds, like flavonoids or phenolic acids [11,12,160,180,193]. There are four main valuable groups of ingredients found in hops: soft and hard resins, essential oils, and polyphenols [11]. Xanthohumol represents the main compound found within the hard hop resins, and is accompanied by 13 other compounds also belonging to the prenylated chalcones, but in up to 1–100 fold lower concentrations compared to xanthohumol [142]. It was also confirmed that xanthohumol can be found only in hops, while other prenylflavonoids can be found in some other plant families [142,143]. In regard to the polyphenols present in hop cones, there are flavanols (32–191 mg/100 g air dry hops), proanthocyanidins (91–599 mg/100 g air dry hops) [194], flavanol glycosides (quercetin: 0.092 mg/100 g, kaempferol 0,12 mg/100 g) [195], and phenolic acids (hydroxycinnamic acids 59–288 mg/100 g air dry hops, hydroxybenzoic acids: <1–10 mg/100 g air dry hops) [194]. Jerkovic and Collin [196], investigated the total content of *trans*-resveratrol and *trans*-piceid, which ranged between 0.5 to 11.7 mg/kg in 40 samples of hop cones. They also concluded that harvest year strongly influenced the content of stilbenoids in hops, as well as that hop varieties with a lower content of α-acid usually contain a higher content of stilbenoids [196,197]. It was found that the vintage, i.e., the harvest year and date of the harvest influence, to a large extent, the quantity and quality of polyphenols in hop cones. Inui et al. [198] found that the content of polyphenols increases if the harvest was performed earlier, but the development of specific polyphenolic compounds differed. Kavalier et al. [199] showed that the content of some terpenophenolics increased during hop ripening, but the tendency of flavanols, flavonols, and phenolic acids was not clearly defined. It was also determined that the content of proanthocyanidins in hops is influenced by the growing conditions as well as by the hop variety [200]. Besides the influence of hop variety and growing conditions, i.e., the harvest year, the content of polyphenols in hops also depends on the type of hop product [180,201]. Mainly, hop is processed into the hop extracts, hop pellets, and isomerized products. During production of hop pellets, the raw hop goes through processes such as drying and grinding, which cause slight loss of polyphenol content, while the process of pelletization did not significantly influenced the content of polyphenolic compounds [202]. It has also been shown that the time length of storage influences the concentration of polyphenols in hop, and Mikyška and Krofta [203] showed that after twelve months of storage the polyphenols content had decreased significantly. In regard to hop extracts it was shown that the type of polyphenolic compounds and their concentrations depend on extraction solvent. In this respect, hop extract obtained using supercritical CO_2 is used as a source of xanthohumol, because flavanones and other prenylated chalcones are insoluble in this kind of solvent [204]. Quiet recently, production of hop polyphenol, as well as tannin extract, has been used in order to improve light stability and the storage of beer [205,206].

After malt production and prior alcoholic fermentation, the next step in beer making is wort production and its boiling with hop addition. During these brewing processes, changes of the total and individual polyphenol content occurs constantly. Therefore, all important influencing parameters should be considered. First, before malt undergone mashing, it has to be milled. There are two types of milling, dry, and wet with water addition. It was shown that the total phenolic concentration and ferulic acid content decreased during wet milling [207]. Mashing technology can be performed in two ways, one in which the whole mash is treated with heating steps that are aligned with the activity of enzymes, and another in which part of the mash is separated and heated in another kettle, and after this part has undergone boiling in order to inactivate enzymes it is returned to the main mash, at the same time increasing the temperature in the main tank. It has been shown that temperature, mashing-in time, the thickness of mash, as well the grist coarseness influence release of phenolic acids. Vanbeneden et al. [208] found that the optimal temperature of cinnamoyl esterase is 30 °C, while optimal temperature for ferulic acid release is 40 °C. Other researchers also confirmed that temperature of 40 to 45 °C is optimal for releasing ferulic acid [209,210]. Vanbeneden et al. [208] showed that more ferulic acid was released using finer grists, and longer mashing time gave wort with a higher level of ferulic acid [209]. Overall, it was shown that only small part of hydroxycinnamic acids had been transferred into wort during the mashing process, the majority was left in the consumed grain. Zhao et al. [211] found that the total phenolic content decreased during the phase of enzyme inactivation. After mashing, the following step is wort separation from the consumed grain, using a mash filter or a lauter tun, the special type of vessel for the purpose of filtering the spent grain. Higher reduction of the phenolic compounds content was found when lauter tun was used [212]. These finding are not in accordance with the results of Pascoe et al. [213], who found that during the lautering phase the concentration of total polyphenols increased, due to the extraction from phenol rich spelt material. After the wort is separated the next phase is its boiling and hop addition, and within these phases reaction of polymerization occurs, as well as precipitation and interactions between proteins and polyphenols. According to Forster and Gahr [214], the transfer rate of the total polyphenols from hop during boiling is from 50 to 70% and this rate is different for different groups of polyphenolic compounds, depending on their polarity as well as on their affinity to interact with the proteins from wort. Hop can be added at the beginning of wort boiling in order to obtain the desired bitterness, or at the end of wort boiling, or even during whirlpool rest, and in that way will influence final beer aroma. The whirlpool rest is the operation that follows wort boiling and hop addition, in order to get clear wort separated from hot trub, in which are left insoluble proteins, bitter and organic substances, and ash. During this process, a significant decrease of phenolic substances occurs due to their adsorption into the hot trub. Late hop addition was proved as useful for better oxidative beer stability [40,41]. Wietstock et al. [215] found that a modified dose of hop improved oxidative beer stability, and a lower content of staling aldehydes were determined after storage compared to single hop dosing at the beginning of wort boiling. Mikyška et al. [201] did not observe significant differences between the beer hopped at the beginning of wort boiling, with hop pellets (type 90), or hop CO_2 extract. However, the early addition of hop and longer wort boiling resulted in a higher depletion of phenolic compounds [11,206]. Higher content of total polyphenols, un-isomerized α-acids, and anthocyanogens (flavan-3,4-diols) were achieved when a late hop addition regime was applied [40,41,44,45]. Wannemacher et al. [12] researched the impact of different hop products and hoping regimes on the concentration of total phenolic and antioxidant potential in beer and found that besides higher content of total polyphenols, the higher content of individual phenolic compounds was also observed in beer with second hop addition during whirlpool rest. Furthermore, in their research higher sensory scores were given to the beers in which hop was added during whirlpool rest. Recently, Mikyška [110] investigated the influence of different hopping regimes during wort boiling, and concluded that flavanols content (epicatechin, catechin) in hopped wort and beer mainly depends on the hop raw material, and in the case of the addition of an aromatic and polyphenols rich hop, that two thirds of the polyphenol level in beer is influenced by hops. They also found that the flavonols kaempferol, quercetin, and multifidol are present in wort and beer in their glycosidic form, and the origin of these compounds is solely from the hops. During wort boiling it takes 15–30 min to release the flavonoids from hop into wort, and the dose of hop and low-pressure boiling technology did not influence significantly the polyphenol content. The addition of hops with a higher content of polyphenols turned out to be a better source of these antioxidants, compared to the addition of hop extract in beer [110].

3.2. Changes during Alcoholic Fermentation

There have been more studies in regards to the changes of polyphenols content during alcoholic fermentation in wine comparing to the beer. The main reason for this is that during the alcoholic fermentation of grapes, particularly red, maceration, i.e., extraction of polyphenolic compounds from grapes, occurs. Beside the chosen grape variety, this process is affected by many microbiological and technological parameters, such as yeast, enzymes, temperature, and the applied vinification techniques. During alcoholic fermentation, polyphenolic compounds in wine increase, while the concentration of total phenolic substances during fermentation in beer decreases [212].

Brandolini et al. [216] have shown that the yeast strain used for alcoholic fermentation, besides having an important role in the sensory quality of wine, also influences the polyphenol content. In the research of Kostadinović et al. [217], it was shown that by using dissimilar yeast strains it was possible to influence the stilbenes concentrations and antioxidant activity in Merlot and Vranac wines. Similar results were obtained for the wines Albariño [218], Pinot Noir [219], and Gaglioppo [220]. Investigating singular yeast strains, confirmed that the used starters have the ability to influence polyphenolic composition of wine. The importance of inoculation with commercial yeast starters, in order to modulate the content of total polyphenols in wine, was also highlighted [111]. Recently, Grieco et al. [27] investigated the influence of autochthonous yeasts isolated from the grapes of varieties Primitivo and Negroamaro grown in the Apulia region, and they determined significantly higher total concentrations of stilbenes in wines of both varieties fermented with autochthonous yeast (18.40–67.78 mg/L) than with commercial ones (3.70–18.83 mg/L). In their research the sum of determined phenolic acids (caftaric, caffeic, and *p*-coumaric acid), in wines fermented using commercial yeast ranged from 350.90 to 677.80 mg/L, while in wines fermented with autochthonous yeast strains, it ranged from 731.00 to 1976.70 mg/L. Similarly, the sum of determined flavonols (myricetin, quercetin, and kaempferol) varied up to 106.1 mg/L in wines fermented with an autochthonous yeast strain, and up to 49.80 mg/L in wines obtained utilizing commercial yeast. Only the content of identified flavanols (catechin and epicatechin) did not show significant differences, varying from 16.07–20.09 mg/L and 17.05–22.50 mg/L for wines fermented with commercial and autochthonous yeast, respectively. Total phenolic content and antioxidant activity in wines that utilized native yeast strains, obtained up to 1569.3 ± 7.6 mgGAE/L and 96.4 ± 1.5 mmol TE/100 mL, while the highest value of total phenolic content of wines fermented with commercial yeast was 1221.9 ± 7.6 mgGAE/L, and for antioxidant activity that value was 76.60 ± 2.1 mmol TE/100 mL. Their results indicated that the use of native yeast can considerably affect the composition of polyphenolic compounds. The significant influence of different commercial yeasts on the concentration of total, and some individual, phenolic compounds resulted in an increased of total phenolic compounds, and particularly stilbenes [70]. Besides the use of yeast, i.e., performing the traditional fermentation or using commercial selected dry yeast or isolated native yeast, it is also possible to vary other parameters. It was determined that higher content of phenolic compounds was observed performing alcoholic fermentation in fermenters comparing to traditional vinification in PVC barrels, and the addition of grape tannins, enzyme, and oak chips increased the content of total polyphenols, total anthocyanins, and total flavan-3-ols [33]. Recently, Generalić Mekinić et al. [28] investigated the impact of two different commercial pectolytic enzymes, which were based on polygalacturonase, with the activity of 7500 and 7600 units/g, on the phenolics extraction during maceration as well on the antioxidant activity of the analyzed wines. In their research, the use of commercial pectolytic enzymes had a slightly negative influence on the content of total phenolic compounds and antioxidant features of the wine. In the control sample, without enzyme addition, the content of determined phenolic acids varied from 0.91 ± 0.05 mg/L for protocatechuic acid, to 28.63 ± 0.04 mg/L for gallic acid; the content of flavonoids varied from 1.27 ± 0.03 mg/L for quercetin, to 82.60 ± 0.18 mg/L for catechin; the resveratrol content was 0.70 ± 0.02 mg/L, and anthocyanins varied from 0.36 ± 0.01 mg/L (petunidin-3-(6-*O*-coumaryoyl)glucoside), to 50.49 ± 0.15 mg/L (malvidin-3-*O*-glucoside). While in wines treated with enzyme (7500 polygalacturonase units/g) the content of determined phenolic acids varied from 1.67 ± 0.02 mg/L for protocatechuic acid, to 65.45 ± 0.02 mg/L for gallic acid; the content of flavonoids varied from 2.18 ± 0.01 mg/L for quercetin, to 91.99 ± 0.10 mg/L for catechin; the resveratrol content was 0.51 ± 0.02 mg/L, and anthocyanins varied from 0.29 ± 0.00 mg/L (cyanidin-3-(6-*O*-coumaryoyl)glucoside), to 39.29 ± 0.10 mg/L (malvidin-3-*O*-glucoside). In wines treated with enzyme (7600 polygalacturonase units/g) the content of determined phenolic acids varied from 1.78 ± 0.02 mg/L for protocatechuic acid, to 45.04 ± 0.19 mg/L for gallic acid; the content of flavonoids varied from 5.19 ± 0.12 mg/L for quercetin, to 55.49 ± 3.93 mg/L for catechin; the resveratrol content was 1.07 ± 0.13 mg/L, and anthocyanins varied from 0.21 ± 0.00 mg/L (cyanidin-3-(6-*O*-coumaryoyl)glucoside), to 21.63 ± 0.09 mg/L (malvidin-3-*O*-glucoside). Lower content of total phenolic compounds, particularly of monomeric anthocyanins in wines treated with enzymes, was probably caused by polymerization reactions, or due to glycosidase activity causing hydrolysis of these compounds [28]. Furthermore, the time of maceration also influenced the anthocyanin content, as well the wine color; low anthocyanin content and weak color was obtained in wines with a short period of maceration, while the prolonged maceration time resulted in unstable and poor color wine characteristics [221]. There are many studies regarding the use of different winemaking techniques and enzymes, and the obtained results are contradictory, but they all improve the knowledge, with the aim of choosing appropriate vinification techniques [28,221–225].

As was mentioned, knowledge in regard to the phenolic compound change during alcoholic fermentation in beer is incomplete, and the influence of different yeast strains and technological variations should be considered. However, by now the decrease of phenolic compounds during fermentation, warm rest, and chill-lagering is confirmed [212]. It was also found that these processing phases did not have a significant impact on the catechin and phenolic acids, except for ferulic acid, whose concentration decreased by 35% during warm rest at the end of fermentation [213]. Contrarily, Coghe et al. [209], noticed increase of ferulic acid during fermentation, which they attributed to the activity of enzyme in yeast feruloyl esterase.

3.3. Maturation, Aging, and Storage

Maturation and aging represent very important steps in wine and beer production. Changes in wine during the process of maturation and aging reflect, first, on the wine color and its sensory properties, in terms of the harmonizing astringency. Color change is associated with a decrease in anthocyanins content in aged wines, and changes depend on wine pH and SO_2 content. Degradation reactions of grape-derived anthocyanins occur in an oxygen excess, forming the insoluble complexes of brown compounds [226]. Additionally, during wine aging, anthocyanins bind covalently with other compounds in wine, such as flavan-3-ols, then they form pyranoanthocyanin, and undergo the polymerization reactions; all these new compounds improve the wine stability, and there is less bleaching in the presence of SO_2 [227,228]. The structure of formed compounds varies according to the molecular weight, from low flavanyl-vinylpyranoanthocyanins [229], and pyranoanthocyanins [230], and to the large molecules like tannin–anthocyanin adducts [231]. As one of the important factors also influencing the change of anthocyanin loss is the temperature of storage [231]. The choice of vessel for maturation and the aging time, beside the wine style, also influence sensory characteristics and the content of polyphenolic composition in wine. For wine maturation, many types of vessels can be used, with different size and materials, such are stainless steel and wood. Stainless steel is good due to its permeability to oxygen, easier temperature control, and is mainly used for keeping the non-expensive wines prior to bottling. On the other hand, high quality wines usually age in wooden barrels, and during the time spent in wooden barrels wine is exposed to controlled oxygenation and wood compounds are transferred to the wine. Barrel size and temperature influence the wine maturation, a smaller size of barrel, means maturation will be quicker, and if the temperature is lower it will slow the maturation [232]. Macromolecules, which are present in oak wood, belong to the class of polysaccharides (hemicellulose and cellulose) and to the class of polyphenols (lignin), while among other components ellagitannins are the most abundant in oak, there are also some low molecular weight polyphenols and volatile substances. These compounds are extracted into the wine during maturation, and this is the main reason for the choice of this aging technology. During wine maturation in oak barrels, ellagitannins are transferred into the wine, and give aged wine astringency and bitterness sensations [233–235], and due to their ability to consume the oxygen they act as antioxidants [236,237]. Ellagitannins also react with anthocyanins forming complexes that are much more stable compared to the grape derived anthocyanins [230,238], they can be also found associated with flavonoids, forming derivatives that have been determined in aged wines [239], and which are also interesting due to their antioxidant activity [240]. The most used, and the most traditional, oak wood belongs to the *Quercus* species from France (*Q. robur* and *Q. petraea*) and from the USA (*Q. alba*) [13]. The concentrations of compounds in oak depend on the type of the oak, i.e., on its drying and toasting conditions, and on the origin of the oak, and there is great variability between the examined species and between the forests [13]. Furthermore, Jeremic et al. [14] tested the antioxidant ability of oenological tannins (procyanidins from grape seed and skin), ellagitannins from oak wood, and gallotannins from gallnut in commercial Chianti red wine, and in a model solution. They concluded that the rate of O_2 consumption was the highest when ellagitannins were added, representing an effective tool in winemaking when there is a need for instant protection against oxidation, but the effect of ellagitannins on the consumption of O_2 in wine decreases rapidly with time, which is a limiting factor for their use. While tannins from skin and seed had more consistent reactivity and lower rate of oxygen consumption, gallotannins showed low performance in protection against the oxygen exposure. Besides the addition of tannins into wine in order to improve the concentration of phenolic compounds and antioxidant activity, there are studies that report that the addition of some by-products that appear during winemaking, or derived during making wood barrels, can also be used. The addition of seed from white grape by-products increased the antioxidant activity, phenolic content, and color stabilization of red wine [241–244]. Jara-Palacios et al. [244], studied the influence of winemaking by-products (seed, stems, skin, and pomace) on the wine antioxidant activity and copigmentation, and concluded that addition of these by-products could improve the wine color and its bioactivity. Escudero-Gilete et al. [245] evaluated the potential use of wood shavings, by-products that appear during wooden barrel production, in order to improve red wine color and antioxidant activity. They used two types of shavings, American and Ukrainian, and concluded that these kinds of cooperage products represent a natural source of copigments and antioxidants, and that Ukrainian shavings provided better color stability.

In beer production, after fermentation, maturation, and lagering, it was found that the content of phenolic compounds was lost by 17%, because of the adsorption to cold trub and yeast [162]. In biological terms beer represents a stable product, but its shelf life is not unlimited, because of its flavor and colloidal stability changes. Vanderhaegen et al. [246] observed that during beer aging, a typical aging aroma will appear, and bitterness sensation decreases. Storage conditions (light and temperature) play an important role in beer aging, and it was noted that these factors influenced the significant decrease of α-acids after 5 months of storage [247]. Besides change in the content of α-acids during beer aging, the content of phenolic compounds also was changed with beer aging. In the research of Li et al. [248] it was determined that the substantial decrease of phenolic compounds occurred within the first three months of aging. Total phenolic content was decreased in examined samples from 16 to 23% within six months of storage, and the antioxidant activity of examined samples behaved in the same way [248]. Even in the 80s of the last century, researchers investigated the changes of polyphenolic compounds during beer production. It was found that after storage of six months, concentrations of the flavanols group ((+)-catechin and (−)-epicatechin), and prodelphinidin and proanthocyanidin B3, decreased [249]. Higher stability of monomeric flavanols was also observed. As was mentioned, during wine aging due to acidic wine conditions and the reaction that occurs among procyanidins, large sized molecules are formed that afterwards precipitate, leading to the decrease of astringency. An important role in these reactions is played by the presence of oxygen [14]. Conditions in beer a medium are less acidic, therefore these kinds of reactions can be performed more slowly, and beer is not exposed to the same levels of oxygen as wine, it is unlikely that procyanindin polymerization will occur during beer storage. Considerable decrease of small flavonoid molecules (monomer to trimers) was observed after the storage of one year at 20 °C [250]. It was also determined that prenylated flavonoids are distinguished by high stability in beer during storage, even after 10 years beer aging at 20 °C, determined concentration of prenylated flavonoids was not significantly different to the concentrations in fresh beer [251]. Heuberger et al. [252] observed that xanthohumol concentration decreased during beer storage, while on the contrary, the concentration of its isomer isoxanthohumol increased. The changes of free trans-piceid and trans-resveratrol, during one year of beer aging at different temperatures, 4 and 10 °C, were investigated by Jerkovic et al. [253], and it was shown that trans-piceid content was not changed, while the content of trans-resveratrol decreased.

Recently, new methods for increasing the content of bioactive compounds in beer, and in the same way of increasing its antioxidant potential have been reported [10,113,254,255]. Particularly, the global rapid growth of the craft beer industry has achieved a huge success, even in the countries that are not recognized as traditional beer producers. Uniqueness and the additional value of the craft beer are due to the addition of innovative raw materials, along with constant main ingredients (malt, hop, water, and yeast). Special hibiscus ale beers, in which different extract concentrations were added, and compared by antioxidant activity, and the content of total phenolic compounds besides other physicochemical changes, during a forced aging process, were analyzed [113]. Analyzing antioxidant capacity and total phenolic content during period of 7 days storage at 45 °C, decrease of both parameters were observed, but it was shown that hibiscus extract is an important source of anthocyanins and phenolic compounds with antioxidant features. Eggplant peel extract was added at the end of the maturation process in order to obtain a high value-added beer, and it was observed that total flavonoid content was stable during whole tested period, while a slight decrease of total phenolic content was noticed after seven days of storage, but even with this slight decrease total phenolic content was significantly higher in beer enriched with eggplant peel extract compared to the control beer. Regarding antioxidant activity, it was found that it rises with addition of eggplant peel extract [10]. These results were in agreement with results obtained by Đorđević et al. [254] who used different extracts of medicinal plants in lager beers, and by Ulloa et al. [255] who investigated the addition of propolis in lager beers. Veljović et al. [256] have shown that it is possible to produce a pleasant, special type of beer using a fermenting mixture of grape must and wort, with higher content of total phenolic compounds in comparison to commercial light beers. In their research, total polyphenolic content in the special types of beer made from grapes, depending on grape variety, different yeast strains, and different wort to grape ratio, was also investigated. They observed the significant impact of grape varieties, content of their addition to wort, and also of the yeast strains, on the total polyphenolic compounds in analyzed samples. For alcoholic fermentation, wine strain *Saccharomyces cerevisiae* and a strain used for brewing *Saccharomyces pastorianus*, were used and higher contents of polyphenols were obtained using wine yeast *S. cerevisiae*. Beer without grape addition, fermented with brewing yeast obtained a mean value of the content of total polyphenols of 95.94 mg/L, while beer with the addition of 30% of Cabernet Sauvignon grapes, which utilized yeast from wine industry, obtained 754.4 mg/L. While among investigated grape varieties (Prokupac, Pinot Noir, and Cabernet Sauvignon), the highest polyphenols content was determined when Cabernet Sauvignon was used, and the lowest in case of Prokupac beer sample; with increasing the amount of added crushed grapes, phenolic content also significantly increased. A recent study by Lasanta et al. [257] investigated the use of five different strains, all of which belonged to *Saccharomyces cerevisiae*, but two for bottom and three for top alcoholic fermentation, varying also the temperatures. Polyphenolic content was higher when lower temperature was applied (12 °C), which was explained by longer fermentation, and at same time longer maceration of these compounds from the used raw material. In addition, the influence of the yeast strain was not shown as significant, particularly at applied lower temperatures. Moreover, recently, the addition of different kinds of fruits (cherry, peach, apricot, raspberry, plum, grape, apple, and orange), and the influence on polyphenolic content and antioxidant activity in beer was analyzed [258]. Most fruit beers obtained a higher antioxidant activity, and total flavonoids and polyphenols content, particularly for beer with the addition of cherries, followed by beers with addition of grape, plum, and orange. All fruit beers have shown enrichment in the content of catechin and quercetin. Polyphenolic content and antioxidant activity were also investigated in beer with addition of mango fruit, and polyphenolic content in mango beer rose, up to even 44% compared to the control beer, which was also in accordance with higher antioxidant activity [259]. These studies indicated improvements in polyphenolic content and antioxidant activity with the addition of different kinds of fruits, and also confirmed better organoleptic features [259].

4. Conclusions

Phenolic compounds present in beer and wine have shown high antioxidant and anti-inflammatory features, and in the last decades beneficial effects on human health due to moderate beer and wine consumption have been indicated by many research studies. In this review, a comparative overview of qualitative and quantitative phenolic compound profiles in wine and beer was evaluated. As was expected, due to the different used raw material and technological processes, there are differences between wine and beer in the presence, as well in the concentrations, of phenolic substances. It was shown that some polyphenol classes can only be found in beer (chalcones and flavanones) and other are mainly found in wine (stilbenes, proanthocyanidins), while flavanols and flavan-3-ols are found in similar concentrations in both beverages. Both beverages represent natural fermented products, and minor changes within the growth of raw material and clime conditions, as well as within the used technology, will impact the final chemical composition of these products. Considering the literature data, the obtained results favor wine as the beverage with a higher content of bioactive compounds, particularly phenolic compounds, and as was expected, with higher antioxidant activity. Overall, in order to decide which of these two alcoholic beverages represents the better choice as a functional drink, a lot of parameters should be considered (social occasion, quantity, individual tolerance, etc.). As it was mentioned, drinking pattern is very important, only light-moderate drinking is recommended, and it should not be forgotten that the choice firstly depends on individual preference.

Nowadays, the brewing industry and winemakers put a lot of effort in obtaining a final product that will be unique, with more potent antioxidant activity, and with satisfying sensory characteristics, to attract consumers, who are now more aware of alcoholic beverages influence on human health. Winemakers began even from the vineyard, applying new additives that would improve phenolic composition in the grape, and afterwards, taking care through every step to the final product. After all, from the winemaker's point of view, the aim is to produce wine that will satisfy all required safety conditions, and with this added value, and at the same time attempting not to increase the costs. A good solution to this, is the use of all by-products, which occur during grape processing and winemaking. In the brewing industry, besides changing hopping regimes and influencing other technology phases, craft breweries that have expanded rapidly all over the world, are doing their best to produce authentic beer, in terms of flavor. Research on using different kinds of fruit in order to obtain special beer with added value, with a focus on sensory improvement and differentiation, has been performed. Both industries should consider changes in clime conditions, and research on new modified technologies is always an open issue, like the use of some by-products and additives within the production.

Author Contributions: S.R. and T.K. designed the review; V.M., S.R., and J.R. analysed the bibliography; S.R. and T.K. wrote the paper. All authors have read and agreed to the published version of the manuscript.

Funding: This research did not receive any specific grant from funding agencies in the public, commercial, or not-for-profit sectors.

References

1. Li, H.; Wang, H.; Li, H.; Goodman, S.; van der Leed, P.; Xu, Z.; Fortunato, A.; Yanga, P. The worlds of wine: Old, new and ancient. *Wine Econ. Pol.* **2018**, *7*, 178–182. [CrossRef]

2. Wang, J.; Liu, L.; Ball, T.; Yu, L.; Li, Y.; Xing, F. Revealing a 5000-y-old beer recipe in china. *Proc. Natl. Acad. Sci. USA* **2016**, *113*, 6444–6448. [CrossRef] [PubMed]

3. Roerecke, M.; Rehm, J. Alcohol consumption, drinking patterns, and ischemic heart disease: A narrative review of meta-analyses and a systematic review and meta-analysis of the impact of heavy drinking occasions on risk for moderate drinkers. *BMC Med.* **2014**, *12*, 182. [CrossRef] [PubMed]

4. Gromes, R.; Zeuch, M.; Piendl, A. Further investigations into the dietary fibre content of beers. *Brau. Int.* **2000**, *18*, 24–28.

5. Powell, J.J.; McNaughton, S.A.; Jugdaohsingh, R.; Anderson, S.H.; Dear, J.; Khot, F.; Mowatt, L.; Gleason, K.L.; Sykes, M.; Thompson, R.P.; et al. A provisional database for the silicon content of foods in the United Kingdom. *Br. J. Nutr.* **2005**, *94*, 804–812. [CrossRef]

6. Castaldo, L.; Narváez, A.; Izzo, L.; Graziani, G.; Gaspari, A.; Di Minno, G.; Ritieni, A. Red wine consumption and cardiovascular health. *Molecules* **2019**, *24*, 3626. [CrossRef]

7. Di Pietro, M.B.; Bamforth, C.W. A comparison of the antioxidant potential of wine and beer. *J. Inst. Brew.* **2011**, *117*, 547–555. [CrossRef]

8. Martinez-Gomez, A.; Caballero, I.; Blanco, C.A. Phenols and melanoidins as natural antioxidants in beer. structure, reactivity and antioxidant activity. *Biomolecules* **2020**, *10*, 400. [CrossRef]

9. Habschied, K.; Lončarić, A.; Mastanjević, K. Screening of polyphenols and antioxidative activity in industrial beers. *Foods* **2020**, *9*, 238. [CrossRef]

10. Horincar, G.; Enachi, E.; Bolea, C.; Râpeanu, G.; Aprodu, I. Value-added lager beer enriched with eggplant (*Solanum melongena* l.) peel extract. *Molecules* **2020**, *25*, 731. [CrossRef]

11. Wannenmacher, J.; Gastl, M.; Becker, T. Phenolic substances in beer: Structural diversity, reactive potential and relevance for brewing process and beer quality. *Compr. Rev. Food Sci.* **2018**, *17*, 953–988. [CrossRef]

12. Wannenmacher, J.; Cotterchio, C.; Schlumberger, M.; Reuber, V.; Gastl, M.; Becker, T. Technological influence on sensory stability and antioxidant activity of beers measured by ORAC and FRAP. *J. Sci. Food Agric.* **2019**, *99*, 6628–6637. [CrossRef] [PubMed]

13. Martínez-Gil, A.; del Alamo-Sanza, M.; Sánchez-Gómez, R.; Nevares, I. Alternative woods in enology: Characterization of tannin and low molecular weight phenol compounds with respect to traditional oak woods. A Review. *Molecules* **2020**, *25*, 1474. [CrossRef] [PubMed]

14. Jeremic, J.; Vongluanngam, I.; Ricci, A.; Parpinello, G.P.; Versari, A. The oxygen consumption kinetics of commercial oenological tannins in model wine solution and chianti red wine. *Molecules* **2020**, *25*, 1215. [CrossRef]

15. Afroz, R.; Tanvir, E.; Little, P. Honey-derived flavonoids: Natural products for the prevention of atherosclerosis and cardiovascular diseases. *Clin. Exp. Pharmacol.* **2016**, *6*, 1–4. [CrossRef]

16. Mozaffarian, D.; Rosenberg, I.; Uauy, R. History of modern nutrition science—Implications for current research, dietary guidelines, and food policy. *BMJ* **2018**, *361*, 2392. [CrossRef]

17. Cheng, C.K.; Luo, J.; Lau, C.W.; Chen, Z.; Tian, X.Y.; Huang, Y. Pharmacological basis and new insights of resveratrol action in the cardiovascular system. *Br. J. Pharmacol.* **2019**, *177*, 1258–1277. [CrossRef]

18. De Gaetano, G.; Cerletti, C.; Alkerwi, A.; Iacoviello, L.; Badimon, L.; Costanzo, S.; Pounis, G.; Trevisan, M.; Panico, S.; Stranges, S.; et al. Effects of moderate beer consumption on health and disease: A consensus document. *Nutr. Metab. Cardiovasc. Dis.* **2016**, *26*, 443–467. [CrossRef]

19. Redondo, N.; Nova, E.; Díaz-Prieto, L.E.; Marcos, A. Effects of moderate beer consumption on health. *Nutr. Hosp.* **2018**, *35*, 41–44. [CrossRef]

20. Chiva-Blanch, G.; Magraner, E.; Condines, X.; Valderas-Martínez, P.; Roth, I.; Arranz, S.; Casas, R.; Navarro, M.; Hervas, A.; Sisó, A.; et al. Effects of alcohol and polyphenols from beer on atherosclerotic biomarkers in high cardiovascular risk men: A randomized feeding trial. *Nutr. Metab. Cardiovasc. Dis.* **2014**, *25*, 36–45. [CrossRef]

21. Lugasi, A. Polyphenol content and antioxidant properties of beer. *Acta Aliment.* **2003**, *32*, 181–192. [CrossRef]

22. Arranz, S.; Valderas-Martínez, P.; Chiva-Blanch, G.; Medina-Remón, A.; Lamuela-Raventós, R.M.; Estruch, R. Wine, beer, alcohol and polyphenols on cardiovascular disease and cancer. *Nutrients* **2012**, *4*, 759–781. [CrossRef] [PubMed]

23. Quifer, P.; Martínez, M.; Jáuregui, O.; Estruch, R.; Lamuela, R.; Chiva, G.; Vallverdú-Queralt, A. A comprehensive characterisation of beer polyphenols by high resolution mass spectrometry (LC–ESI-LTQ-Orbitrap-MS). *Food Chem.* **2014**, *169*, 336–343. [CrossRef]

24. Čechovská, L.; Konečný, M.; Velíšek, J.; Cejpek, K. Effect of Maillard reaction on reducing power of malts and beers. *Czech J. Food Sci.* **2012**, *30*, 548–556. [CrossRef]

25. Capece, A.; Romaniello, R.; Pietrafesa, A.; Siesto, G.; Pietrafesa, R.; Zambuto, M.; Romano, P. Use of *Saccharomyces cerevisiae* var. *boulardii* in co-fermentations with *S. cerevisiae* for the production of craft beers with potential healthy value-added. *Int. J. Food Microbiol.* **2018**, *284*, 22–30. [CrossRef]

26. Lingua, M.S.; Neme Tauil, R.M.; Batthyány, C.; Wunderlin, D.A.; Baroni, M.V. Proteomic analysis of Saccharomyces cerevisiae to study the effects of red wine polyphenols on oxidative stress. *J. Food Sci. Technol.* **2019**, *56*, 4129–4138. [CrossRef]

27. Grieco, F.; Carluccio, M.A.; Giovinazzo, G. Autochthonous *Saccharomyces cerevisiae* starter cultures enhance polyphenols content, antioxidant activity, and anti-inflammatory response of Apulian red wines. *Foods* **2019**, *8*, 453. [CrossRef] [PubMed]

28. Generalić Mekinić, I.; Skračić, Ž.; Kokeza, A.; Soldo, B.; Ljubenkov, I.; Banović, M.; Skroza, D. Effect of winemaking on phenolic profile, colour components and antioxidants in Crljenak kaštelanski (sin. Zinfandel, Primitivo, Tribidrag) wine. *J. Food Sci. Technol.* **2019**, *56*, 1841–1853. [CrossRef] [PubMed]

29. Mitić, M.; Kostić, D.; Pavlović, A. The phenolic composition and the antioxidant capacity of Serbian red wines. *Adv. Technol.* **2014**, *3*, 16–22. [CrossRef]

30. Mitić, M.; Kostić, D.; Pavlović, A.; Micić, R.; Stojanović, B.; Paunović, D.; Dimitrijević, D. Antioxidant activity and polyphenol profile of Vranac red wines from Balkan region. *Hem. Ind.* **2016**, *70*, 265–275. [CrossRef]

31. Kondrashov, A.; Ševčík, R.; Benáková, H.; Koštířová, M.; Štípek, S. The key role of grape variety for antioxidant capacity of red wines. *e-SPEN Eur. e-J. Clin. Nutr. Metab.* **2009**, *4*, 41–46. [CrossRef]

32. Di Majo, D.; La Guardia, M.; Giammanco, S.; La Neve, L.; Giammanco, M. The antioxidant capacity of red wine in relationship with its polyphenolic constituents. *Food Chem.* **2008**, *111*, 45–49. [CrossRef]

33. Raičević, D.; Božinović, Z.; Petkov, M.; Ivanova-Petropulos, V.; Kodžulović, V.; Mugoša, M.; Šućur, S.; Maraš, V. Polyphenolic content and sensory profile of Montenegrin Vranac wines produced with different oenological products and maceration. *Maced. J. Chem. Chem. Eng.* **2017**, *36*, 229–238. [CrossRef]

34. Callemien, D.; Collin, S. Structure, organoleptic properties, quantification methods, and stability of phenolic compounds in beer-a review. *Food Rev. Int.* **2010**, *26*, 1–84. [CrossRef]

35. Gerhäuser, C. Beer constituents as potential cancer chemopreventive agents. *Eur. J. Cancer* **2005**, *41*, 1941–1954. [CrossRef]

36. Shahidi, F.; Nazk, M. *Phenolics in Food and Nutraceuticals*, 2nd ed.; CRC Press LLC: Boca Raton, FL, USA, 2004. [CrossRef]

37. Andersen, M.L.; Gundermann, M.; Danielsen, B.P.; Lund, M.N. Kinetic models for the role of protein thiols during oxidation in beer. *J. Agric. Food Chem.* **2017**, *65*, 10820–10828. [CrossRef]

38. Andersen, M.L.; Outtrup, H.; Skibsted, L.H. Potential antioxidants in beer assessed by ESR spin trapping. *J. Agric. Food Chem.* **2000**, *48*, 3106–3111. [CrossRef]

39. Lindenmeier, M.; Faist, V.; Hofmann, T. Structural and functional characterization of pronyl-lysine, a novel protein modification in bread crust melanoidins showing in vitro antioxidative and phase I/II enzyme modulating activity. *J. Agric. Food Chem.* **2002**, *50*, 6997–7006. [CrossRef]

40. Maillard, M.N.; Billaud, C.; Chow, Y.N.; Ordonaud, C.; Nicolas, J. Free radical scavenging, inhibition of polyphenoloxidase activity and copper chelating properties of model Maillard systems. *LWT Food Sci. Technol.* **2007**, *40*, 1434–1444. [CrossRef]

41. Kunz, T.; Frenzel, J.; Wietstock, P.C.; Methner, F.J. Possibilities to improve the antioxidative capacity of beer by optimized hopping regimes. *J. Inst. Brew.* **2014**, *120*, 415–425. [CrossRef]

42. Wietstock, P.; Kunz, T.; Shelhammer, T.; Schon, T.; Methner, F.J. Behaviour of antioxidants derived from hops during wort boiling. *J. Inst. Brew.* **2010**, *116*, 157–166. [CrossRef]

43. Ting, P.L.; Lusk, L.; Refling, J.; Kay, S.; Ryder, D. Identification of antiradical hop compounds. *J. Am. Soc. Brew. Chem.* **2008**, *66*, 116–126. [CrossRef]

44. Leopoldini, M.; Russo, N.; Toscano, M. The molecular basis of working mechanism of natural polyphenolic antioxidants. *Food Chem.* **2011**, *125*, 288–306. [CrossRef]

45. Szwajgier, D. Content of individual phenolic acids in worts and beers and their possible contribution to the antiradical activity of beer. *J. Inst. Brew.* **2009**, *115*, 243–252. [CrossRef]

46. Piazzon, A.; Forte, M.; Nardini, M. Characterization of phenolics content and antioxidant activity of different beer types. *J. Agric. Food Chem.* **2010**, *58*, 10677–10683. [CrossRef]

47. Snopek, L.; Mlcek, J.; Sochorova, L.; Baron, M.; Hlavacova, I.; Jurikova, T.; Kizek, R.; Sedlackova, E.; Sochor, J. Contribution of red wine consumption to human health protection. *Molecules* **2018**, *23*, 1684. [CrossRef]

48. Osorio-Paz, I.; Brunauer, R.; Alavez, S. Beer and its non-alcoholic compounds in health and disease. *Crit. Rev. Food Sci. Nutr.* **2019**, *1*, 1–14. [CrossRef]

49. Di Renzo, L.; Marsella, L.T.; Carraro, A.; Valente, R.; Gualtieri, P.; Gratteri, S.; Tomasi, D.; Gaiotti, F.; De Lorenzo, A. Changes in LDL oxidative status and oxidative and inflammatory gene expression after red wine intake in healthy people: A randomized trial. *Mediat. Inflamm.* **2015**, *2015*, 317348:1–317348:13. [CrossRef]

50. Annunziata, G.; Maisto, M.; Schisano, C.; Ciampaglia, R.; Narciso, V.; Hassan, S.T.; Tenore, G.C.; Novellino, E. Effect of grape pomace polyphenols with or without pectin on TMAO serum levels assessed by LC/MS-based assay: A preliminary clinical study on overweight/obese subjects. *Front. Pharmacol.* **2019**, *10*, 575:1–575:11. [CrossRef]

51. Nova, E.; San Mauro-Martín, I.; Díaz-Prieto, L.E.; Marcos, A. Wine and beer within a moderate alcohol intake is associated with higher levels of HDL-c and adiponectin. *Nutr. Res.* **2019**, *63*, 42–50. [CrossRef]

52. Golan, R.; Gepner, Y.; Shai, I. Wine and health–new evidence. *Eur. J. Clin. Clin. Nutr.* **2018**, *72*, 55–59. [CrossRef] [PubMed]

53. Torres, A.; Cachofeiro, V.; Millán, J.; Lahera, V.; Nieto, M.; Martin, R.; Bello, E.; Alvarez-Sala, L.; Nieto, M. Red wine intake but not other alcoholic beverages increases total antioxidant capacity and improves pro-inflammatory profile after an oral fat diet in healthy volunteers. *Rev. Clín. Esp.* **2015**, *215*, 486–494. [CrossRef] [PubMed]

54. Lippi, G.; Franchini, M.; Favaloro, E.J.; Targher, G. Moderate red wine consumption and cardiovascular disease risk: Beyond the "French paradox". In *Seminars in Thrombosis and Hemostasis*; Favaloro, E.J., Levi, M., Lisman, T., Kwaan, H.C., Schulman, S., Eds.; Thieme Medical Publishers: Stuttgart, Germany, 2010; pp. 59–70.
55. Lim, S.S.; Vos, T.; Flaxman, A.D.; Danaei, G.; Shibuya, K.; Adair-Rohani, H.; Amann, M.; Anderson, H.R.; Andrews, K.G.; Aryee, M.; et al. A comparative risk assessment of burden of disease and injury attributable to 67 risk factors and risk factor clusters in 21 regions, 1990–2010: A systematic analysis for the global burden of disease study 2010. *Lancet* **2012**, *380*, 2224–2260. [CrossRef]
56. De Salvo, K.B.; Olson, R.; Casavale, K.O. Dietary guidelines for Americans. *JAMA* **2016**, *315*, 457–458. [CrossRef]
57. Giovinazzo, G.; Ingrosso, I.; Paradiso, A.; De Gara, L.; Santino, A. Resveratrol biosynthesis: Plant metabolic engineering for nutritional improvement of food. *Plant Foods Hum. Nutr.* **2012**, *67*, 191–199. [CrossRef]
58. Calabriso, N.; Scoditti, E.; Massaro, M.; Pellegrino, M.; Storelli, C.; Ingrosso, I.; Giovinazzo, G.; Carluccio, M.A. Multiple anti-inflammatory and anti-atherosclerotic properties of red wine polyphenolic extracts: Differential role of hydroxycinnamic acids, flavonols and stilbenes on endothelial inflammatory gene expression. *Eur. J. Nutr.* **2016**, *55*, 477–489. [CrossRef]
59. Giovinazzo, G.; Grieco, F. Functional properties of grape and wine polyphenols. *Plant Foods Hum. Nutr.* **2015**, *70*, 454–462. [CrossRef]
60. Lesjak, M.; Beara, I.; Simin, N.; Pintać, D.; Majkić, T.; Bekvalac, K.; Orčić, D.; Mimica-Duk, N. Antioxidant and anti-inflammatory activities of quercetin and its derivatives. *J. Funct. Foods* **2018**, *40*, 68–75. [CrossRef]
61. Wang, P.; Sang, S. Metabolism and pharmacokinetics of resveratrol and pterostilbene. *BioFactors* **2018**, *44*, 16–25. [CrossRef]
62. Singh, M.; Kaur, M.; Silakari, O. Flavones: An important scaffold for medicinal chemistry. *Eur. J. Med. Chem.* **2014**, *84*, 206–239. [CrossRef] [PubMed]
63. Ghosh, D. Tannins from foods to combat diseases. *Int. J. Pharma Res. Rev.* **2015**, *4*, 40–44.
64. Rufián Henares, J.; Morales, F. Functional properties of melanoidins: In vitro antioxidant, antimicrobial and antihypertensive activities. *Food Res. Int.* **2013**, *40*, 995–1002. [CrossRef]
65. Samuels, J.; Shashidharamurthy, R.; Rayalam, S. Novel anti-obesity effects of beer hops compound xanthohumol: Role of AMPK signalling pathway. *Nutr. Metab.* **2018**, *15*, 42:1–42:11. [CrossRef] [PubMed]
66. Liu, M.; Hansen, P.E.; Wang, G.; Qiu, L.; Dong, J.; Yin, H.; Qian, Z.; Yang, M.; Miao, J. Pharmacological profile of xanthohumol, a prenylated flavonoid from hops (*Humulus lupulus*). *Molecules* **2015**, *20*, 754–779. [CrossRef] [PubMed]
67. Rivero, D.; Pérez, S.; González, M.L.; Valls, V.; Codoñer, P.; Muñiz, P. Inhibition of induced DNA oxidative damage by beers: Correlation with the content of polyphenols and melanoidins. *J. Agric. Food Chem.* **2005**, *53*, 3637–3642. [CrossRef] [PubMed]
68. Zhao, H.; Li, H.; Sun, G.; Yang, B.; Zhao, M. Assessment of endogenous antioxidative compounds and antioxidant activities of lager beers. *J. Sci. Food Agric.* **2013**, *93*, 910–917. [CrossRef] [PubMed]
69. Huang, W.Y.; Liu, Y.M.; Wang, J.; Wang, X.N.; Li, C.Y. Anti-inflammatory effect of the blueberry anthocyanins malvidin-3-glucoside and malvidin-3-galactoside in endothelial cells. *Molecules* **2014**, *19*, 12827–12841. [CrossRef]
70. Radonjić, S.; Košmerl, T.; Ota, A.; Prosen, H.; Maraš, V.; Demšar, L.; Polak, T. Technological and microbiological factors affecting polyphenolic profile of Montenegrin red wines. *Chem. Ind. Chem. Eng. Q.* **2019**, *25*, 309–319. [CrossRef]
71. Pajović-Šćepanović, R.; Wendelin, S.; Eder, R. Phenolic composition and varietal discrimination of Montenegrin red wines (*Vitis vinifera* var. Vranac, Kratošija, and Cabernet Sauvignon). *Eur. Food Res. Technol.* **2018**, *244*, 2243–2254. [CrossRef]
72. Lima, A.; Oliveira, C.; Santos, C.; Campos, F.M.; Couto, J.A. Phenolic composition of monovarietal red wines regarding volatile phenols and its precursors. *Eur. Food Res. Technol.* **2018**, *244*, 1985–1994. [CrossRef]
73. Bartolomé, B.; Peña-Neira, A.; Gómez-Cordovés, C. Phenolics and related substances in alcohol-free beers. *Eur. Food Res. Technol.* **2000**, *210*, 419–423. [CrossRef]
74. Dvořáková, M.; Hulín, P.; Karabín, M.; Dostálek, P. Determination of polyphenols in beer by an effective method based on solid-phase extraction and high performance liquid chromatography with diode-array detection. *Czech J. Food Sci.* **2007**, *25*, 182–188. [CrossRef]

75. Jandera, P.; Škeříková, V.; Řehová, L.; Hájek, T.; Baldriánová, L.; Škopová, G.; Kellner, V.; Horna, A. RP-HPLC analysis of phenolic compounds and flavonoids in beverages and plant extracts using a CoulArray detector. *J. Sep. Sci.* **2005**, *28*, 1005–1022. [CrossRef]

76. McMurrough, I.; Roche, G.P.; Cleary, K.G. Phenolic acids in beers and worts. *J. Inst. Brew.* **1984**, *90*, 181–187. [CrossRef]

77. Zhao, H.; Chen, W.; Lu, J.; Zhao, M. Phenolic profiles and antioxidant activities of commercial beers. *Food Chem.* **2010**, *119*, 1150–1158. [CrossRef]

78. Floridi, S.; Montanari, L.; Marconi, O.; Fantozzi, P. Determination of free phenolic acids in wort and beer by coulometric array detection. *J. Agric. Food Chem.* **2003**, *51*, 1548–1554. [CrossRef]

79. Marques, D.; Cassis, M.; Quelhas, J.; Bertozzi, J.; Visentainer, J.; Oliveira, C.; Monteiro, A. Characterization of craft beers and their bioactive compounds. *Chem. Eng. Trans.* **2017**, *57*, 1747–1752.

80. Marova, I.; Parilova, K.; Friedl, Z.; Obruca, S.; Duronova, K. Analysis of phenolic compounds in lager beers of different origin: A contribution to potential determination of the authenticity of Czech beer. *Chromatographia* **2011**, *73*, 83–95. [CrossRef]

81. Alonso Garcıa, A.; Cancho Grande, B.; Simal Gandara, J. Development of a rapid method based on solid-phase extraction and liquid chromatography with ultraviolet absorbance detection for the determination of polyphenols in alcohol-free beers. *J. Chromatogr. A* **2004**, *1054*, 175–180. [CrossRef]

82. Kellner, V.; Jurková, M.; Čulík, J.; Horák, T.; Čejka, P. Some phenolic compounds in Czech hops and beer of pilsner type. *Brew. Sci.* **2007**, *60*, 31–37.

83. Socha, R.; Pająk, P.; Fortuna, T.; Buksa, K. Antioxidant activity and the most abundant phenolics in commercial dark beers. *Int. J. Food Prop.* **2017**, *20*, 1–15. [CrossRef]

84. Nardini, M.; Ghiselli, A. Determination of free and bound phenolic acids in beer. *Food Chem.* **2004**, *84*, 137–143. [CrossRef]

85. Lamuela-Raventos, R.M.; Romero-Perez, A.I.; Waterhouse, A.L.; Carmen de la Torre-Borona, M. Direct HPLC Analysis of *cis*- and *trans*-resveratrol and piceid isomers in Spanish red *Vitis vinifera* wines. *J. Agric. Food Chem.* **1995**, *43*, 281–283. [CrossRef]

86. Chiva-Blanch, G.; Urpi-Sarda, M.; Rotchés-Ribalta, M.; Zamora-Ros, R.; Llorach, R.; Lamuela-Raventós, R.M.; Andrés-Lacueva, C. Determination of resveratrol and piceid in beer matrices by solid-phase extraction and liquid chromatography-tandem mass spectrometry. *J. Chromatogr. A* **2011**, *1218*, 698–705. [CrossRef]

87. Paulo, L.; Domingues, F.; Queiroz, J.A.; Gallardo, E. Development and validation of an analytical method for the determination of *trans*- and *cis*-resveratrol in wine: Analysis of its contents in 186 Portuguese red wines. *J. Agric. Food Chem.* **2011**, *59*, 2157–2168. [CrossRef]

88. Nour, V.; Trandafir, I.; Muntean, C. Ultraviolet irradiation of trans-resveratrol and HPLC determination of *trans*-resveratrol and *cis*-resveratrol in Romanian red wines. *J. Chromatogr. Sci.* **2012**, *50*, 920–927. [CrossRef]

89. Zoechling, A.; Reiter, E.; Eder, R.; Wendelin, S.; Leiber, F.; Jungbauer, A. The flavonoid kaempferol is responsible for majority of estrogenic activity in red wine. *Am. J. Enol. Vitic.* **2009**, *60*, 223–232.

90. Basalekou, M.; Kyraleou, M.; Pappas, C.; Tarantilis, P.; Kotseridis, Y.; Kallithraka, S. Proanthocyanidin content as an astringency estimation tool and maturation index in red and white winemaking technology. *Food Chem.* **2019**, *299*, 125–135. [CrossRef]

91. Stark, T.; Wollmann, N.; Wenker, K.; Lösch, S.; Glabasnia, A.; Hofmann, T. Matrix-calibrated LC-MS/MS quantitation and sensory evaluation of oak ellagitannins and their transformation products in red wines. *J. Agric. Food Chem.* **2010**, *58*, 6360–6369. [CrossRef]

92. García-Estévez, I.; Escribano-Bailón, M.T.; Rivas-Gonzalo, J.C.; Alcalde-Eon, C. Validation of a mass spectrometry method to quantify oak ellagitannins in wine samples. *J. Agric. Food Chem.* **2012**, *60*, 1373–1379. [CrossRef]

93. Jourdes, M.; Michel, J.; Saucier, C.; Quideau, S.; Teissedre, P.L. Identification, amounts, and kinetics of extraction of C-glucosidic ellagitannins during wine aging in oak barrels or in stainless steel tanks with oak chips. *Anal. Bioanal. Chem.* **2011**, *401*, 1531–1539. [CrossRef] [PubMed]

94. Ma, W.; Guo, A.; Zhang, Y.; Wang, H.; Liu, Y.; Li, H.J. A review on astringency and bitterness perception of tannins in wine. *Trends Food Sci. Technol.* **2014**, *40*, 6–19. [CrossRef]

95. Niculescu, V.C.; Paun, N.; Ionete, R.E. The evolution of polyphenols from grapes to wines. In *Grapes and Wines—Advances in Production, Processing, Analysis and Valorization*; Jordão, A.M., Cosme, F., Eds.; Intech Open: London, UK, 2017. [CrossRef]

96. Barnaba, C.; Dellacassa, E.; Nicolini, G.; Nardin, T.; Malacarne, M.; Larcher, R. Identification and quantification of 56 targeted phenols in wines, spirits, and vinegars by online solid-phase extraction—Ultrahigh-performance liquid chromatography—Quadrupole-orbitrap mass spectrometry. *J. Chromatogr. A* **2015**, *1423*, 124–135. [CrossRef] [PubMed]

97. Cabrera-Banegil, M.; Hurtado-sánchez, M.C.; Galeano-Díaz, T.; Durán-Merás, I. Front-face fluorescence spectroscopy combined with second-order multivariate algorithms for the quantification of polyphenols in red wine samples. *Food Chem.* **2016**, *220*, 168–176. [CrossRef]

98. Kustrin, S.; Hettiarachchi, C.; Morton, D.; Razic, S. Analysis of phenolics in wine by high performance thin-layer chromatography with gradient elution and high resolution plate imaging. *J. Pharm. Biomed. Anal.* **2014**, *102*, 93–99. [CrossRef]

99. Justesen, U.; Knuthsen, P.; Leth, T. Quantitative analysis of flavonols, flavones, and flavanones in fruits, vegetables and beverages by high-performance liquid chromatography with photo-diode array and mass spectrometric detection. *J. Chromatogr. A* **1998**, *799*, 101–110. [CrossRef]

100. Achilli, G.; Cellerino, G.P.; Gamache, P.H. Identification and determination of phenolic constituents in natural beverages and plant extracts by means of a coulometric electrode array system. *J. Chromatogr. A* **1993**, *632*, 111–117. [CrossRef]

101. Martinez-Periñan, E.; Hernández-Artiga, M.P.; Palacios-Santander, J.M.; ElKaoutit, M.; Naranjo-Rodriguez, I.; Bellido-Milla, D. Estimation of beer stability by sulphur dioxide and polyphenol determination. Evaluation of a laccase-sonogel-carbon biosensor. *Food Chem.* **2011**, *127*, 234–239. [CrossRef]

102. Sýs, M.; Metelka, R.; Vytřas, K. Comparison of tyrosinase biosensor based on carbon nanotubes with DPPH spectrophotometric assay in determination of TEAC in selected Moravian wines. *Monatsh. Chem. Chem. Mon.* **2015**, *146*, 813–817. [CrossRef]

103. Martysiak-Żurowska, D.; Wenta, W. A Comparison of ABTS and DPPH methods for assessing the total antioxidant capacity of human milk. *Acta Sci. Pol. Technol. Aliment.* **2012**, *1*, 83–89.

104. Özyürek, M.; Güçlü, K.; Tütem, E.; Sözgen-Başkan, K.; Erçağ, E.; Çelik, S.E.; Baki, S.; Yildiz, L.; Karaman, Ş.; Apak, R. A comprehensive review of CUPRAC methodology. *Anal. Methods* **2011**, *3*, 2439–2453. [CrossRef]

105. Šeruga, M.; Novak, I.; Jakobek, L. Determination of polyphenols content and antioxidant activity of some red wines by differential pulse voltammetry, HPLC and spectrophotometric methods. *Food Chem.* **2011**, *124*, 1208–1216. [CrossRef]

106. Đorđević, N.O.; Pejin, B.; Novaković, M.M.; Stanković, D.M.; Mutić, J.J.; Pajović, S.B.; Tešević, V.V. Some chemical characteristics and antioxidant capacity of novel Merlot wine clones developed in Montenegro. *Sci. Hortic.* **2017**, *225*, 505–511. [CrossRef]

107. Ricci, A.; Teslic, N.; Petropolus, V.; Parpinello, G.P.; Versari, A. Fast analysis of total polyphenol content and antioxidant activity in wines and oenological tannins using a flow injection system with tandem diode array and electrochemical detections. *Food Anal. Methods* **2019**, *12*, 347–354. [CrossRef]

108. Minkova, S.; Hristova-Avakumova, N.; Traykov, T.; Nikolova, K.; Hadjimitova, V. Antioxidant scavenging capacity of Bulgarian red wines by chemiluminescent and stable free radical methods. *Bulg. Chem. Commun.* **2019**, *51*, 299–303.

109. García-Guzmán, J.J.; López-Iglesias, D.; Cubillana-Aguilera, L.; Lete, C.; Lupu, S.; Palacios-Santander, J.M.; Bellido-Milla, D. Assessment of the polyphenol indices and antioxidant capacity for beers and wines using a tyrosinase-based biosensor prepared by sinusoidal current method. *Sensors* **2018**, *19*, 66. [CrossRef] [PubMed]

110. Mikyška, A.; Dušek, M. How wort boiling process affects flavonoid polyphenols in beer. *Kvas. Prum.* **2019**, *65*, 192–200. [CrossRef]

111. Šućur, S.; Maraš, V.; Kodžulović, V.; Raičević, J.; Mugoša, M.; Jug, T.; Košmerl, T. The impact of different commercial yeasts on quality parameters of Montenegrin red wine—Vranac and Kratošija. *Biol. Eng. Med.* **2016**, *1*. [CrossRef]

112. Mattivi, F.; Nicolini, G. Analysis of polyphenols and resveratrol in Italian wines. *BioFactors* **1997**, *6*, 445–448. [CrossRef]

113. Martínez, A.; Vegara, S.; Herranz-López, M.; Martí, N.; Valero, M.; Micol, V.; Saura, D. Kinetic changes of polyphenols, anthocyanins and antioxidant capacity in forced aged hibiscus ale beer. *J. Inst. Brew.* **2017**, *123*, 58–65. [CrossRef]

114. Chen, O.; Blumberg, J. Flavonoids in Beer and Their Potential Benefit on the Risk of Cardiovascular Disease. In *Beer in Health and Disease Prevention*, 1st ed.; Preedy, V.R., Ed.; Elsevier: Amsterdam, The Netherlands, 2010; pp. 831–841.

115. Galarce, O.; Pavon, J.; Aranda, M.; Novoa, L.; Henriquez, K. Chemometric Optimization of QuEChERS extraction method for polyphenol determination in beers by liquid chromatography with ultraviolet detection. *Food Anal. Methods* **2018**, *12*, 448–457. [CrossRef]

116. Lentz, M. The impact of simple phenolic compounds on beer aroma and flavor. *Fermentation* **2018**, *4*, 20. [CrossRef]

117. Pihlava, J.M.; Kurtelius, T.; Hurme, T. Total hordatine content in different types of beers. *J. Inst. Brew.* **2016**, *122*, 212–217. [CrossRef]

118. Ferreira-Lima, N.; Vallverdú-Queralt, A.; Meudec, E.; Pinasseau, L.; Verbaere, A.; Bordignon-Luiz, M.T.; Le Guernevé, C.; Cheynier, V.; Sommerer, N. Quantification of hydroxycinnamic derivatives in wines by UHPLC-MRM-MS. *Anal. Bioanal. Chem.* **2018**, *410*, 3483–3490. [CrossRef] [PubMed]

119. Kallitraka, S.; Tsoutsouras, E.; Tzourou, E.; Lanaridis, P. Principal phenolic compounds in Greek red wines. *Food Chem.* **2006**, *99*, 784–793. [CrossRef]

120. Mitić, S.; Paunović, D.; Pavlović, A.; Tošić, S.; Stojković, M.; Mitić, M. Phenolic profiles and total antioxidant capacity of marketed beers in Serbia. *Int. J. Food Prop.* **2014**, *17*, 908–922. [CrossRef]

121. Moura, N.; Cazaroti, T.; Dias, N.; Fernandes, P.; Monteiro, M.; Perrone, D.; Guedes, A. Phenolic compounds of Brazilian beers from different types and styles and application of chemometrics for modeling antioxidant capacity. *Food Chem.* **2016**, *199*, 105–113. [CrossRef]

122. Waterhouse, A.L. Wine phenolics. *Ann. N. Y. Acad. Sci.* **2002**, *957*, 21–36. [CrossRef]

123. Baur, J.A.; Pearson, K.J.; Price, N.L.; Jamieson, H.A.; Lerin, C.; Kalra, A.; Prabhu, V.V.; Allard, J.S.; Lopez-Lluch, G.; Lewis, K.; et al. Resveratrol improves health and survival of mice on a high-calorie diet. *Nature* **2006**, *444*, 337–342. [CrossRef]

124. Kalantari, H.; Das, D.K. Physiological effects of resveratrol. *BioFactors* **2010**, *36*, 401–406. [CrossRef]

125. Wu, J.M.; Hsieh, T.C. Resveratrol: A cardioprotective substance. *Ann. N. Y. Acad. Sci.* **2011**, *1215*, 16–21. [CrossRef] [PubMed]

126. Sahebkar, A. Effects of resveratrol supplementation on plasma lipids: A systematic review and meta-analysis of randomized controlled trials. *Nutr. Rev.* **2013**, *71*, 822–835. [CrossRef]

127. Penumathsa, S.V.; Maulik, N. Resveratrol: A promising agent in promoting cardioprotection against coronary heart disease. *Can. J. Physiol. Pharmacol.* **2009**, *87*, 275–286. [CrossRef]

128. Sakata, Y.; Zhuang, H.; Kwansa, H.; Koehler, R.C.; Doré, S. Resveratrol protects against experimental stroke: Putative neuroprotective role of heme oxygenase 1. *Exp. Neurol.* **2010**, *224*, 325–329. [CrossRef]

129. Pandey, K.B.; Rizvi, S.I. Anti-oxidative action of resveratrol: Implications for human health. *Arab. J. Chem.* **2011**, *4*, 293–298. [CrossRef]

130. Hussein, M.A. A convenient mechanism for the free radical scavenging activity of resveratrol. *Int. J. Phytomed.* **2011**, *3*, 459–469.

131. Godelmann, R.; Fang, F.; Humpfer, E.; Schütz, B.; Bansbach, M.; Schäfer, H.; Spraul, M. Targeted and Nontargeted Wine Analysis by 1H NMR spectroscopy combined with multivariate statistical analysis. diferentiation of important parameters: Grape variety, geographical origin, year of vintage. *J. Agric. Food Chem.* **2013**, *61*, 5610–5619. [CrossRef] [PubMed]

132. Cheiran, K.P.; Raimundo, V.P.; Manfroi, V.; Cheiran, K.P.; Rodrigues, E.; Anzanello, M.J.; Raimundo, V.P.; Frazzon, J.; Kahmann, A. Simultaneous identification of low-molecular weight phenolic and nitrogen compounds in craft beers by HPLC-ESI-MS/MS. *Food Chem.* **2019**, *286*, 113–122. [CrossRef] [PubMed]

133. Molina, L.; Ruiz, A.; Luisa, F.M. A novel multicommuted fluorimetric optosensor for determination of resveratrol in beer. *Talanta* **2010**, *83*, 850–856. [CrossRef] [PubMed]

134. Jerkovic, V.; Collin, S. Fate of resveratrol and piceid through different hop processings and storage times. *J. Agric. Food Chem.* **2008**, *56*, 584–590. [CrossRef]

135. Gambini, J.; Inglés, M.; Olaso, G.; Lopez-Grueso, R.; Bonet-Costa, V.; Gimeno-Mallench, L.; Mas-Bargues, C.; Abdelaziz, K.M.; Gomez-Cabrera, M.C.; Vina, J.; et al. Properties of resveratrol: In vitro and in vivo studies about metabolism, bioavailability, and biological effects in animal models and humans. *Oxid. Med. Cell. Longev.* **2015**, *2015*, 837042:1–837042:13. [CrossRef]

136. Chira, K.; Pacella, N.; Jourdes, M.; Teissedre, P.-L. Chemical and sensory evaluation of Bordeaux wines (Cabernet-Sauvignon and Merlot) and correlation with wine age. *Food Chem.* **2011**, *126*, 1971–1977. [CrossRef] [PubMed]

137. Markoski, M.M.; Garavaglia, J.; Oliveira, A.; Olivaes, J.; Marcadenti, A. Molecular properties of red wine compounds and cardiometabolic benefits. *Nutr. Metab. Insights* **2016**, *9*, 51–57. [CrossRef]

138. Feng, W.; Hao, Z.; Li, M. Isolation and Structure Identification of Flavonoids. In *Flavonoids—From Biosynthesis to Human Health*; Justino, G.C., Ed.; Intech Open: London, UK, 2017. [CrossRef]

139. Escobar-Cévoli, R.; Castro-Espín, C.; Béraud, V.; Buckland, G.; Zamora-Ros, R.; Béraud, G.B.V. An overview of global flavonoid intake and its food sources. In *Flavonoids—From Biosynthesis to Human Health*; Justino, G.C., Ed.; Intech Open: London, UK, 2017. [CrossRef]

140. Elrod, S.M. Xanthohumol and the medicinal benefits of beer. In *Polyphenols: Mechanisms of Action in Human Health and Disease*; Elsevier: Amsterdam, The Netherlands, 2018; pp. 19–32.

141. Andrés-Iglesias, C.; Blanco, C.A.; Blanco, J.; Montero, O. Mass spectrometry-based metabolomics approach to determine differential metabolites between regular and non-alcohol beers. *Food Chem.* **2014**, *157*, 205–212. [CrossRef] [PubMed]

142. Stevens, J.F.; Page, J.E. Xanthohumol and related prenylflavonoids from hops and beer: To your good health! *Phytochemistry* **2004**, *65*, 1317–1330. [CrossRef] [PubMed]

143. Venturelli, S.; Burkard, M.; Biendl, M.; Lauer, U.M.; Frank, J.; Busch, C. Prenylated chalcones and flavonoids for the prevention and treatment of cancer. *Nutrition* **2016**, *32*, 1171–1178. [CrossRef]

144. Back, W.; Zurcher, A.; Wunderlich, S. Verfahren zur Herstellung eines xanthohumolhaltigen Getränkes aus Malz- und/oder Rohfruchtwürze sowie derart hergestelltes Getränk. *DE10256166A1* **2004**, *A1*, 1–5.

145. Collin, S.; Jerkovic, V.; Bröhan, M.; Callemien, D. Polyphenols and beer quality. In *Natural Products*, 1st ed.; Ramawat, K.G., Mérillon, J.M., Eds.; Springer: Heidelberg, Germany, 2013; pp. 2334–2353.

146. Garrido, J.; Borges, F. Wine and grape polyphenols—A chemical perspective. *Food Res. Int.* **2013**, *54*, 1844–1858. [CrossRef]

147. Dadic, M.; Belleau, G. Polyphenols and beer flavor. *Proc. Am. Soc. Brew. Chem.* **1973**, *31*, 107–114. [CrossRef]

148. D'Archivio, M.; Filesi, C.; Varì, R.; Scazzocchio, B.; Masella, R. Bioavailability of the polyphenols: Status and controversies. *Int. J. Mol. Sci.* **2010**, *11*, 1321–1342. [CrossRef]

149. Gerhäuser, C.; Becker, H. Phenolic Compounds in Beer. In *Beer in Health and Disease Prevention*; Elsevier: Amsterdam, The Netherlands, 2008; Chapter 12; ISBN 9780080453521.

150. de Pascual-Teresa, S.; Santos-Buelga, C.; Rivas-Gonzalo, J. Quantitative analysis of flavan-3-ols in Spanish foodstuffs and beverages. *J. Agric. Food Chem.* **2000**, *48*, 5331–5337. [CrossRef] [PubMed]

151. Koponen, J.M.; Happonen, A.M.; Mattila, P.H.; Törrönen, A.R. Contents of anthocyanins and ellagitannins in selected foods consumed in Finland. *J. Agric. Food Chem.* **2007**, *55*, 1612–1619. [CrossRef]

152. Dixon, R.A.; Xie, D.Y.; Sharma, S.B. Proanthocyanidins—A final frontier in flavonoid research? *New Phytol.* **2005**, *165*, 9–28. [CrossRef] [PubMed]

153. Santos-Buelga, C.; Scalbert, A. Proanthocyanidins and tannin–like compounds—Nature, occurrence, dietary intake and effects on nutrition and health. *J. Sci. Food Agric.* **2000**, *80*, 1094–1117. [CrossRef]

154. McCullough, M.L.; Peterson, J.J.; Patel, R.; Jacques, P.F.; Shah, R.; Dwyer, J.T. Flavonoid intake and cardiovascular disease mortality in a prospective cohort of US adults. *Am. J. Clin. Nutr.* **2012**, *95*, 454–464. [CrossRef] [PubMed]

155. Gu, L.; Kelm, M.A.; Hammerstone, J.F.; Beecher, G.; Holden, J.; Haytowitz, D.; Gebhardt, S.; Prior, R.L. Concentrations of proanthocyanidins in common foods and estimations of normal consumption. *J. Nutr.* **2004**, *134*, 613–617. [CrossRef]

156. Cheynier, V.; Dueñas-Paton, M.; Salas, E.; Maury, C.; Souquet, J.M.; Sarni-Manchado, P.; Fulcrand, H. Structure and properties of wine pigments and tannins. *Am. J. Enol. Vitic.* **2006**, *57*, 298–305.

157. Gerhäuser, C.; Alt, A.P.; Klimo, K.; Knauft, J.; Frank, N.; Becker, H. Isolation and potential cancer chemopreventive activities of phenolic compounds of beer. *Phytochem. Rev.* **2002**, *1*, 369–377. [CrossRef]

158. Oregon State University Home Page. Available online: https://lpi.oregonstate.edu/mic/dietary-factors/phytochemicals/flavonoids (accessed on 13 October 2020).

159. Perez-Vizcaino, F.; Duarte, J. Flavonols and cardiovascular disease. *Mol. Asp. Med.* **2010**, *31*, 478–494. [CrossRef]

160. Forster, A.; Beck, B.; Schmidt, R.; Jansen, C.; Mellenthin, A. On the composition of low molecular polyphenols in different varieties of hops and from two growing areas. *Mon. Brauwiss.* **2002**, *55*, 98–108.

161. Gangopadhyay, N.; Rai, D.K.; Brunton, N.P.; Gallagher, E.; Hossain, M.B. Antioxidant-guided isolation and mass spectrometric identification of the major polyphenols in barley (*Hordeum vulgare*) grain. *Food Chem.* **2016**, *210*, 212–220. [CrossRef] [PubMed]

162. Stevens, J.F.; Taylor, A.W.; Deinzer, M.L. Quantitative analysis of xanthohumol and related prenylfiavonoids in hops and beer by liquid chromatography-tandem mass spectrometry. *J. Chromatogr. A* **1999**, *832*, 97–107. [CrossRef]

163. Moritz Seliger, J.; Misuri, L.; Maser, E.; Hintzpeter, J. The hop-derived compounds xanthohumol, isoxanthohumol and 8-prenylnaringenin are tight-binding inhibitors of human aldo-keto reductases 1B1 and 1B10. *J. Enzyme Inhib. Med. Chem.* **2018**, *33*, 607–614. [CrossRef]

164. Lee, Y.-M.; Yoon, Y.; Yoon, H.; Park, H.-M.; Song, S.; Yeum, K.-J.J.N. Dietary anthocyanins against obesity and inflammation. *Nutrients* **2017**, *9*, 1089. [CrossRef]

165. Cassidy, A.; Rimm, E.B.; O'Reilly, É.J.; Logroscino, G.; Kay, C.; Chiuve, S.E.; Rexrode, K.M. Dietary flavonoids and risk of stroke in women. *Stroke* **2012**, *43*, 946–951. [CrossRef]

166. Cassidy, A.; Mukamal, K.J.; Liu, L.; Franz, M.; Eliassen, A.H.; Rimm, E.B. High anthocyanin intake is associated with a reduced risk of myocardial infarction in young and middle-aged women. *Circulation* **2013**, *127*, 188–196. [CrossRef] [PubMed]

167. Cassidy, A.; Bertoia, M.; Chiuve, S.; Flint, A.; Forman, J.; Rimm, E.B. Habitual intake of anthocyanins and flavanones and risk of cardiovascular disease in men. *Am. J. Clin. Nutr.* **2016**, *104*, 587–594. [CrossRef] [PubMed]

168. Castillo-Munoz, N.; Winterhalter, P.; Weber, F.; Gomez, M.V.; Gomez-Alonso, S.; Garcia- Romero, E.; Hermosin-Gutierrez, I. Structure elucidation of peonidin 3,7-O-β-diglucoside isolated from Garnacha Tintorera (*Vitis vinifera* L.) grapes. *J. Agric. Food Chem.* **2010**, *58*, 11105–11111. [CrossRef]

169. Hannah, L.; Roehrdanz, P.R.; Ikegami, M.; Shepard, A.V.; Shaw, M.R.; Tabor, G.; Zhi, L.; Marquet, P.A.; Hijmans, R.J. Climate change, wine, and conservation. *Proc. Natl. Acad. Sci. USA* **2013**, *110*, 6907–6912. [CrossRef]

170. Jones, G.V.; White, M.A.; Cooper, O.R.; Storchmann, K. Climate change and global wine quality. *Clim. Chang.* **2005**, *73*, 319–343. [CrossRef]

171. Bogićević, M.; Maraš, V.; Mugoša, M.; Kodžulović, V.; Raičević, J.; Šućur, S.; Failla, O. The effect of early leaf removal and cluster thinning treatments on berry growth and grape composition in cultivars Vranac and Cabernet Sauvignon. *Chem. Biol. Technol. Agric.* **2015**, *2*, 13. [CrossRef]

172. Kemp, B.S.; Harrison, R.; Creasy, G.L. Effect of mechanical leaf removal and its timing on flavan-3-ol composition and concentrations in Vitis vinifera L. cv. Pinot noir wine. *Aust. J. Grape Wine Res.* **2011**, *17*, 270–279. [CrossRef]

173. Gatti, M.; Bernizzoni, F.; Civardi, S.; Poni, S. Effects of cluster thinning and pre-flowering leaf removal on growth and grape composition in cv. Sangiovese. *Am. J. Enol. Vitic.* **2012**, *63*, 325–332. [CrossRef]

174. Lee, J.; Skinkis, P.A. Oregon 'Pinot noir' grape anthocyanin enhancement by early leaf removal. *Food Chem.* **2013**, *139*, 893–901. [CrossRef]

175. Košmerl, T.; Bertalanič, L.; Maraš, V.; Kodžulović, V.; Šućur, S.; Abramovič, H. Impact of Yield on Total polyphenols, Anthocyanins, Reducing Sugars and Antioxidant potential in White and Red Wines Produced from Montenegrin Autochthonous Grape Varieties. *Food Sci. Technol.* **2013**, *1*, 7–15. [CrossRef]

176. Vezzulli, S.; Civardi, S.; Ferrari, F.; Bavaresco, L. Methyl jasmonate treatment as a trigger of resveratrol synthesis in cultivated grapevine. *Am. J. Enol. Vitic.* **2007**, *58*, 530–533.

177. Santamaria, A.R.; Mulinacci, N.; Valletta, A.; Innocenti, M.; Pasqua, G. Effects of elicitors on the production of resveratrol and viniferins in cell cultures of *Vitis vinifera* L. cv Italia. *J. Agric. Food Chem.* **2011**, *59*, 9094–9101. [CrossRef]

178. Giacosa, S.; Ossola, C.; Botto, R.; Río Segade, S.; Paissoni, M.A.; Pollon, M.; Gerbi, V.; Rolle, L. Impact of specific inactive dry yeast application on grape skin mechanical properties, phenolic compounds extractability, and wine composition. *Food Res. Int.* **2019**, *116*, 1084–1093. [CrossRef]

179. Narziss, L. Polyphenolgehalt und Polymerisationsindex von Gersten und Kleinmalzen. *Mon. Brauwiss.* **1976**, *29*, 9–19.

180. Almaguer, C.; Schonberger, C.; Gastl, M.; Arendt, E.K.; Becker, T. Humulus lupulus—A story that begs to be told. A review. *J. Inst. Brew.* **2014**, *120*, 289–314. [CrossRef]

181. Holtekjølen, A.K.; Kinitz, C.; Knutsen, S.H. Flavanol and bound phenolic acid contents in different barley varieties. *J. Agric. Food Chem.* **2006**, *54*, 2253–2260. [CrossRef] [PubMed]

182. Penalvo, J.L.; Haajanen, K.M.; Botting, N.; Adlercreutz, H. Quantification of lignans in food using isotope dilution gas chromatography/mass spectrometry. *J. Agric. Food Chem.* **2005**, *53*, 9342–9347. [CrossRef]

183. Kohyama, N.; Ono, H. Hordatine a beta-d-glucopyranoside from ungerminated barley grains. *J. Agric. Food Chem.* **2013**, *61*, 1112–1116. [CrossRef] [PubMed]

184. Carvalho, D.O.; Curto, A.F.; Guido, L.F. Determination of phenolic content in different barley varieties and corresponding malts by liquid chromatography-diode array detection-electrospray ionization tandem mass spectrometry. *Antioxidants* **2015**, *4*, 563–576. [CrossRef] [PubMed]

185. Dvořáková, M.; Douanier, M.; Jurková, M.; Kellner, V.; Dostálek, P. Comparison of antioxidant activity of barley (hordeum vulgare l.) and malt extracts with the content of free phenolic compounds measured by high performance liquid chromatography coupled with coularray detector. *J. Inst. Brew.* **2008**, *114*, 150–159. [CrossRef]

186. Dvořáková, M.; Moreira, M.M.; Dostálek, P.; Skulilova, Z.; Guido, L.F.; Barros, A.A. Characterization of monomeric and oligomeric flavan-3-ols from barley and malt by liquid chromatography-ultraviolet detection-electrospray ionization mass spectrometry. *J. Chromatogr. A* **2008**, *1189*, 398–405. [CrossRef]

187. Madhujith, T.; Shahidi, F. Antioxidative and antiproliferative properties of selected barley (*Hordeum vulgare* L.) cultivars and their potential for inhibition of low-density lipoprotein (LDL) cholesterol oxidation. *J. Agric. Food Chem.* **2007**, *55*, 5018–5024. [CrossRef]

188. Madhujith, T.; Shahidi, F. Antioxidant potential of barley as affected by alkaline hydrolysis and release of insoluble-bound phenolics. *Food Chem.* **2009**, *117*, 615–620. [CrossRef]

189. Leitao, C.; Marchioni, E.; Bergaentzlé, M.; Zhao, M.; Didierjean, L.; Miesch, L.; Holder, E.; Miesch, M.; Ennahar, S. Fate of polyphenols and antioxidant activity of barley throughout malting and brewing. *J. Cereal Sci.* **2012**, *55*, 318–322. [CrossRef]

190. Inns, E.L.; Buggey, L.A.; Booer, C.; Nursten, H.E.; Ames, J.M. Effect of heat treatment on the antioxidant activity, color, and free phenolic acid profile of malt. *J. Agric. Food Chem.* **2007**, *55*, 6539–6546. [CrossRef]

191. Muñoz-Insa, A.; Gastl, M.; Becker, T. Variation of sunstruck flavor-related substances in malted barley, triticale and spelt. *Eur. Food Res. Technol.* **2016**, *242*, 11–23. [CrossRef]

192. Stavri, M.; Schneider, R.; O' Donnell, G.; Lechner, D.; Bucar, F.; Gibbons, S. The antimycobacterial components of hops (*Humulus lupulus*) and their dereplication. *Phytother. Res.* **2004**, *18*, 774–776. [CrossRef] [PubMed]

193. Prencipe, F.B.; Brighenti, V.; Rodolfi, M.; Mongelli, A.; dall'Asta, C.; Ganino, T.; Bruni, R.; Pellati, F. Development of a new high-performance liquid chromatography method with diode array and electrospray ionization-mass spectrometry detection for the metabolite fingerprinting of bioactive compounds in *Humulus lupulus* L. *J. Chromatogr. A* **2014**, *1349*, 50–59. [CrossRef]

194. Forster, A.; Beck, B.; Schmidt, R.; Jansen, C.; Mellenthin, A. U ber die Zusammensetzung von niedermolekularen Polyphenolen in verschiedenen Hopfensorten und zwei Anbaugebieten. *Mon. Brauwiss.* **2002**, *55*, 98–108.

195. McMurrough, I.; Hennigan, G.P.; Loughrey, M.J. Quantitative analysis of hop flavonols using high-performance liquid chromatography. *J. Agric. Food Chem.* **1982**, *30*, 1102–1106. [CrossRef]

196. Jerkovic, V.; Collin, S. Occurrence of resveratrol and piceid in American and European hop cones. *J. Agric. Food Chem.* **2007**, *55*, 8754–8758. [CrossRef]

197. Jerkovic, V.; Callemien, D.; Collin, S. Determination of stilbenes in hop pellets from different cultivars. *J. Agric. Food Chem.* **2005**, *53*, 4202–4206. [CrossRef]

198. Inui, T.; Okumura, K.; Matsui, H.; Hosoya, T.; Kumazawa, S. Effect of harvest time on some in vitro functional properties of hop polyphenols. *Food Chem.* **2007**, *225*, 69–76. [CrossRef]

199. Kavalier, A.R.; Litt, A.; Ma, C.; Pitra, N.J.; Coles, M.C.; Kennelly, E.J.; Matthews, P.D. Phytochemical and morphological characterization of hop (*Humulus lupulus* L.) cones over five developmental stages using high performance liquid chromatography coupled to time-of-flight mass spectrometry, ultrahigh performance liquid chromatography photodiode array detection, and light microscopy techniques. *J. Agric. Food Chem.* **2011**, *59*, 4783–4793. [CrossRef]

200. Olšovska, J.; Kameník, Z.; Čejka, P.; Jurkova, M.; Mikyška, A. Ultra-high-performance liquid chromatography profiling method for chemical screening of proanthocyanidins in Czech hops. *Talanta* **2013**, *116*, 919–926. [CrossRef]

201. Mikyška, A.; Krofta, K.; Haškova, D.; Culık, J.; Čejka, P. The influence of hopping on formation of carbonyl compounds during storage of beer. *J. Inst. Brew.* **2012**, *117*, 47–54. [CrossRef]
202. Krofta, K.; Mikyška, A.; Haškova, D. Antioxidant characteristics of hops and hop products. *J. Inst. Brew.* **2008**, *114*, 160–166. [CrossRef]
203. Mikyška, A.; Krofta, K. Assessment of changes in hop resins and polyphenols during long-term storage. *J. Inst. Brew.* **2012**, *118*, 269–279. [CrossRef]
204. Steenackers, B.; De Cooman, L.; De Vos, D. Chemical transformations of characteristic hop secondary metabolites in relation to beer properties and the brewing process: A review. *Food Chem.* **2015**, *172*, 742–756. [CrossRef] [PubMed]
205. Jaskula-Goiris, B.; Goiris, K.; Syryn, E.; Van Opstaele, F.; De Rouck, G.; Aerts, G.; De Cooman, L. The use of hop polyphenols during brewing to improve flavor quality and stability of pilsner beer. *J. Am. Soc. Brew. Chem.* **2014**, *72*, 175–183. [CrossRef]
206. Muñoz-Insa, A.; Gastl, M.; Becker, T. Use of polyphenol-rich hop products to reduce sunstruck flavor in beer. *J. Am. Soc. Brew. Chem.* **2015**, *73*, 228–235. [CrossRef]
207. Szwajgier, D. Dry and wet milling of malt. A preliminary study comparing fermentable sugar, total protein, total phenolics and the ferulic acid content in non-hopped worts. *J. Inst. Brew.* **2012**, *117*, 569–577. [CrossRef]
208. Vanbeneden, N.; Van Roey, T.; Willems, F.; Delvaux, F.; Delvaux, F.R. Release of phenolic flavour precursors during wort production: Influence of process parameters and grist composition on ferulic acid release during brewing. *Food Chem.* **2008**, *111*, 83–91. [CrossRef]
209. Coghe, S.; Benoot, K.; Delvaux, F.; Vanderhaegen, B.; Delvaux, F.R. Ferulic acid release and 4-vinylguaiacol formation during brewing and fermentation: Indications for feruloyl esterase activity in *Saccharomyces cerevisiae*. *J. Agric. Food Chem.* **2004**, *52*, 602–608. [CrossRef]
210. Schwarz, K.J.; Boitz, L.I.; Methner, F.J. Release of phenolic acids and amino acids during mashing dependent on temperature, pH, time and raw materials. *J. Am. Soc. Brew. Chem.* **2012**, *70*, 290–295. [CrossRef]
211. Zhao, H.; Zhao, M. Effects of mashing on total phenolic contents and antioxidant activities of malts and worts. *Int. J. Food Sci. Technol.* **2012**, *47*, 240–247. [CrossRef]
212. Fumi, M.D.; Galli, R.; Lambri, M.; Donadini, G.; De Faveri, D.M. Effect of full-scale brewing process on polyphenols in italian all-malt and maize adjunct lager beers. *J. Food Compos. Anal.* **2011**, *24*, 568–573. [CrossRef]
213. Pascoe, H.M.; Ames, J.M.; Sachin, C. Critical stages of the brewing process for changes in antioxidant activity and levels of phenolic compounds in ale. *J. Am. Soc. Brew. Chem.* **2003**, *61*, 203–209. [CrossRef]
214. Forster, A.; Gahr, A. On the fate of certain hop substances during dry hopping. *Brew. Sci.* **2013**, *66*, 93–103.
215. Wietstock, P.C.; Kunz, T.; Methner, F.J. Influence of hopping technology on oxidative stability and staling-related carbonyls in pale lager beer. *Brew. Sci.* **2016**, *69*, 73–84.
216. Brandolini, V.; Fiore, C.; Maietti, A.; Tedeschi, P.; Romano, P. Influence of *Saccharomyces cerevisiae* strains on wine total antioxidant capacity evaluated by photochemiluminescence. *World J. Microbiol. Biotechnol.* **2007**, *23*, 581–586. [CrossRef]
217. Kostadinović, S.; Wilkens, A.; Stefova, M.; Ivanova, V.; Vojnovski, B.; Mirhosseini, H.; Winterhalter, P. Stilbene levels and antioxidant activity of Vranec and Merlot wines from Macedonia: Effect of variety and enological practices. *Food Chem.* **2012**, *135*, 3003–3009. [CrossRef]
218. Carrascosa, A.V.; Bartolome, B.; Robredo, S.; Leon, A.; Cebollero, E.; Juega, M.; Nunez, Y.P.; Martinez, M.C.; Martinez-Rodriguez, A.J. Influence of locally-selected yeast on the chemical and sensorial properties of Albariño white wines. *LWT Food Sci. Technol.* **2012**, *46*, 319–325. [CrossRef]
219. Carew, A.L.; Dugald, P.S.; Close, C.; Curtin, C.; Dambergs, R.G. Yeast effects on Pinot noir wine phenolics, color, and tannin composition. *J. Agric. Food Chem.* **2013**, *61*, 9892–9898. [CrossRef]
220. Caridi, A.; Sidari, R.; Solieri, L.; Cufari, A.; Giudici, P. Wine colour adsorption phenotype: An inheritable quantitative trait loci of yeasts. *J. Appl. Microbiol.* **2007**, *103*, 735–742. [CrossRef]
221. Bautista-Ortın, A.; Fernandez-Fernandez, J.I.; Lopez-Roca, J.M.; Gomez-Plaza, E. The effects of enological practices in anthocyanins, phenolic compounds and wine colour and their dependence on grape characteristics. *J. Food Comp. Anal.* **2007**, *20*, 546–552. [CrossRef]
222. Mojsov, K.; Ziberoski, J.; Bozinovic, Z.; Petreska, M. A comparison of effects of three commercial pectolytic enzyme preparations in red winemaking. *Int. J. Pure Appl. Sci. Technol.* **2010**, *1*, 127–136. [CrossRef]

223. Río Segade, S.; Pace, C.; Torchio, F.; Giacosa, S.; Gerbi, V.; Rolle, L. Impact of maceration enzymes on skin softening and relationship with anthocyanin extraction in wine grapes with different anthocyanin profiles. *Food Res. Int.* **2015**, *71*, 50–57. [CrossRef]

224. González-Neves, G.; Favre, G.; Piccardo, D.; Gil, G. Anthocyanin profile of young red wines of Tannat, Syrah and Merlot made using maceration enzymes and cold soak. *Int. J. Food Sci. Technol.* **2016**, *51*, 260–267. [CrossRef]

225. Aguilar, T.; Loyola, C.; de Bruijn, J.; Bustamante, L.; Vergara, C.; von Baer, D.; Mardones, C.; Serra, I. Effect of thermomaceration and enzymatic maceration on phenolic compounds of grape must enriched by grape pomace, vine leaves and canes. *Eur. Food Res. Technol.* **2016**, *242*, 1149–1158. [CrossRef]

226. Waterhouse, A.L.; Sacks, G.L.; Jeffery, D.W. Topics related to aging. In *Understanding Wine Chemistry*; Waterhouse, A.L., Sacks, G.L., Jeffery, D.W., Eds.; John Wiley & Sons, Ltd.: West Sussex, UK, 2016; pp. 294–317. ISBN 978-1-118-73070-6.

227. Sarni-Manchado, P.; Fulcrand, H.; Souquet, J.M.; Cheynier, V.; Moutounet, M. Stability and color of unreported wine anthocyanin-derived pigments. *J. Food Sci.* **1996**, *61*, 938–941. [CrossRef]

228. Asenstorfer, R.E.; Hayasaka, Y.; Jones, G.P. Isolation and structures of oligomeric wine pigments by bisulfite-mediated ion-exchange chromatography. *J. Agric. Food Chem.* **2001**, *49*, 5957–5963. [CrossRef]

229. Mateus, N.; Silva, A.M.; Rivas-Gonzalo, J.C.; Santos-Buelga, C.; De Freitas, V. A new class of blue anthocyanin-derived pigments isolated from red wines. *J. Agric. Food Chem.* **2003**, *51*, 1919–1923. [CrossRef]

230. Remy, S.; Fulcrand, H.; Labarbe, B.; Cheynier, V.; Moutounet, M. First confirmation in red wine of products resulting from direct anthocyanin-tannin reactions. *J. Sci. Food Agric.* **2000**, *80*, 745–751. [CrossRef]

231. He, F.; Mu, L.; Yan, G.L.; Liang, N.N.; Pan, Q.H.; Wang, L.; Reeves, M.J.; Duan, C.Q. Biosynthesis of anthocyanins and their regulation in coloured grapes. *Molecules* **2010**, *15*, 9057–9091. [CrossRef]

232. Grainger, K.; Tattersall, H. *Wine Production: Vine to Bottle*, 2nd ed.; John Wiley & Sons, Ltd.: West Sussex, UK, 2016; pp. 138–143. ISBN 978-1-405-17354-4.

233. Glabasnia, A.; Hofmann, T. Sensory-directed identification of taste-active ellagitannins in American (*Quercus alba* L.) and European oak wood (*Quercus robur* L.) and quantitative analysis in bourbon whiskey and oak-matured red wines. *J. Agric. Food Chem.* **2006**, *54*, 3380–3390. [CrossRef]

234. Rasines-Perea, Z.; Jacquet, R.; Jourdes, M.; Quideau, S.; Teissedre, P.L. Ellagitannins and flavano-ellagitannins: Red wines tendency in different areas, barrel origin and ageing time in barrel and bottle. *Biomolecules* **2019**, *9*, 316. [CrossRef] [PubMed]

235. Michel, J.; Albertin, W.; Jourdes, M.; Le Floch, A.; Giordanengo, T.; Mourey, N.; Teissedre, P.L. Variations in oxygen and ellagitannins, and organoleptic properties of red wine aged in French oak barrels classified by a near infrared system. *Food Chem.* **2016**, *204*, 381–390. [CrossRef]

236. Navarro, M.; Kontoudakis, N.; Giordanengo, T.; Gómez-Alonso, S.; García-Romero, E.; Fort, F.; Canals, J.M.; Hermosín-Gutíerrez, I.; Zamora, F. Oxygen consumption by oak chips in a model wine solution; Influence of the botanical origin, toast level and ellagitannin content. *Food Chem.* **2016**, *199*, 822–827. [CrossRef] [PubMed]

237. García-Estévez, I.; Alcalde-Eon, C.; Martínez-Gil, A.M.; Rivas-Gonzalo, J.C.; Escribano-Bailón, M.T.; Nevares, I.; Del Alamo-Sanza, M. An approach to the study of the interactions between ellagitannins and oxygen during oak wood aging. *J. Agric. Food Chem.* **2017**, *65*, 6369–6378. [CrossRef] [PubMed]

238. Chassaing, S.; Lefeuvre, D.; Jacquet, R.; Jourdes, M.; Ducasse, L.; Galland, S.; Grelard, A.; Saucier, C.; Teissedre, P.L.; Dangles, O.; et al. Physicochemical studies of new anthocyano-ellagitannin hybrid pigments: About the origin of the influence of oak C-glycosidic ellagitannins on wine color. *Eur. J. Org. Chem.* **2010**, *2010*, 55–63. [CrossRef]

239. Zhang, B.; Cai, J.; Duan, C.Q.; Reeves, M.J.; He, F. A review of polyphenolics in oak woods. *Int. J. Mol. Sci.* **2015**, *16*, 6978–7014. [CrossRef]

240. Quideau, S.; Jourdes, M.; Lefeuvre, D.; Montaudon, D.; Saucier, C.; Glories, Y.; Pardon, P.; Pourquier, P. The chemistry of wine polyphenolic C-glycosidic ellagitannins targeting human topoisomerase II. *Chem. A Eur. J.* **2005**, *11*, 6503–6513. [CrossRef]

241. Jara-Palacios, M.J.; Rodríguez-Pulido, F.J.; Hernanz, D.; Escudero-Gilete, M.L.; Heredia, F.J. Determination of phenolic substances of seeds skins and stems from white grape marc by near-infrared hyperspectral imaging. *Aust. J. Grape Wine Res.* **2016**, *22*, 11–15. [CrossRef]

242. Jara-Palacios, M.J.; Hernanz, D.; Escudero-Gilete, M.L.; Heredia, F.J. The use of grape seed by-products rich in flavonoids to improve the antioxidant potential of red wines. *Molecules* **2016**, *21*, 1526. [CrossRef]

243. Rivero, F.J.; Gordillo, B.; Jara-Palacios, M.J.; González-Miret, M.L.; Heredia, F.J. Effect of addition of overripe seeds from white grape by-products during red wine fermentation on wine colour and phenolic composition. *LWT Food Sci. Technol.* **2017**, *84*, 544–550. [CrossRef]

244. Jara-Palacios, M.J.; Gordillo, B.; González-Miret, M.L.; Hernanz, D.; Escudero-Gilete, M.L.; Heredia, F.J. Comparative study of the enological potential of different winemaking Byproducts: Implications in the antioxidant activity and color expression of red wine anthocyanins in a model solution. *J. Agric. Food Chem.* **2014**, *62*, 6975–6983. [CrossRef] [PubMed]

245. Escudero-Gilete, M.L.; Hernanz, D.; Galán-Lorente, C.; Heredia, F.J.; Jara-Palacios, M.J. Potential of cooperage byproducts rich in ellagitannins to improve the antioxidant activity and color expression of red wine anthocyanins. *Foods* **2019**, *8*, 336. [CrossRef]

246. Vanderhaegen, B.; Neven, H.; Verachtert, H.; Derdelinckx, G. The chemistry of beer aging—A critical review. *Food Chem.* **2006**, *95*, 357–381. [CrossRef]

247. Karabín, M.; Rýparová, A.; Jelínek, L.; Kunz, T.; Wietstock, P.; Methner, F.J.; Dostálek, P. Relationship of iso-α-acid content and endogenous antioxidative potential during storage of lager beer. *J. Inst. Brew.* **2014**, *120*, 212–219. [CrossRef]

248. Li, H.; Zhao, M.; Cui, C.; Sun, W.; Zhao, H. Antioxidant activity and typical ageing compounds: Their evolutions and relationships during the storage of lager beers. *Int. J. Food Sci. Technol.* **2016**, *51*, 2026–2033. [CrossRef]

249. Moll, M.; Fonknechten, G.; Carnielo, M.; Flayeux, R. Changes in polyphenols from raw materials to finished beer. *MBAA Tech. Q.* **1984**, *21*, 79–87.

250. Callemien, D.; Collin, S. Involvement of flavanoids in beer color instability during storage. *J. Agric. Food Chem.* **2007**, *55*, 9066–9073. [CrossRef]

251. Intelmann, D.; Haseleu, G.; Dunkel, A.; Lagemann, A.; Stephan, A.; Hofmann, T. Comprehensive sensomics analysis of hop-derived bitter compounds during storage of beer. *J. Agric. Food Chem.* **2011**, *59*, 1939–1953. [CrossRef]

252. Heuberger, A.L.; Broeckling, C.D.; Lewis, M.R.; Salazar, L.; Bouckaert, P.; Prenni, J.E. Metabolomic profiling of beer reveals effect of temperature on non-volatile small molecules during short-term storage. *Food Chem.* **2012**, *135*, 1284–1289. [CrossRef]

253. Jerkovic, V.; Nguyen, F.; Timmermanns, A.; Collin, S. Comparison of procedures for resveratrol analysis in beer: Assessment of stilbenoids stability through wort fermentation and beer aging. *J. Inst. Brew.* **2008**, *114*, 143–149. [CrossRef]

254. Đorđević, S.; Popović, D.; Despotović, S.; Veljović, M.; Atanacković, M.; Cvejić, J.; Nedović, V.; Leskosek-Cukalovik, I. Extracts of medicinal plants as functional beer additives. *Chem. Ind. Chem. Eng. Q.* **2016**, *22*, 301–308. [CrossRef]

255. Ulloa, P.A.; Vidal, J.; Ávila, M.I.; Labbe, M.; Cohen, S.; Salazar, F.N. Effect of the addition of propolis extract on bioactive compounds and antioxidant activity of craft beer. *J. Chem.* **2017**, *2017*, 1–7. [CrossRef]

256. Veljović, M.; Despotović, S.; Pecić, S.; Davidović, S.; Djordjević, R.; Vukosavljević, P.; Leskosek-Cukalovik, I. The influence of raw materials and fermentation conditions on the polyphenol content of grape beer. In Proceedings of the 6th Central European Congress on Food, CEFood 2012, Novi Sad, Serbia, 23–26 May 2012; Lević, J., Nedović, V., Ilić, N., Tumbas, V., Kalušević, A., Eds.; University of Novi Sad, Institute of Food Technology: Novi Sad, Serbia, 2012; pp. 1137–1141.

257. Lasanta, C.; Durán-Guerrero, E.; Díaz, A.B.; Castro, R. Influence of fermentation temperature and yeast type on the chemical and sensory profile of handcrafted beers. *J. Sci. Food Agric.* **2020**, 1–8. [CrossRef]

258. Nardini, M.; Garaguso, I. Characterization of bioactive compounds and antioxidant activity of fruit beers. *Food Chem.* **2020**, *305*. [CrossRef]

259. Gasiński, A.; Kawa-Rygielska, J.; Szumny, A.; Czubaszek, A.; Gąsior, J.; Pietrzak, W. Volatile compounds content, physicochemical parameters, and antioxidant activity of beers with addition of mango fruit (*Mangifera Indica*). *Molecules* **2020**, *25*, 3033. [CrossRef]

Publisher's Note: MDPI stays neutral with regard to jurisdictional claims in published maps and institutional affiliations.

Article

Unraveling the Antioxidant, Binding and Health-Protecting Properties of Phenolic Compounds of Beers with Main Human Serum Proteins: In Vitro and In Silico Approaches [†]

Raja Mohamed Beema Shafreen [1], Selvaraj Alagu Lakshmi [1], Shunmugiah Karutha Pandian [1], Yong Seo Park [2], Young Mo Kim [3], Paweł Paśko [4], Joseph Deutsch [5], Elena Katrich [5] and Shela Gorinstein [5,*]

[1] Department of Biotechnology, Alagappa University, Science Campus, Karaikudi, Tamil Nadu 630003, India; beema.shafreen@gmail.com (R.M.B.S.); lakshmivinay.317@gmail.com (S.A.L.); sk_pandian@rediffmail.com (S.K.P.)

[2] Department of Horticultural Science, Mokpo National University, Muan, Jeonnam 534-729, Korea; ypark@mokpo.ac.kr

[3] Department of Food Nutrition, Gwangju Health University, Gwangsan-gu, Gwangju 506-723, Korea; bliss0816@hanmail.net

[4] Department of Food Chemistry and Nutrition, Jagiellonian University Medical College, Krakow 30-688, Poland; paskopaw@poczta.fm

[5] Institute for Drug Research, School of Pharmacy, Faculty of Medicine, The Hebrew University of Jerusalem, Jerusalem 9112001, Israel; josephd@ekmd.huji.ac.il (J.D.); ekatrich@gmail.com (E.K.)

* Correspondence: shela.gorin@mail.huji.ac.il

[†] This article is dedicated to the memory of my dear brother Prof. Simon Trakhtenberg, who died on 20 November 2011, who encouraged our research group during all his life. He carried out the clinical research on humans and consumption of beer. He was a special our friend, outstanding scientist and long time cooperator. He will always remain in our hearts.

Academic Editor: Mirella Nardini

Received: 20 September 2020; Accepted: 22 October 2020; Published: 27 October 2020

Abstract: Our recently published in vivo studies and growing evidence suggest that moderate consumption of beer possesses several health benefits, including antioxidant and cardiovascular effects. Although beer contains phenolic acids and flavonoids as the major composition, and upon consumption, the levels of major components increase in the blood, there is no report on how these beer components interact with main human serum proteins. Thus, to address the interaction potential between beer components and human serum proteins, the present study primarily aims to investigate the components of beer from different industrial sources as well as their mode of interaction through in silico analysis. The contents of the bioactive compounds, antioxidant capacities and their influence on binding properties of the main serum proteins in human metabolism (human serum albumin (HSA), plasma circulation fibrinogen (PCF), C-reactive protein (CRP) and glutathione peroxidase 3 (GPX3)) were studied. In vitro and in silico studies indicated that phenolic substances presented in beer interact with the key regions of the proteins to enhance their antioxidant and health properties. We hypothesize that moderate consumption of beer could be beneficial for patients suffering from coronary artery disease (CAD) and other health advantages by regulating the serum proteins.

Keywords: beer; phenolic compounds; antioxidants; binding; health properties; docking

---

1. Introduction

Beer is an important beverage, containing high amounts of polyphenols and showing antioxidant activity [1–4]. The phenolic compounds vary in high and low fermented, non-alcoholic and fruit beers [5–7]. It is known from a large number of reports that beer positively influences the health properties of human metabolism for protection from cardiovascular risk, lipid metabolism and antioxidant activity [8–11]. These actions depend on the antioxidant and anti-inflammatory properties of non-alcoholic compounds and slightly on the ethanol-dependent activity of beer [12,13]. Beer represents a source of phenolic compounds that could act synergistically, providing valuable data for moderate dietary beer inclusion studies [14–17]. The antioxidant properties of phenolics are responsible for the inhibition of oxidation of low density lipoprotein cholesterol. Moderate consumption of beverages in cholesterol-containing diets leads to a decrease in the content of total cholesterol in the liver in experiments on laboratory animals and in hypercholesterolemic patients [8,10–12]. Flavonoids could be linked to the beneficial effects of beer, as shown for the first time by our international research group in a number of reports in vitro and in vivo [9,18,19]. Recent studies have suggested that those flavonoids and some phenolic acids, which are abundant in beers, are present in many natural products and show health and binding properties [20,21]. Although numerous human studies have shown consistent effects of beer and other beverages on several intermediate markers for cardiovascular diseases [9,19,22–24], it is still unknown whether their action could be specifically related to polyphenols and especially to main human proteins (human serum albumin (HSA), plasma circulation fibrinogen (PCF), C-reactive protein (CRP), glutathione peroxidase 3 (GPX3)), which are relatively new biomarkers of coronary artery disease (CAD). In connection with the recent information described above, the present study aims to unveil the antioxidant capacities of phenolic compounds (total polyphenols, phenolic acids, flavonoids and flavanols), which are present in commercially available lager alcoholic beers in the context of health promotions, by in silico and in vitro analyses. The binding properties of investigated beers were determined in in vitro studies by fluorescence assays in comparison with main flavonoids and phenolic acids. Interactive behavior of the main serum proteins HSA, PCF, CRP and GPX3 with catechin, epicatechin, quercetin, ferulic and caffeic acids was also studied through molecular docking evaluation.

2. Results and Discussion

2.1. Total Polyphenols, Flavonoids, Flavanols and Phenolic Acids Content of Beers

The amounts of total polyphenols, flavonoids and flavanols in 11 beer samples are shown in Table 1.

On the basis of our published in vivo in results of health properties of moderate beer consumption [9,19,22,23], the main aim of the present study was to determine the functional properties of some individual phenolic compounds by interaction with the main human serum proteins, using fluorescence and molecular docking. Non-selective spectrometric methods were used for determination of several phenolic substances. A correlation was found between the most phenolic compounds, antioxidant and binding properties of beers. The comparison between the advanced analytical methods for determination of phenolic compounds was not the aim of this study, and in the literature there are numerous reports describing the analysis of these compounds and some of them were cited [6,7,16,25]. There are some differences and similarities in the obtained results. Total phenolic contents of low fermented lager beers were slightly lower in comparison with the previous report [5], showing the range of 373–473 mg/L of tyrosol (302–383 mg gallic acid equivalent (GAE)/L) of low fermentation of samples. The results of Amstel beer (Table 1) were higher than previously reported [5]. High fermentation beers showed a slightly higher amount of polyphenols from 453 to 599 mg/L of tyrosol (366.9–485.2 mg GAE/L), and only 'Murphys' showed 915 mg/L of tyrosol (741.2 mg GAE/L). In the Nardini et al. [6] study, the conventional lager beers showed lower polyphenol content (320.6–273.8 mg GAE/L) and total flavonoids (27–64 mg catechin equivalent (CE)/L) than in the investigated samples

(Table 1, 668.3–442.1 mg GAE/L; 35.8–52.5 mg CE/L). The obtained results of total polyphenols were between 464.3 and 539.5 mg/L GAE. Chiva-Blanch et al. [10] evaluated that the amount of polyphenols in Carlsberg (510.2 ± 15.5 mg GAE/L) was higher in comparison with the values shown in Table 1 (450.5 ± 7.5 mg GAE/L). Oppositely, in the report of Mitić et al. [16], polyphenols in Amstel and Heineken beers were 1.46 and 1.11 times lower, respectively, than in Table 1. In the study of Mitić et al. [16], the amount of flavonoids in quercetin equivalent (QE) (103.9–185.3 mg QE/L) were relatively high and did not correlate with the values of the antioxidant activities of Amstel and Heineken beers. These results differed from those presented in the previous report [26], where the total polyphenols of Maccabee beer were 345.1 ± 12.1 mg GAE/L, epicatechin −65.5 mg/L and quercetin −0.95 mg/L. The amount of total flavanols in the presently measured samples did not show direct correlation between total polyphenols and antioxidant activities (Table 1). Beer contains a complex mixture of phenolic compounds (hydroxybenzoic acids (gallic acid), hydroxycinnamic acids (ferulic acid) and flavonoids (catechin)) that have expressed high antioxidant activity [3]. It was shown as well that caffeic acid is found in the lowest concentrations than other phenolic acids, and ferulic acid and some flavonoids were the most abundant [1,3,16,17], and therefore, in investigated beers, individual phenolic compounds were determined (Table 2). The correlation between the highest (Goldstar (GOLD), Kamenitza (KAM), Rostocker (ROST)), average (Maccabee (MACC), Heineken (HEIN), Oranjeboom (ORJB), Amstel (AMST), Żywiec (ŻYW)) and the lowest (Carlsberg (CARL), Miller Genuine Draft (MGD), Corona (COR)) concentrations of total polyphenols and flavonoids (Table 1) and the amounts of caffeic and ferulic acids, catechin, epicatechin and quercetin (Table 2) was found.

Table 1. Antioxidant properties of beer samples.

Beer Code	Style	Country of Production	Alcohol Strength % Vol	Total Pol., mg GAE/L	Total Flavonoids, mg CE/L	Total Flavanols, mg CE/L	β-Carot, % AA	ABTS, mM TE
MACC	Pale lager	Israel	5.0	510.1 ± 10.1 [b]	45.1 ± 0.5 [b]	40.3 ± 1.8 [a]	28.1 ± 0.8 [b]	2.06 ± 0.01 [b]
GOLD	Dark lager	Israel	4.9	552.6 ± 9.6 [a,b]	48.9 ± 0.8 [a,b]	23.3 ± 1.4 [c]	30.7 ± 1.2 [a,b]	2.21 ± 0.02 [a,b]
HEIN	Pale lager	Israel	5.0	466.3 ± 6.2 [c]	41.3 ± 0.7 [c]	21.9 ± 1.5 [d]	25.2 ± 0.7 [c]	1.88 ± 0.02 [c]
CARL	Pale lager	Israel	5.0	450.5 ± 5.5 [c,d]	40.1 ± 0.6 [c,d]	21.2 ± 0.9 [d]	24.6 ± 0.8 [d]	1.82 ± 0.01 [c,d]
MGD	Pale lager	USA	4.6	456.7 ± 7.2 [c,d]	40.8 ± 0.9 [c,d]	16.3 ± 0.4 [e]	25.1 ± 1.1 [c]	1.85 ± 0.01 [c,d]
COR	Pale lager	Mexico	4.5	442.1 ± 4.3 [d]	35.8 ± 0.5 [d]	19.2 ± 0.8 [d,e]	24.2 ± 0.7 [d]	1.79 ± 0.03 [d]
ORJB	Pale lager	Netherlands	5.0	482.3 ± 6.8 [b,c]	42.9 ± 0.9 [b,c]	29.6 ± 1.5 [b]	26.7 ± 1.0 [b,c]	1.95 ± 0.01 [b,c]
AMST	Pale lager	Netherlands	4.1	501.3 ± 7.5 [b,c]	44.3 ± 1.3 [b,c]	21.6 ± 0.9 [d]	27.6 ± 1.1 [b,c]	2.02 ± 0.01 [b]
KAM	Pale lager	Bulgaria	4.4	647.4 ± 11.3 [a]	51.1 ± 1.4 [a]	25.4 ± 1.1 [b,c]	33.6 ± 1.3 [a]	2.61 ± 0.05 [a]
ROST	Golden lager	Germany	4.9	668.3 ± 13.3 [a]	52.5 ± 1.5 [a]	26.4 ± 1.2 [b,c]	34.5 ± 1.2 [a]	2.68 ± 0.03 [a]
ŻYW	Blond lager	Poland	5.6	471.3 ± 7.2 [c]	41.5 ± 0.9 [c]	22.2 ± 1.0 [c,d]	26.6 ± 1.0 [b,c]	1.90 ± 0.02 [b,c]

Values are means ± SD of 5 measurements; Means within a column with the different superscripts are statistically different ($p < 0.05$; Student's *t*-test). Abbreviations: Maccabee (MACC); Goldstar (GOLD); Heineken (HEIN); Carlsberg (CARL); Miller Genuine Draft (MGD); Corona (COR); Oranjeboom (ORJB); Amstel (AMST); Kamenitza (KAM); Rostocker (ROST); Żywiec (ŻYW); gallic acid equivalent (GAE); catechin equivalent (CE); total polyphenols (Total Pol.); 2,2′-azino-bis (3-ethylbenzothiazoline-6-sulfonic acid) assay (ABTS); Trolox equivalent (TE).

Table 2. Individual phenolic compounds of beer samples (mg/L).

Beer Code	Caffeic Acid	Ferulic Acid	Catechin	Epicatechin	Quercetin
MACC	2.17 ± 0.08 [b,c]	14.10 ± 0.39 [b]	3.03 ± 0.06 [b]	1.09 ± 0.08 [a,b]	1.40 ± 0.08 [b]
GOLD	2.34 ± 0.07 [b]	15.22 ± 0.54 [a,b]	3.27 ± 0.09 [a,b]	1.17 ± 0.07 [a,b]	1.52 ± 012 [a,b]
HEIN	1.97 ± 0.07 [c,d]	12.92 ± 0.45 [c,d]	2.78 ± 0.07 [b,c]	0.99 ± 0.05 [b,c]	1.24 ± 0.07 [c]
CARL	1.91 ± 0.04 [c,d]	12.48 ± 0.32 [d]	2.69 ± 0.09 [c]	0.96 ± 0.09 [b,c]	1.28 ± 0.08 [b,c]
MGD	1.94 ± 0.05 [c,d]	12.62 ± 0.35 [c,d]	2.69 ± 0.08 [c]	0.97 ± 0.06 [b,c]	1.25 ± 0.11 [b,c]
COR	1.87 ± 0.06 [d]	12.23 ± 0.44 [d]	2.64 ± 0.08 [c]	0.94 ± 0.07 [c]	1.21 ± 0.07 [c]
ORJB	2.07 ± 0.08 [c]	13.31 ± 0.54 [c]	2.83 ± 0.05 [b,c]	1.02 ± 0.07 [b]	1.32 ± 0.13 [b]
AMST	2.12 ± 0.06 [b,c]	13.89 ± 0.48 [b,c]	2.99 ± 0.09 [b,c]	1.07 ± 0.07 [b]	1.37 ± 0.01 [b]
KAM	2.73 ± 0.06 [a]	17.87 ± 0.61 [a]	3.82 ± 0.12 [a]	1.37 ± 0.08 [a]	1.77 ± 0.12 [a]
ROST	2.83 ± 0.08 [a]	18.47 ± 0.51 [a]	3.98 ± 0.15 [a]	1.42 ± 0.09 [a]	1.83 ± 0.07 [a]
ŻYW	2.08 ± 0.07 [c]	13.01 ± 0.36 [c]	2.26 ± 0.08 [d]	1.01 ± 0.07 [b]	1.29 ± 0.09 [b,c]

Values are means ± SD of 5 measurements; Means within a column with the different superscripts are statistically different ($p < 0.05$; Student's *t*-test). Abbreviations: Maccabee (MACC); Goldstar (GOLD); Heineken (HEIN); Carlsberg (CARL); Miller Genuine Draft (MGD); Corona (COR), Oranjeboom (ORJB); Amstel (AMST); Kamenitza (KAM); Rostocker (ROST); Żywiec (ŻYW).

The obtained results differ from other reports, where slightly lower estimations of ferulic acid (0.85–2.16 mg/L), catechin (0.57–1.21 mg/L) and epicatechin (0.08–0.39 mg/L) were reported [16,17,25] than determined (Table 2). Most of the reports showed that among the different phenolic acids, ferulic and gallic acids are the most copious in commercial beers (around 14 and 6 mg/mL, respectively), followed by sinapic, vanillic, caffeic, *p*-coumaric, syringic and 4-hydroxyphenylacetic acids (between 0.5 and 4.2 mg/mL) [2,6]. Gallic and ferulic acids were more than 50% of the total content of individual phenolic compounds found during beer studies and are the most reported phenolics in beer [1,3]. The comparison of the same type of beer, but produced in different countries, showed differences because of the modifications in the technological processes, raw materials and conditions of the extraction of the main components. According to the data presented in the report of Szwajgier [17], the total amounts of phenolic acids in Heineken and Corona beers were 6.78 ± 0.39 and 6.13 ± 0.43, respectively. These results differ from the ones presented in Table 2. As was shown in the same report [17], vanillic and ferulic acids exerted a lower share of total antiradical activity against free radicals than the minor phenolic acids; therefore, caffeic acid was determined in all investigated beer samples (Table 2).

2.2. Beer Antioxidant Activities

The antioxidant activities of investigated beers are presented in Table 1. The obtained results were higher than reported by Nardini and Foddai [6], where the antioxidant activities of lager beers varied and showed values by 2-azino-bis (3-ethyl-benzothiazoline-6-sulfonic acid) diammonium salt (ABTS) assay in the range of 1.5–1.8 mM Trolox equivalent (TE) in comparison with the data in Table 1 (1.8–2.7 mM TE). The present results were in accordance with the published report of Habschied et al. [25], where three different kinds of lager beers (4.7–5.2% (*v/v*) of alcohol content) showed corresponding values of ABTS tests of 1.29–2.03 mM TE/L. Low antioxidant values of beer samples such as 0.21–0.23 mM TE were reported by Mitić et al. [16]. It can be concluded that the antioxidant activities measured in conventional beers varied, but were consistent with our previous results and with the published data [2,9,26–29]. The results of the β-carotene test were in correlation with the values of the ABTS assay (Table 1). The obtained results of some investigated beers can be compared with the report of Wang et al. [28]. In this report, Heineken beer showed the amount of total polyphenols of 393.9 mg GAE/L and the corresponding ability to scavenge free radicals by 1,1-diphenyl-2-picrylhydrazyl (DPPH) assay, of 27%. The same type of beer (Table 1) produced in another country showed the amount of total polyphenols of 466.3 mg GAE/L and the ability to scavenge free radicals, using β-carotene assay, with scavenging activity of 25%. Corona beer with total polyphenols of 285 mg GAE/L and the ability to scavenge free radicals (DPPH scavenging activity of 21%) can be compared with the same type of beer in which the amount of total polyphenols was 442 mg GAE/L. The ability to scavenge free radicals by the β-carotene test was similar to the published results [28] and showed scavenging activity of 24% (Table 1). The values of the polyphenols, flavonoids and the two antioxidant assays, and the expression of the units of antioxidant activities, did not change the correlation of the presented indices in all investigated industrial beer samples. The correlation with ABTS assay was slightly higher than with the β-carotene test, suggesting that ABTS test is based on hydrogen-donating ability. This fact underlines that phenolic compounds mainly influence the antioxidant properties of beer. It is also suggested that the beer samples with relatively high ABTS values can stabilize active oxygen radicals and have better flavor stability [3,6].

2.3. Binding Properties of Beers and Some Phenolic Compounds with Main Human Proteins

The binding properties of beer samples and some individual phenolic compounds were compared in interaction with human serum albumin and plasma circulation fibrinogen (Table 3).

Table 3. Binding properties of beer samples, standard flavonoids and phenolic acids with human serum proteins.

Beer Code	λ_{em} (nm)	FI (A.U.)	Binding to HSA (%)	λ_{em} (nm)	FI (A.U.)	Binding to PCF (%)
MACC	349	731.8 ± 2.1 [c,d]	24.1 ± 2.5 [b]	347	674.9 ± 2.8 [d]	14.1 ± 0.9 [c]
GOLD	350	713.6 ± 2.6 [d]	26.0 ± 2.8 [a,b]	348	666.3 ± 1.2 [e]	15.1 ± 1.2 [b]
HEIN	347	750.9 ± 3.8 [b,c]	22.2 ± 2.7 [c]	346	684.3 ± 2.7 [c]	12.8 ± 0.8 [d]
CARL	346	759.4 ± 3.5 [b,c]	21.3 ± 2.6 [c,d]	346	683.6 ± 2.8 [c]	12.9 ± 1.1 [d]
MGD	346	755.3 ± 4.2 [b,c]	21.7 ± 1.9 [c,d]	346	685.9 ± 2.1 [c]	12.6 ± 0.9 [d]
COR	345	763.9 ± 4.3 [b,c]	20.8 ± 1.5 [d]	346	686.7 ± 2.7 [c]	12.5 ± 1.1 [d]
ORJB	348	745.1 ± 5.8 [c]	22.8 ± 1.9 [c]	346	680.0 ± 3.0 [c]	13.3 ± 1.2 [c,d]
AMST	349	735.9 ± 5.5 [c,d]	23.7 ± 1.3 [b,c]	346	676.5 ± 4.1 [d]	13.8 ± 1.1 [c,d]
KAM	350	671.3 ± 6.3 [e]	30.4 ± 1.4 [a]	350	646.7 ± 4.3 [f]	17.6 ± 1.5 [a]
ROST	351	663.2 ± 7.3 [e]	31.3 ± 1.5 [a]	350	642.4 ± 4.2 [f]	18.1 ± 1.3 [a]
ŻYW	347	748.8 ± 4.2 [c]	22.4 ± 0.9 [c]	346	682.8 ± 4.0 [c]	13.0 ± 0.9 [c,d]
EtOH	344	934.8 ± 2.9 [a]	3.1 ± 0.2 [f]	343	764.3 ± 5.9 [a,b]	2.6 ± 0.1 [f]
Catechin	348	743.8 ± 4.4 [c]	22.9 ± 1.9 [c]	344	736.9 ± 5.7 [b]	6.1 ± 0.7 [e]
Epicatechin	348	745.7 ± 4.8 [c]	22.7 ± 2.0 [c]	344	738.7 ± 5.4 [b]	6.9 ± 0.9 [e]
Quercetin	347	754.4 ± 4.7 [b,c]	21.8 ± 2.1 [c,d]	344	743.2 ± 6.1 [b]	5.3 ± 0.5 [e,f]
Caffeic acid	345	820.1 ± 3.2 [a,b]	14.9 ± 1.5 [e]	360	667.9 ± 4.9 [e]	14.9 ± 1.1 [c]
Ferulic acid	345	803.6 ± 5.2 [b]	16.7 ± 1.5 [d,e]	360	663.9 ± 4.9 [e]	15.4 ± 0.9 [b]
HSA/buffer	343	964.7 ± 3.8 [a]	-	-	-	-
PCF/buffer	-	-	-	344	784.8 ± 4.9 [a]	-

Values are means ± SD of 5 measurements; Means within a column with the different superscripts are statistically different ($p < 0.05$; Student's *t*-test). Abbreviations: Maccabee (MACC); Goldstar (GOLD); Heineken (HEIN); Carlsberg (CARL); Miller Genuine Draft (MGD); Corona (COR); Oranjeboom (ORJB); Amstel (AMST); Kamenitza (KAM); Rostocker (ROST); Żywiec (ŻYW); human serum albumin (HSA); maximum emission peak (λ_{em}); fluorescence intensity (FI); arbitral units (A.U.); plasma circulation fibrinogen (PCF).; Binding to HSA (%) and binding to PCF (%) is the % decrease of fluorescence emission of the fractions of the binding sites of the proteins occupied by the ligand.

HSA had a strong fluorescence emission peak at 343 nm, when excited with a wavelength of 280 nm. The addition of beer samples and pure phenolic compounds caused a gradual decrease in the fluorescence intensity of HSA, and the emission maximum had a red-shift of 8 nm. The principles of such measures and the obtained results (Table 3) are documented in the report of Poloni et al. [20], who used the classic indirect method of fluorescent quenching of tryptophan residues for the binding of polyphenols with porcine LDL and BSA, and where the binding data were obtained by titration of the proteins with increasing amounts of phenolic ligands. In this way, Stern Volmer plots have been obtained, and this allowed the measurement of binding constants and determination of the static nature of quenching, and the inner filter effects were negligible at the phenols concentrations used. In the present report, we have used a simplified measure to show only the decrease of fluorescence emission after the addition of a single concentration of ligands. This can be regarded as a relative measure of binding, providing that the inner filter is similarly negligible within the series of ligands. Thus, the % decrease of fluorescence represents the fraction of the binding sites of the protein occupied by the ligand, rather than the fraction of the total ligand bound to the protein (Table 3). The same report [20] showed the results of the experimental binding study using fluorescent quenching for quercetin and its 3-O-glucuronide. The albumin binding site for polyphenols had been previously identified by Dufour and Dangles [21]. Pattanayak et al. [30] and Latruffe et al. [31] reported binding properties between ellagic acid, resveratrol and other polyphenols, where phenolic acids and flavonoids effectively quenched the intrinsic fluorescence of HSA by static quenching. Leontowicz et al. [32] and Kim et al. [33] showed the binding properties of polyphenols from kiwi fruit and persimmons with HSA. All the above studies, including the present one, showed the evaluation of transport and releasing efficiency at the target site in the human physiological system since HSA is the most important carrier protein in blood serum. Our explanation of the obtained data was based on the interaction of the polyphenols and flavonoids with the main serum protein HSA. Oppositely, Poloni et al. [20] and Tung et al. [24] found that competition studies between serum albumin and LDL showed that substantial lipoprotein binding occurs even in the presence of a great molar excess of albumin, the major blood protein. The excitation of fibrinogen gave an emission maximum at 344 nm, which had a shift of 6 nm with the binding of phenolic acids and some beer samples. As can be seen from the results (Table 3), the obtained evaluation is in agreement with recent reports about the influence of ethanol with HSA and fibrinogen interaction, where the binding was in the range of 2.6–3.1%. According to some reports [32,33], the ethanolic extracts showed quenching of HSA in comparison with water extracts of about 2.9%. Ethanol has a low influence on the quenching of HSA, but in different samples of investigated beers, having high amounts of total polyphenols and flavonoids (Table 1), increasing binding percentages appeared (Table 3, binding of ROST about 31%), and corresponded to higher antioxidant activity of the product. These results are in agreement with the data reported in [34], where the efficiency of flavonoids as free radical scavengers was proved. The obtained results (Table 3) on quenching of fibrinogen with investigated samples (12.5–18.1%) are in line with other reports, where the absorption peak at about 351 nm was measured at the interaction of resveratrol, and the fluorescence intensity exhibited a decrease [35–37]. The comparison of the obtained results of quenching of HSA with the investigated samples showed about 1.7 times higher quenching than with fibrinogen, especially with flavonoids, and this is in agreement with a recent report [37]. As it was shown in Tables 1–3, there is a correlation between polyphenols, antioxidant activities and binding properties of the investigated beer samples. The low values of ethanol binding with HSA and fibrinogen once more supports the hypothesis that it is not alcohol that prevents coronary artery disease, but rather the non-alcoholic composition of beer which contains a high amount of phenolic substances [38]. In the study of Sierksma et al. [38], plasma C-reactive protein and fibrinogen levels were decreased after three weeks' consumption of beer, as compared to non-alcohol beer consumption. The conclusion of this report is that moderate alcohol consumption significantly decreased these two indices, based on anti-inflammatory action of alcohol in the protection of coronary artery disease. These conclusions are opposite to the present report and to our previous results in in vitro and in vivo studies [19,22,26]. In the present study results showed that binding properties of main human serum proteins with ethanol are not the main components in beer. Oppositely, the non-alcoholic substances prevent CAD, which was proved also by molecular docking evaluation. It is impossible to compare the pure compounds found in beer with real beer samples. The obtained binding properties of the pure standards and the beer samples are not equal (Table 3). The results are dependent on synergism of the bioactive substances in the product. From another point of view [24], it was found that low plasma concentrations make polyphenols and their metabolites poor plasma antioxidants. The concentration of these compounds in lipoproteins and cells is sufficient for polyphenols to act in the protection of heart diseases using their antioxidant properties.

2.4. Molecular Docking of Beer Components with Serum Proteins

Molecular docking studies with CRP revealed that flavonoids have achieved high dock score >58 compared to the phenolic acids. The flavonoids, epicatechin and quercetin, with the dock score of 60.268 and 58.609, respectively, have shown a similar binding pattern to CRP (Table 4).

Table 4. Molecular docking results are indicated with dock score for the flavonoids and phenolic acids against different serum proteins.

Ligand Name	Dock Score	Bond Formation	Chain	Interacting Amino Acids
Human C-Reactive Protein				
Catechin	62.693	2(H-bond), 1(Pi-sigma)	Chain A	ALA92, VAL94, ASP112
Epicatechin	60.268	3(H),1(Pi-alkyl), 1(carbon-H)	Chain A	PHE39, THR41, SER44, TYR49, TRP67, TYR73, THR90, VAL91, ALA92, VAL94, ASP112, VAL111
Ferulic acid	55.343	6(H),1(Pi-sigma), 1(Pi-alkyl), 2(carbon-H)	Chain A	TYR49, TYR73, ALA92, VAL94, ASP112
Caffeic acid	53.062	2(H),1(Pi-alkyl), 2(carbon-H)	Chain A	TYR73, VAL89, ALA92, VAL94, ASP112
Quercetin	58.609	6(H),2(Pi-Pi), 1(Pi-alkyl),1(carbon-H)	Chain A	PHE39, THR41, SER44, TYR49, TRP67, THR90, GLU88, VAL94, ASP112
Human Serum Albumin				
Catechin	53.679	2(H),1(Pi-Pi), 2(Pi-alkyl),3(carbon-H)	Chain A	ILE142,HIS146,PHE157,TYR161,ARG186, GLY189
Epicatechin	53.033	2(H),2(Pi-Pi), 1(Pi-alkyl),1(Pi-sigma)	Chain A	ILE142, HIS146, PHE149, TYR161, ARG186, GLY189, LEU115
Ferulic acid	39.165	1(Pi-Pi),1(Pi-alkyl)	Chain A	ILE142, PHE157
Caffeic acid	36.825	3(Pi-Pi),1(Pi-alkyl), 1(Van der Waals)	Chain A	ILE142, PHE157, HIS146, GLY189, LYS190
Quercetin	51.170	2(H),1(Pi-Pi),1(Pi-alkyl), 1(carbon-H), 1(Pi-sigma)	Chain A	ILE142, HIS146, PHE149, TYR161, ARG186, GLY189
Human glutathione peroxidase 3 (GPX3)				
Epicatechin	103.364	2(H),1(Pi-Pi), 3(Pi-alkyl)	Chain A	LEU46, TYR53, GLN86, ALA90
Ferulic acid	82.449	1(Pi-Pi),1(Pi-alkyl), 1(carbon-H)	Chain A	TYR53, ALA90, ASN131
Caffeic acid	83.956	1(Pi-Pi),1(Pi-alkyl), 1(carbon-H)	Chain A	TYR53, ALA90, ASN131
Quercetin	102.459	1(amide-Pi),2(Pi-alkyl), 1(Pi-lone), 1(Van der Waals)	Chain A	ALA90, ASN131, LEU46, PHE132, GLN133
Human Fibrinogen				
Ferulic acid	95.517	2(H),1(Pi-amide), 1(Pi-S), 1(carbon-H)	Chain J (α), Chain I (β), chain L (γ)	CYS19, PRO20, THR21, THR22, CYS45
Caffeic acid	95.095	3(H),1(Pi-amide), 1(Pi-S), 1(carbon-H),1(Van der Waals)	Chain J (α), Chain H (β), Chain L (γ)	CYS19, THR22, CYS45, THR78, PRO77

The most common amino acids showing interaction with epicatechin and quercetin are PHE39, THR41, SER44, TYR49, TRP67, THR90, VAL94 and ASP112 (Figure 1).

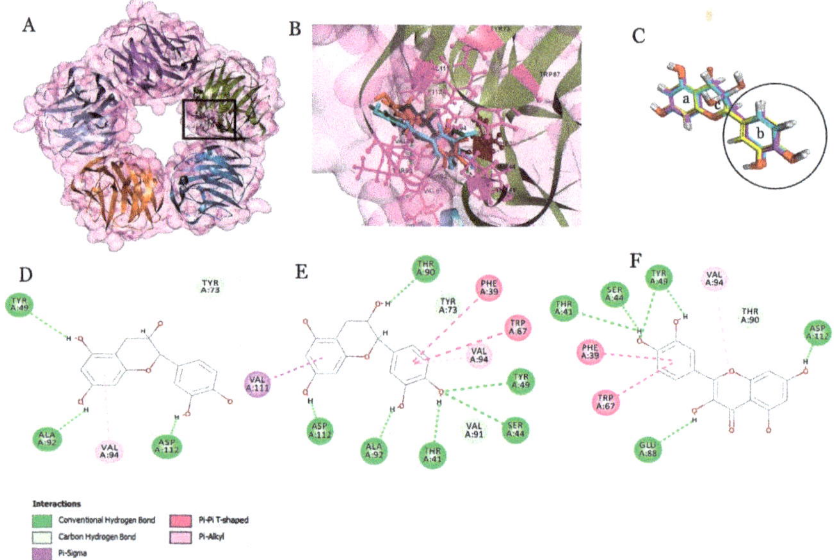

Figure 1. Molecular docking studies with C-reactive protein (CRP). (**A**) Interaction of the ligands into binding pocket (black box) of the pentameric protein; (**B**) expanded view of the binding pocket shows the interacting amino acids (ball and stick model) with the ligands; (**C**) molecular overlay of the flavonoids—catechin, epicatechin and quercetin (a, b) aromatic ring and c is the heterocyclic ring. Circle represents the aromatic ring (b) which has favored the interaction with CRP. (**D–F**) represent the 2D view for interaction of catechin, epicatechin and quercetin with CRP.

From HSA docking analysis, both flavonoids and phenolic acids had interactions in domain I, the major drug binding pocket of HSA. The crucial residues involved in binding are ILE142, HIS146, PHE149, TYR161, ARG186, GLY189 and LEU115. Among these, TYR161 are the crucial residues involved in drug recognition (Figure 2). All the three flavonoids investigated in the study have achieved the highest dock score >50. Catechin, epicatechin and quercetin have also shown consistent interaction with the key residues (TYR161). However, the phenolic acids have shown a dock score of >35. In the case of GPX3, epicatechin exhibited the highest dock score of 103.36, followed by quercetin with a dock score of 102.45 (a score identical to the dock score of epicatechin). Phenolic acids have similar dock scores, 82.448 and 83.955 for ferulic and caffeic acids, respectively.

Overall, flavonoids show the highest dock score compared to the phenolic compounds and the residues implicated in the interactions were LEU46, TYR53, GLN86, ALA90, ASN131, PHE132 and GLN133 (Figure 3).

Fibrinogen interaction with beer components revealed the possible interactions with phenolic acids with the dock score of 95.517 and 95.094 for ferulic acid and caffeic acid, respectively. The residues such as CYS19, PRO20, THR21, THR22, CYS45, THR78 and PRO77 are involved in interactions (Figure 4).

As it was mentioned previously, the secondary metabolites such as flavonoids and phenolic acids, which are found as well in beer, are investigated widely as antioxidants to prevent oxidative damage responsible for many diseases such as cancer, atherosclerosis, dyslipidaemia, chronic inflammation and other diseases [39–42]. Flavonoids and phenolic acids are well-known for their therapeutic benefits but as candidates their effectiveness still remains unclear. In the present study, the flavonoids and phenolic acids from the beer were investigated for their interactive behavior with serum proteins such as C-reactive protein (CRP), human serum albumin (HSA), GPX3 and fibrinogen through molecular docking studies.

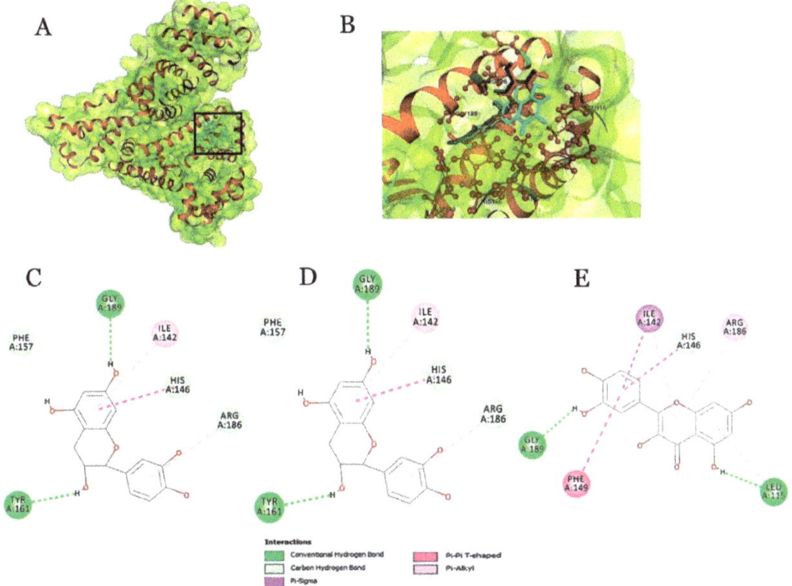

Figure 2. In silico docking of HSA. (**A**) Interaction of the ligands into binding pocket (black box) of HSA; (**B**) Expanded view representing ligands interacting with the amino acids of the receptors (ball and stick model); 2D plot for interaction of catechin (**C**), epicatechin (**D**) and quercetin (**E**) with HSA.

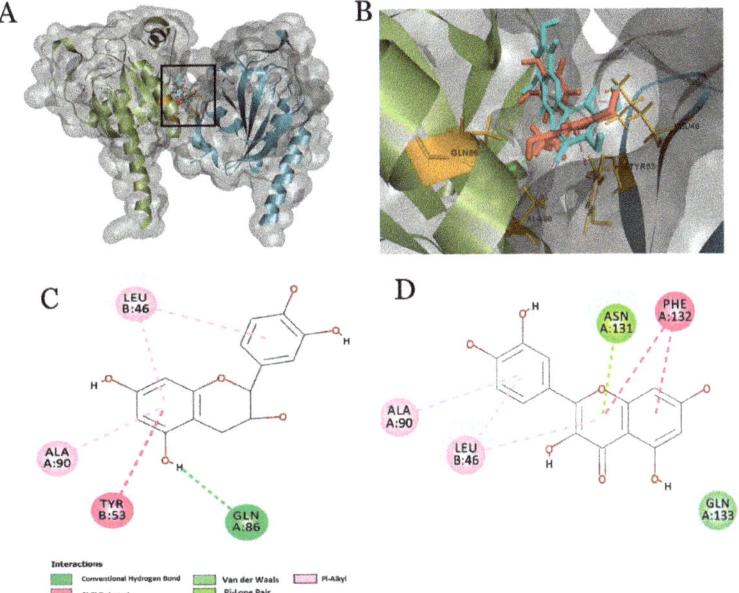

Figure 3. Interaction analysis with glutathione peroxidase 3 (GPX3) protein. (**A**) The black box represents the binding pocket of the tetrameric protein. (**B**) Expanded view shows ligand interaction with the amino acids in the binding pocket; 2D plot representing the amino acid interaction with epicatechin (**C**) and quercetin (**D**).

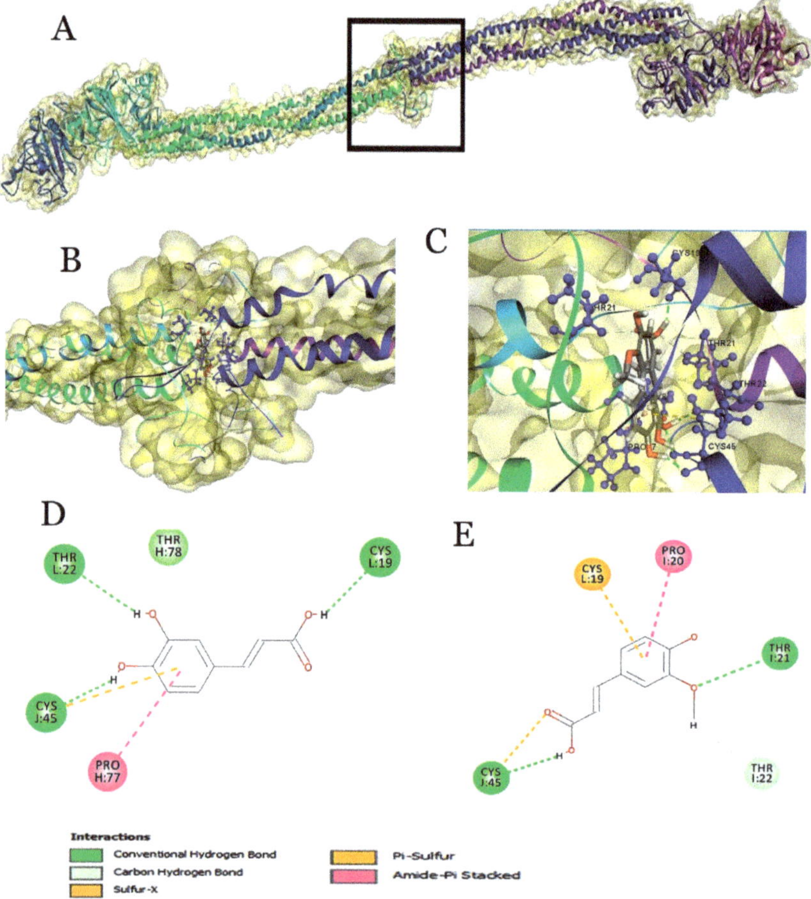

Figure 4. Molecular docking of ligands with fibrinogen. (**A**) Surface view of fibrinogen representing the central nodule (black box) present in the E region; (**B**) Expanded view of the central nodule. (**C**) Interaction of ligands with the binding pocket amino acids of fibrinogen; 2D plot showing interaction with caffeic acid (**D**) and ferulic acid (**E**).

Among the target proteins, CRP is a known biomarker detected in the human serum during inflammation as well as classified as a putative pattern recognition receptor (PPR) of the innate immune system, which indicates the invasion of the pathogens and removal of dead cells by eliciting the innate complement pathway [43–45]. CRP activates the macrophages and induces oxidative stress damage; therefore, CRP is also regarded to be itself a risk factor for cardiovascular diseases. The amount of CRP determines the risk levels of different diseases and is an indicator for cardiovascular disease (CVD), rheumatoid arthritis (RA), lupus nephritis and chronic inflammation. Moua et al. [46] reported that coffee, containing bioactive compounds, may reduce CRP levels as a biomarker of chronic inflammation. Mangnus et al. [47] showed that moderate alcohol consumption is protective against RA development and associated with lower levels of systemic inflammation in RA and with lower levels of CRP. However, autoantibodies are produced against 35–47 amino acids of CRP which is associated with the severity of the disease [48]. Thus, the epitope interacts with CRP after undergoing a conformational change.

Additionally, the residues covering from 35–47 amino acids are considered to be the important residues for therapeutics and diagnostics studies. Among them, LEU37, PHE39, TYR40 and LEU43 amino acids were found buried in the native protein and exposed only when the protein underwent conformational change of the monomeric form as epitope. Epicatechin and quercetin interacting with key residues of CRP is determined to be significant for therapeutic studies. Though catechin and epicatechin are under the same class flavan-3-ol, the binding pattern with CRP is completely different. The difference in catechin and epicatechin is mainly due to the presence of the hydroxyl groups in the β and α position of the C3, respectively (Figure 1C). Quercetin is a flavanol, and its chemical structure completely lacks the OH group in the C3 position. However, quercetin has shown similar interaction as epicatechin (flavan-3-ol) with CRP. In addition, epicatechin and quercetin were reported [30,34,44] as important dietary flavonoids with strong antioxidant properties and were investigated for their preventive role against CVD. HSA is an important biomarker which is synthesized in liver and found circulating in the blood. HSA has an indispensable role as an important antioxidant of blood and maintains the blood pH level. Besides, HSA is regarded as an important carrier for exogenous and endogenous substances. HSA also plays an important role in pharmaceuticals by binding to the drug and preventing the oxidation of the drug. However, a low level of HSA indicates the risk level of cardiovascular disease [49]. On the contrary, HSA is associated with an anti-inflammatory role, but the mechanism is unclear.

HSA is a 67 kDa protein with 585 amino acid residues. It consists of three identical domains (5–190, 191–383 and 384–585) with two drug binding sites, I and II. Site I appears at the second domain while site II appears at the third domain. Interestingly, docking of HSA with beer components revealed the interactions at the rearmost end of the first domain.

GPX3, a selenium containing glutathione peroxidase 3, is synthesized in the kidney and actively expressed in plasma. It protects the cells from oxidative stress by catalyzing the hydrogen peroxide into alcohol [50]. It has already been reported that flavonoids, in particular quercetin, interacts with GPX through in vitro studies. Besides the antioxidant potential of the quercetin–GPX complex, it has also been reported for cytotoxicity effect. However, there are no clear reports on how flavonoids bind to GPX3 at the molecular level. Here, docking with beer components revealed that flavonoids such as quercetin and epicatechin have a higher affinity toward GPX3 (than polyphenolic acids) and this observation was well-consistent with the previous in vitro report of Nagata et al. [51], wherein they have shown the interaction of endogenous GPX with flavonoids in rat BL9 (hepatocyte) cells through in vitro assays.

Additionally, they have shown that synergistic interaction of flavonoids and GPX are critical factors for enhancing their antioxidant activities. Here, flavonoids exhibited good interactions with GPX3. Nonetheless, the residues that make interactions with GPX3 partially differ among them which indicates that each flavonoid may have a differential binding region. In the case of phenolic acids, the residues implicated in binding are similar, suggesting that phenolic acids may interact with GPX3 in a similar fashion. Surprisingly, the overall interactions were observed at the adjacent region to the active site residue Seu-73 of GPX3. But how these flavonoids binding alter the conformations of GPX3 to activate the enzymes requires a comprehensive study. Overall, our results suggest that these flavonoids interact in the distal region of the GPX3 active site and account for antioxidant potential in the plasma. Moreover, high levels of such interactions with GPX protein can enhance the GPX3 activity. This enhancement has a beneficial role in reducing the risk of cardiovascular and chronic kidney diseases [52]. It also delays the aging process as aging occurs mainly due to the decline in GPX3 [53]. Fibrinogen is a glycoprotein, which circulates in blood plasma and is synthesized by the liver [54–56]. It comprises two sets of non-identical polypeptide chains termed α, β and γ ($\alpha_2\beta_2\gamma_2$). An enzymatic conversion of fibrinogen into fibrin by thrombin is one of the critical steps for maintaining the homeostasis of blood. Here, ferulic and caffeic acids made interactions with all three α, β and γ chains (Table 4). When compared to flavonoids, phenolic acids (ferulic and caffeic acids) exerted higher affinity towards the central nodule of the E region of fibrinogen. In general, hydroxycinnamic acids are illustrious for antioxidant property. Thus, based on our results, we postulate that phenolic acids have an important role in interacting with fibrinogen than flavonoids.

The present results are in good agreement with the fluorescence measurements and literature report of Luo et al. [55], in which three type II phenolic acids (caffeic, *p*-hydroxycinnamic and ferulic acids) were used to synthesize a total of 18 phenolic acid derivatives. With molecular docking for molecule design and the evaluation of haemostatic and anticoagulant activities with blood assays, the data of Luo et al. [55] indicated that caffeic acid derivatives showed certain anticoagulant or procoagulant activities and that two other series contained compounds with the best anticoagulant activities (Table 4). The interaction of fibrinogen with investigated flavonoids and their docking is in line with other reports [56], where six compounds, including quercetin, catechin and epicatechin, were examined for the inhibition of thrombin amidolytic activity. Quercetin, catechin and epicatechin caused the inhibition of thrombin amidolytic activity and only quercetin from the three mentioned above changed thrombin proteolytic activity. It is possible that these compounds can change the activity of thrombin. From another point of view, most phenolic substances are not stable in vivo and their bioavailability in the digestive tract is relatively low. From a number of previous and present experiments (Table 3), it was proved that polyphenol compounds can also bind with many components of blood plasma (mainly by human serum albumin) and the real effect of these compounds on coagulation may be mediated also by a different mechanism than their action on thrombin [56].

As was mentioned previously, the present study was aimed at investigating the interaction of individual components with different serum proteins that are responsible for the health benefits. Overall, the study indicates that beer components such as flavonoids and phenolic acids interact with the key regions of the proteins to enhance their antioxidant and binding properties. Among them, flavonoids have a significant role in enhancing the beneficial properties. It has already been reported that consumption of beer increases the flavonoid and phenolic acid content in the plasma and thereby promotes the cardiovascular health benefits [8–12,18,19,22,24].

3. Materials and Methods

3.1. Materials

Caffeic and ferulic acids, catechin, epicatechin, quercetin, Trolox, human serum albumin, fibrinogen, sodium nitrite, aluminum chloride, potassium peroxodisulfate and 2,2′-azino-bis(3-ethylbenzothiazoline-6-sulfonic acid) diammonium salt (ABTS), were from Sigma (St. Louis, MO, USA). Standard phenolics were dissolved in methanol (1 mg/mL), stored at −80 °C.

3.2. Samples

Commercial beer bottles were purchased at markets and beer shops and were investigated in this study. The eleven beers were common lager beers from different countries of production (Maccabee (MACC); Goldstar (GOLD); Heineken (HEIN); Carlsberg (CARL); Miller Genuine Draft (MGD); Corona (COR); Oranjeboom (ORJB); Amstel (AMST); Kamenitza (KAM); Rostocker (ROST); Żywiec (ŻYW)). Every sample was bought in triplicate, from the same batch and with the identical shelf life. The sample set included craft and mainstream beer varieties with alcohol by volume ranging from 4.1 to 5.6%. Four beer samples were produced in Israel. Beer bottles were stored in the dark and analyzed immediately after opening. All beer samples were first degasified and then pH was adjusted to 7.0 before analysis with additions of an appropriate amount of 0.1 M sodium hydrogen phosphate solution. Separate samples from the same bottle were frozen at −80 °C for antioxidant status and bioactivity.

3.3. Analyses of Bioactive Compounds

The total polyphenols were determined by the Folin–Ciocalteu method [57], where beer samples were diluted with distilled water till 1 mL, then 0.1 mL of Folin–Ciocalteu's reagent was added. After 5 min, 0.2 mL sodium carbonate (35% *w/v*) was added. Final volume was adjusted to 2 mL with distilled water. After 1 h in the dark, absorbance at 765 nm was measured against an appropriate blank reagent. The results were expressed as milligrams of gallic acid equivalents per liter of beer.

Total flavonoids were determined in 0.05 mL aliquots of the sample using the spectrophotometric method [58], where beer samples were diluted with distilled water to a final volume of 1.5 mL, and then 0.075 mL of 5% $NaNO_2$ solution was added. After 6 min, 0.15 mL of 10% $AlCl_3$ hexahydrate was added and allowed to stand for an additional 5 min, before 0.5 mL 1 M NaOH was added. The volume was adjusted to 2.5 mL with distilled water, mixed, and absorbance at 510 nm was measured immediately. The results are expressed as milligrams of catechin equivalents per liter of beer.

Total flavanols (TFLs) were estimated using the *p*-dimethylaminocinnamaldehyde (DMACA) method, where 0.2 mL of beer was introduced into a 1.5 mL Eppendorf tube, and 1 mL of DMACA solution was added. The mixture was vortexed and allowed to react at room temperature for 10 min. The absorbance at 640 nm was then read against a blank prepared similarly without DMACA. The presence of flavanols on the nuclei with subsequent staining with the DMACA reagent resulted in an intense blue coloration in beer [59].

Some phenolic acids (ferulic and caffeic) and flavonoids (catechin, epicatechin and quercetin) were determined with a HPLC system [3,6,13,60]. The phenolic compounds from beer samples were extracted according to the procedures, described by Nardini and Foddai [6], Bartolomé et al. [13] and Pozo-Bayon et al. [60]. A volume of 50 mL of each of 11 beer samples was extracted three times with 25 mL of diethyl ether and then three times with 25 mL of diethyl acetate, and the organic fractions were combined. After 30 min of drying with anhydrous Na_2SO_4, the extract was filtered through a Whatman-40 filter and evaporated to dryness in a rotary evaporator. The residue was dissolved in 2 mL of methanol/water (1:1, *v/v*) and analyzed by high-performance liquid chromatography (HPLC), according to the conditions described in the Bartolomé et al. [13] report. A Waters (Milford, MA, USA) chromatograph equipped with a 600-MS controller, a 717 plus autosampler and a 996 photodiode-array detector was used. A gradient of solvent A (water/acetic acid, 98:2, *v/v*) and solvent B (water/acetonitrile/acetic acid, 78:20:2, *v/v/v*) was applied to a reverse-phase Nova-pack C18 column (30 cm × 3.9 mm Internal Diameter (I. D.)), as following as follows: 0–55 min, 80% B linear, 1.1 mL/min; 55–57 min, 90% B linear, 1.2 mL/min; 57–70 min, 90% B isocratic, 1.2 mL/min; 70–80 min, 95% B linear, 1.2 mL/min; 80–90 min, 100% B linear, 1.2 mL/min; 90–120 min. For HPLC analysis, an aliquot (50 μL) was injected onto the column and eluted at the temperature of 20 °C. Samples were prepared and analyzed in duplicate.

3.4. Determination of Antioxidant Activities

The total antioxidant activity of beers was evaluated by the ABTS radical cation decolorization (ABTS) assay [61] and β-carotene bleaching assay on 0.01 mL of beer samples. The ABTS radical cation was formed by ABTS solution (7 mM) with potassium persulfate (2.45 mM) in distilled water at room temperature, for 16 h before use. A working solution (ABTS reagent) was diluted to obtain absorbance values of 0.7 at 734 nm and equilibrated at 30 °C. After addition of ABTS solution, the absorbance reading was taken 1 min after initial mixing, and up to 6 min percentage inhibition of absorbance was calculated with reference to a Trolox calibration curve and evaluated as mM Trolox equivalent/L of beer.

In an antioxidant assay using a β-carotene linoleate model system, 4 mL of emulsion containing β-carotene (0.2 mg) in 0.2 mL of chloroform, linoleic acid (20 mg) and Tween-40 (200 mg) was mixed, and then the chloroform was removed at 40 °C under vacuum. The resulting mixture was diluted with 10 mL of water. To this emulsion was added 40 mL of oxygenated water. The emulsion (4 mL) was added to the investigated sample. The absorbance at 470 nm was taken for 120 min at an interval of 20 min. The synthetic antioxidant butylated hydroxyanisole (BHA) in EtOH was used for comparative purposes and added to the sample. The antioxidant activity (AA) of the samples was evaluated in terms of bleaching of the β-carotene [62].

3.5. Fluorimetric Measurements

Two-dimensional fluorescence (2D-FL) measurements for all beer samples were recorded on a model FP-6500, Jasco spectrofluorometer, serial N261332, Tokyo, Japan, equipped with 1.0 cm quartz cells and a thermostat bath. The 2D-FL measurements were taken at emission wavelengths from 310 to 500 nm and at excitation of 295 nm. For comparison of the obtained results, caffeic and ferulic acids, catechin, epicatechin and quercetin were used [31]. The solutions for the reaction were in the following concentrations: 1.0×10^{-5} mol/L HSA; 0.05 mol/L Tris HCl buffer with 0.1 mol/L NaCl, pH 7.4. Fibrinogen stock solution was made by dissolving in phosphate buffer (10 mM, pH 7.4) to obtain a concentration of 20 µM. The initial fluorescence intensities of HSA and PCF were measured before the interaction with the investigated samples and pure substances and after interaction with the samples (quenching of fluorescence emission of proteins in our case of HSA and fibrinogen) upon addition of pure phenolic compounds or samples from beer. The differences of the measured fluorescence intensities were used for calculation of the relative binding properties, because the ligands were used only in one concentration, and the decrease of fluorescence represents the fraction of the binding sites of the protein by the ligand [20,32,33].

3.6. Molecular Docking Studies Using Main Human Serum Proteins

Crystal structures of human C-reactive protein (CRP) (PDB ID: 1B09), human serum albumin (HSA) (PDB ID: 1H9Z), human glutathione peroxidase 3 (GPX3) (PDB ID: 2R37) and human fibrinogen (PDB ID: 3GHG) with a resolution of 2.5 Å, 2.5 Å, 1.85 Å and 2.9 Å, respectively, was obtained in PDB format from the PDB database. Similarly, flavonoids and phenolic acids reported in the study were downloaded from the PubChem database in SDF format. The protein and ligand structures were minimized by applying a CHARMM force field and the spherical cut-off radius of 13.0 Å was set for non-bonded interaction. All other parameters were set to their defaults. The potential binding site region of the target proteins were determined using 'Define and edit binding site' protocol. The active site for CRP, HSA, GPX3 and fibrinogen was determined at site 1 (grid box, X: 142.694, Y: 153.060, Z: 30.358), site 1 (grid box, X: 37.172, Y: 10.895, Z: 13.554), site 1 (grid box, X: 21.467, Y: −1.923, Z: −13.674000) and site 5 (grid box, X: 103.163, Y: −40.380, Z: −92.422), respectively. The LigandFit module from Discovery Studio 2.5 (DS2.5) was used for performing the docking studies. Based on the scoring functions, the top scoring ligands resulting with best pose were extracted and analyzed through BIOVIA-DS 17 R2 client [63].

3.7. Statistical Analysis

All results were calculated as the mean with standard deviations. Comparison of the mean values was performed using Duncan's Multiple Range Test. All analyses were performed in five replicates.

4. Conclusions

As the health promoting advantages of beer was mentioned previously in various reports, the present study was aimed at investigating the interaction of individual components of beer with different serum proteins that are responsible for health benefits. Overall, the study unveiled that beer components enriched with flavonoids and phenolic acids interact with the key regions of the serum proteins to enhance their antioxidant and binding properties. Among them, flavonoids had a significant role in enhancing the beneficial properties. It has already been reported that consumption of beer increases the flavonoid and phenolic acid content in the plasma and thereby promotes the cardiovascular health benefits. Our study unveiled that phenolic acids and flavonoids might exert an appreciable health benefit by making contact with serum proteins and significantly contribute to maintain the endogenous redox homeostasis in host. However, in the case of excessive beer consumption, how it exerts deleterious effects needs an elaborate study.

Author Contributions: Conceptualization, visualization, writing—original draft, R.M.B.S., S.K.P. and S.G.; investigation: J.D., E.K. and S.A.L.; data curation, formal analysis, Y.S.P., Y.M.K. and P.P.; software, validation, resources: Y.S.P., Y.M.K. and P.P.; methodology: J.D., E.K. and S.A.L.; supervision, S.G.; reviewing and editing, R.M.B.S., S.K.P. and S.G. All authors have read and agreed to the published version of the manuscript.

Funding: This research receives no external funding.

Acknowledgments: The authors R.M.B.S., S.A. and S.K.P. sincerely acknowledge the computational and bioinformatics facility provided by the Bioinformatics Infrastructure Facility (funded by DBT, GOI; File No. BT/BI/25/012/2012, BIF). R.M.B.S is thankful to RUSA 2.0 (F.24-51/2014-U, Policy (TN Multi-Gen), Dept. of Edn., GoI) for supporting her as RUSA-PDF. The authors are thankful to Judy Siegel-Itzkovich, Breaking Israel News health and science senior reporter, award-winning health & science journalist and translator, for helping in English-style corrections.

Conflicts of Interest: The authors declare that they have no known competing financial interests or personal relationships that could have appeared to influence the work reported in this paper.

References

1. Nardini, M.; Ghiselli, A. Determination of free and bound phenolic acids in beer. *Food Chem.* **2004**, *84*, 137–143. [CrossRef]

2. Piazzon, A.; Forte, M.; Nardini, M. Characterization of Phenolics Content and Antioxidant Activity of Different Beer Types. *J. Agric. Food Chem.* **2010**, *58*, 10677–10683. [CrossRef] [PubMed]

3. Zhao, H.; Chen, W.; Lu, J.; Zhao, M. Phenolic profiles and antioxidant activities of commercial beers. *Food Chem.* **2010**, *119*, 1150–1158. [CrossRef]

4. Wannenmacher, J.; Gastl, M.; Becker, T. Phenolic Substances in Beer: Structural Diversity, Reactive Potential and Relevance for Brewing Process and Beer Quality. *Compr. Rev. Food Sci. Food Saf.* **2018**, *17*, 953–988. [CrossRef]

5. Cerrato-Alvarez, M.; Bernalte, E.; Bernalte-García, M.J.; Pinilla-Gil, E. Fast and direct amperometric analysis of polyphenols in beers using tyrosinase-modified screen-printed gold nanoparticles biosensors. *Talanta* **2019**, *193*, 93–99. [CrossRef] [PubMed]

6. Nardini, M.; Foddai, M.S. Phenolics Profile and Antioxidant Activity of Special Beers. *Molecules* **2020**, *25*, 2466. [CrossRef] [PubMed]

7. Nardini, M.; Garaguso, I. Characterization of bioactive compounds and antioxidant activity of fruit beers. *Food Chem.* **2020**, *305*, 125437. [CrossRef]

8. Nardini, M.; Natella, F.; Scaccini, C.; Ghiselli, A. Phenolic acids from beer are absorbed and extensively metabolized in humans. *J. Nutr. Biochem.* **2006**, *17*, 14–22. [CrossRef]

9. Gorinstein, S.; Caspi, A.; Libman, I.; Leontowicz, H.; Leontowicz, M.; Tashma, Z.; Katrich, E.; Jastrzebski, Z.; Trakhtenberg, S. Bioactivity of beer and its influence on human metabolism. *Int. J. Food Sci. Nutr.* **2007**, *58*, 94–107. [CrossRef]

10. Chiva-Blanch, G.; Magraner, E.; Condines, X.; Valderas-Martínez, P.; Roth, I.; Arranz, S.; Casas, R.; Navarro, M.; Hervas, A.; Sisó, A.; et al. Effects of alcohol and polyphenols from beer on atherosclerotic biomarkers in high cardiovascular risk men: A randomized feeding trial. *Nutr. Metab. Cardiovasc. Dis.* **2015**, *25*, 36–45. [CrossRef]

11. Spaggiari, G.; Cignarelli, A.; Sansone, A.; Baldi, M.; Santi, D. To beer or not to beer: A meta-analysis of the effects of beer consumption on cardiovascular health. *PLoS ONE* **2020**, *15*, e0233619. [CrossRef]

12. Osorio-Paz, I.; Brunauer, R.; Alavez, S. Beer and its non-alcoholic compounds in health and disease. *Crit. Rev. Food Sci. Nutr.* **2019**, 1–14. [CrossRef] [PubMed]

13. Bartolomé, B.; Peña-Neira, A.; Gómez-Cordovés, C. Phenolics and related substances in alcohol-free beers. *Eur. Food Res. Technol.* **2000**, *210*, 419–423. [CrossRef]

14. Humia, B.V.; Santos, K.S.; Barbosa, A.M.; Sawata, M.; Mendonça, M.C.; Padilha, F.F. Beer Molecules and Its Sensory and Biological Properties: A Review. *Molecules* **2019**, *24*, 1568. [CrossRef] [PubMed]

15. Boronat, A.; Soldevila-Domenech, N.; Rodríguez-Morató, J.; Martínez-Huélamo, M.; Lamuela-Raventós, R.; De La Torre, R. Beer Phenolic Composition of Simple Phenols, Prenylated Flavonoids and Alkylresorcinols. *Molecules* **2020**, *25*, 2582. [CrossRef] [PubMed]

16. Mitić, S.S.; Paunović, D.Đ.; Pavlović, A.N.; Tošić, S.B.; Stojković, M.B.; Mitić, M. Phenolic Profiles and Total Antioxidant Capacity of Marketed Beers in Serbia. *Int. J. Food Prop.* **2013**, *17*, 908–922. [CrossRef]

17. Szwajgier, D. Content of Individual Phenolic Acids in Worts and Beers and their Possible Contribution to the Antiradical Activity of Beer. *J. Inst. Brew.* **2009**, *115*, 243–252. [CrossRef]

18. Gasowski, B.; Leontowicz, M.; Leontowicz, H.; Katrich, E.; Lojek, A.; Číž, M.; Trakhtenberg, S.; Gorinstein, S. The influence of beer with different antioxidant potential on plasma lipids, plasma antioxidant capacity, and bile excretion of rats fed cholesterol-containing and cholesterol-free diets. *J. Nutr. Biochem.* **2004**, *15*, 527–533. [CrossRef]

19. Gorinstein, S.; Zemser, M.; Berliner, M.; Goldstein, R.; Libman, I.; Trakhtenberg, S.; Caspi, A. Moderate beer consumption and positive biochemical changes in patients with coronary atherosclerosis. *J. Intern. Med.* **1997**, *242*, 219–224. [CrossRef]

20. Poloni, D.M.; Dangles, O.; Vinson, J.A. Binding of Plant Polyphenols to Serum Albumin and LDL: Healthy Implications for Heart Disease. *J. Agric. Food Chem.* **2019**, *67*, 9139–9147. [CrossRef]

21. Dufour, C.; Dangles, O. Flavonoid–serum albumin complexation: Determination of binding constants and binding sites by fluorescence spectroscopy. *Biochim. Biophys. Acta* **2005**, *1721*, 164–173. [CrossRef]

22. Gorinstein, S.; Caspi, A.; Zemser, M.; Libman, I.; Goshev, I.; Trakhtenberg, S. Plasma circulating fibrinogen stability and moderate beer consumption. *J. Nutr. Biochem.* **2003**, *14*, 710–716. [CrossRef] [PubMed]

23. Gorinstein, S.; Caspi, A.; Rosen, A.; Goshev, I.; Zemser, M.; Weisz, M.; Añon, M.C.; Libman, I.; Lerner, H.T.; Trakhtenberg, S. Structure characterization of human serum proteins in solution and dry state. *J. Peptide Res.* **2002**, *59*, 71–78. [CrossRef]

24. Tung, W.-C.; Rizzo, B.; Dabbagh, Y.; Saraswat, S.; Romanczyk, M.; Codorniu-Hernández, E.; Rebollido-Rios, R.; Needs, P.W.; Kroon, P.A.; Rakotomanomana, N.; et al. Polyphenols bind to low density lipoprotein at biologically relevant concentrations that are protective for heart disease. *Arch. Biochem. Biophys.* **2020**, *694*, 108589. [CrossRef] [PubMed]

25. Habschied, K.; Lončarić, A.; Mastanjević, K. Screening of Polyphenols and Antioxidative Activity in Industrial Beers. *Foods* **2020**, *9*, 238. [CrossRef] [PubMed]

26. Gorinstein, S.; Caspi, A.; Zemser, M.; Trakhtenberg, S. Comparative contents of some phenolics in beer, red and white wines. *Nutr. Res.* **2000**, *20*, 131–139. [CrossRef]

27. Dvorakova, M.; Hulin, P.; Karabin, M.; Dostálek, P. Determination of polyphenols in beer by an effective method based on solid-phase extraction and high performance liquid chromatography with diode-array detection. *Czech J. Food Sci.* **2008**, *25*, 182–188. [CrossRef]

28. Wang, C.; Xie, Y.; Wang, H.; Bai, Y.; Dai, C.; Li, C.; Xu, X.; Zhou, G. Phenolic compounds in beer inhibit formation of polycyclic aromatic hydrocarbons from charcoal-grilled chicken wings. *Food Chem.* **2019**, *294*, 578–586. [CrossRef]

29. Zhao, H.; Li, H.; Sun, G.; Yang, B.; Zhao, M. Assessment of endogenous antioxidative compounds and antioxidant activities of lager beers. *J. Sci. Food Agric.* **2012**, *93*, 910–917. [CrossRef]

30. Pattanayak, R.; Basak, P.; Sen, S.; Bhattacharyya, M. An insight to the binding of ellagic acid with human serum albumin using spectroscopic and isothermal calorimetry studies. *Biochem. Biophys. Rep.* **2017**, *10*, 88–93. [CrossRef]

31. Latruffe, N.; Menzel, M.; Delmas, D.; Buchet, R.; Lançon, A. Compared Binding Properties between Resveratrol and Other Polyphenols to Plasmatic Albumin: Consequences for the Health Protecting Effect of Dietary Plant Microcomponents. *Molecules* **2014**, *19*, 17066–17077. [CrossRef]

32. Leontowicz, H.; Leontowicz, M.; Latocha, P.; Jesion, I.; Park, Y.S.; Katrich, E.; Barasch, D.; Nemirovski, A.; Gorinstein, S. Bioactivity and nutritional properties of hardy kiwifruit *Actinidia arguta* in comparison with *Actinidia deliciosa* 'Hayward' and *Actinidia eriantha* 'Bidan'. *Food Chem.* **2016**, *196*, 281–291. [CrossRef] [PubMed]

33. Kim, Y.M.; Park, Y.S.; Park, Y.-K.; Ham, K.-S.; Kang, S.-G.; Shafreen, R.M.B.; Lakshmi, S.A.; Gorinstein, S. Characterization of Bioactive Ligands with Antioxidant Properties of Kiwifruit and Persimmon Cultivars Using In Vitro and in Silico Studies. *Appl. Sci.* **2020**, *10*, 4218. [CrossRef]

34. Qian, J. The efficiency of flavonoids in polar extracts of Lycium chinense Mill fruits as free radical scavenger. *Food Chem.* **2004**, *87*, 283–288. [CrossRef]

35. Gonçalves, S.; Santos, N.C.; Martins-Silva, J.; Saldanha, C. Fluorescence spectroscopy evaluation of fibrinogen–β-estradiol binding. *J. Photochem. Photobiol. B Biol.* **2007**, *86*, 170–176. [CrossRef]

36. Zhang, J.; Dai, X.-F.; Huang, J.-Y. Resveratrol Binding to Fibrinogen and its Biological Implication. *Food Biophys.* **2011**, *7*, 35–42. [CrossRef]

37. Gligorijević, N.; Radomirović, M.; Rajkovic, A.; Nedić, O.; Cirkovic-Velickovic, T. Fibrinogen Increases Resveratrol Solubility and Prevents it from Oxidation. *Foods* **2020**, *9*, 780. [CrossRef]

38. Sierksma, A.; Van Der Gaag, M.S.; Kluft, C.; Hendriks, H.F.J. Moderate alcohol consumption reduces plasma C-reactive protein and fibrinogen levels; a randomized, diet-controlled intervention study. *Eur. J. Clin. Nutr.* **2002**, *56*, 1130–1136. [CrossRef]

39. Saxena, M.; Saxena, J.; Pradhan, A. Flavonoids and phenolic acids as antioxidants in plants and human health. *Int. J. Pharm. Sci. Rev. Res.* **2012**, *16*, 130–134.

40. Ghiselli, A.; Natella, F.; Guidi, A.; Montanari, L.; Fantozzi, P.; Scaccini, C. Beer increases plasma antioxidant capacity in humans. *J. Nutr. Biochem.* **2000**, *11*, 76–80. [CrossRef]

41. Arranz, S.; Chiva-Blanch, G.; Valderas-Martínez, P.; Remon, A.M.; Raventós, R.M.L.; Estruch, R. Wine, Beer, Alcohol and Polyphenols on Cardiovascular Disease and Cancer. *Nutrients* **2012**, *4*, 759–781. [CrossRef] [PubMed]

42. Padro, T.; Muñoz-García, N.; Vilahur, G.; Chagas, P.; Deyà, A.; Antonijoan, R.M.; Badimon, L. Moderate Beer Intake and Cardiovascular Health in Overweight Individuals. *Nutrients* **2018**, *10*, 1237. [CrossRef] [PubMed]

43. Kumaresan, P.R.; Devaraj, S.; Huang, W.; Lau, E.Y.; Liu, R.; Lam, K.S.; Jialal, I. Synthesis and Characterization of a Novel Inhibitor of C-Reactive Protein–Mediated Proinflammatory Effects. *Metab. Syndr. Relat. Disord.* **2013**, *11*, 177–184. [CrossRef] [PubMed]

44. Morrison, M.; Van Der Heijden, R.; Heeringa, P.; Kaijzel, E.; Verschuren, L.; Blomhoff, R.; Kooistra, T.; Kleemann, R. Epicatechin attenuates atherosclerosis and exerts anti-inflammatory effects on diet-induced human-CRP and NFκB in vivo. *Atherosclerosis* **2014**, *233*, 149–156. [CrossRef]

45. Nagai, T.; Anzai, T.; Kaneko, H.; Mano, Y.; Anzai, A.; Maekawa, Y.; Takahashi, T.; Meguro, T.; Yoshikawa, T.; Fukuda, K. C-reactive protein overexpression exacerbates pressure overload-induced cardiac remodeling through enhanced inflammatory response. *Hypertension* **2011**, *57*, 208–215. [CrossRef]

46. Moua, E.D.; Hu, C.; Day, N.; Hord, N.G.; Takata, Y. Coffee Consumption and C-Reactive Protein Levels: A Systematic Review and Meta-Analysis. *Nutrients* **2020**, *12*, 1349. [CrossRef]

47. Mangnus, L.; Van Steenbergen, H.W.; Nieuwenhuis, W.P.; Reijnierse, M.; Mil, A.H.M.V.D.H.-V. Moderate use of alcohol is associated with lower levels of C reactive protein but not with less severe joint inflammation: A cross-sectional study in early RA and healthy volunteers. *RMD Open* **2018**, *4*, e000577. [CrossRef]

48. Li, Q.-Y.; Li, H.-Y.; Fu, G.; Yu, F.; Wu, Y.; Zhao, M.-H. Autoantibodies against C-Reactive Protein Influence Complement Activation and Clinical Course in Lupus Nephritis. *J. Am. Soc. Nephrol.* **2017**, *28*, 3044–3054. [CrossRef]

49. Arques, S. Human serum albumin in cardiovascular diseases. *Eur. J. Intern. Med.* **2018**, *52*, 8–12. [CrossRef]

50. Chung, S.S.; Kim, M.; Youn, B.S.; Lee, N.S.; Park, J.W.; Lee, I.K.; Lee, Y.S.; Kim, J.B.; Cho, Y.M.; Lee, H.K.; et al. Glutathione peroxidase 3 mediates the antioxidant effect of peroxisome proliferator-activated receptor gamma in human skeletal muscle cells. *Mol. Cell Biol.* **2009**, *29*, 20–30. [CrossRef]

51. Nagata, H.; Takekoshi, S.; Takagi, T.; Honma, T.; Watanabe, K. Antioxidative action of flavonoids, quercetin and catechin, mediated by the activation of glutathione peroxidase. *Tokai J. Exp. Clin. Med.* **1999**, *24*, 1–11. [PubMed]

52. Pang, P.; Abbott, M.; Abdi, M.; Fucci, Q.-A.; Chauhan, N.; Mistri, M.; Proctor, B.; Chin, M.; Wang, B.; Yin, W.; et al. Pre-clinical model of severe glutathione peroxidase-3 deficiency and chronic kidney disease results in coronary artery thrombosis and depressed left ventricular function. *Nephrol. Dial. Transplant.* **2017**, *33*, 923–934. [CrossRef]

53. Pastori, D.; Pignatelli, P.; Farcomeni, A.; Menichelli, D.; Nocella, C.; Carnevale, R.; Violi, F. Aging-Related Decline of Glutathione Peroxidase 3 and Risk of Cardiovascular Events in Patients with Atrial Fibrillation. *J. Am. Hear. Assoc.* **2016**, *5*, e003682. [CrossRef] [PubMed]

54. Domingues, M.M.; Macrae, F.L.; Mcpherson, H.R.; Bridge, K.I.; Ajjan, R.A.; Ridger, V.C.; Connell, S.D.; Philippou, H.; Ari, R.A.S. Thrombin and fibrinogen, impact clot structure by marked effects on intra fibrillar structure and proto fibril packing. *Blood J. Am. Soc. Hematol.* **2019**, *127*, 487–496.

55. Luo, X.; Du, C.; Cheng, H.; Chen, J.; Lin, C. Study on the Anticoagulant or Procoagulant Activities of Type II Phenolic Acid Derivatives. *Molecules* **2017**, *22*, 2047. [CrossRef] [PubMed]

56. Bijak, M.; Ziewiecki, R.; Saluk, J.; Ponczek, M.; Pawlaczyk, I.; Krotkiewski, H.; Wachowicz, B.; Nowak, P. Thrombin inhibitory activity of some polyphenolic compounds. *Med. Chem. Res.* **2013**, *23*, 2324–2337. [CrossRef]

57. Singleton, V.L.; Rossi, J.A. Colorimetry of Total Phenolics with Phosphomolybdic-Phosphotungstic Acid Reagents. *Am. J. Enol. Vitic.* **1965**, *16*, 144–158.
58. Zhishen, J.; Mengcheng, T.; Jianming, W. The determination of flavonoid contents in mulberry and their scavenging effects on superoxide radicals. *Food Chem.* **1999**, *64*, 555–559. [CrossRef]
59. Feucht, W.; Polster, J. Nuclei of Plants as a Sink for Flavanols. *Zeitschrift für Naturforschung C* **2001**, *56*, 479–482. [CrossRef]
60. Pozo-Bayón, M.Á.; Hernández, M.T.; Martín-Álvarez, A.P.J.; Polo, M.C. Study of Low Molecular Weight Phenolic Compounds during the Aging of Sparkling Wines Manufactured with Red and White Grape Varieties. *J. Agric. Food Chem.* **2003**, *51*, 2089–2095. [CrossRef]
61. Re, R.; Pellegrini, N.; Proteggente, A.; Pannala, A.; Yang, M.; Rice-Evans, C. Antioxidant activity applying an improved ABTS radical cation decolorization assay. *Free. Radic. Biol. Med.* **1999**, *26*, 1231–1237. [CrossRef]
62. Singh, R.P.; Murthy, K.N.C.; Jayaprakasha, G.K. Studies on the Antioxidant Activity of Pomegranate (Punicagranatum) Peel and Seed Extracts Using In Vitro Models. *J. Agric. Food Chem.* **2002**, *50*, 81–86. [CrossRef] [PubMed]
63. Dassault Systemes. *BIOVIA, Discovery Studio Modeling Environment*; Release 4.5; Dassault Systemes: San Diego, CA, USA, 2015.

Publisher's Note: MDPI stays neutral with regard to jurisdictional claims in published maps and institutional affiliations.

Article

Association of Moderate Beer Consumption with the Gut Microbiota and SCFA of Healthy Adults

Natalia González-Zancada, Noemí Redondo-Useros, Ligia E. Díaz, Sonia Gómez-Martínez, Ascensión Marcos and Esther Nova *

Immunonutrition Group, Department of Metabolism and Nutrition, Institute of Food Science, Technology and Nutrition (ICTAN), Spanish National Research Council (CSIC), C/Jose Antonio Novais, 28040 Madrid, Spain; nataliagonzalez_91@hotmail.com (N.G.-Z.); noemi_redondo88@hotmail.com (N.R.-U.); ldiaz@ictan.csic.es (L.E.D.); sgomez@ictan.csic.es (S.G.-M.); amarcos@ictan.csic.es (A.M.)
* Correspondence: enova@ictan.csic.es; Tel.: +34-915492300

Received: 30 September 2020; Accepted: 15 October 2020; Published: 17 October 2020

Abstract: Fermented alcoholic drinks' contribution to the gut microbiota composition is mostly unknown. However, intestinal microorganisms can use compounds present in beer. This work explored the associations between moderate consumption of beer, microbiota composition, and short chain fatty acid (SCFA) profile. Seventy eight subjects were selected from a 261 healthy adult cohort on the basis of their alcohol consumption pattern. Two groups were compared: (1) abstainers or occasional consumption (ABS) ($n = 44$; <1.5 alcohol g/day), and (2) beer consumption ≥70% of total alcohol (BEER) ($n = 34$; 200 to 600 mL 5% vol. beer/day; <15 mL 13% vol. wine/day; <15 mL 40% vol. spirits/day). Gut microbiota composition (16S rRNA gene sequencing) and SCFA concentration were analyzed in fecal samples. No differences were found in α and β diversity between groups. The relative abundance of gut bacteria showed that *Clostridiaceae* was lower ($p = 0.009$), while *Blautia* and *Pseudobutyrivibrio* were higher ($p = 0.044$ and $p = 0.037$, respectively) in BEER versus ABS. In addition, *Alkaliphilus*, in men, showed lower abundance in BEER than in ABS ($p = 0.025$). Butyric acid was higher in BEER than in ABS ($p = 0.032$), and correlated with *Pseudobutyrivibrio* abundance. In conclusion, the changes observed in a few taxa, and the higher butyric acid concentration in consumers versus non-consumers of beer, suggest a potentially beneficial effect of moderate beer consumption on intestinal health.

Keywords: alcohol; butyric acid; fiber; polyphenols; drinking pattern

---

1. Introduction

Alcoholic beverage consumption, and its effects on health, is nowadays a controversial topic, with no clear-cut and widely accepted recommendations readily available for all circumstances, even on a population group basis. While scientific evidence on regular and moderate consumption of wine and beer have shown benefits for the risk of cardiovascular disease [1], health organizations claim that the possible benefits do not outweigh the risks, and avoiding alcohol is the best choice for those who are not habitual consumers. Thus, investigations on the impact of these beverages on health are certainly needed. Beer is the most widely consumed alcoholic beverage throughout the world, and contains a multitude of different compounds, many of which are produced during the fermentation process. Some minerals and vitamins such as fluoride, silicon, choline, and folate are present in significant amounts, so that two cans might provide 10% of the recommended dietary allowance (RDA) [1]. Polyphenols from malt and hop are important active compounds in beer that confer antioxidant activity, and can act synergistically with dietary constituents. Beer is also a source of dietary fiber, mainly composed of β-glucans from barley and arabinoxylo-olygosaccharides (AXOS) [2]. These compounds or their combination, i.e., hydrolysable polyphenols which are non-extractable with aqueous organic solvents [3], are mainly conducted undigested to the lower gut and are metabolized by gut bacteria. The symbiotic relationship between the gastrointestinal tract and the gut microbiota has a key role for human health, and many factors can influence this relationship such as lifestyle, environmental factors, the ageing process, etc. While the intestine provides nutrients and good conditions for the gut microorganisms to thrive [4], these microorganisms participate in energy extraction from food and the synthesis of vitamins and aminoacids, in addition to building defensive barriers against pathogens [5,6]. Furthermore, disruption of the normally stable microbial communities, known as gut dysbiosis, is accompanied by variable inflammatory conditions.

Diet is a particularly important factor for intestinal homeostasis and health, and affects greatly the composition and abundance of the microbial community. Regarding alcoholic drinks, separating the effects of ethanol from those of the raw plant components (e.g., polyphenols and fiber) and those formed in the technological and maturation processes of beer making seems a useful research strategy. Evidence in the literature demonstrates that alcohol consumption can lead to quantitative and qualitative dysbiosis in the intestine of rodents and humans [6]. In general, chronic alcohol consumption is associated with bacterial overgrowth, a decrease of *Bacteroidetes,* and an increase of *Proteobacteria* and *Fusobacteria,* endotoxin translocation, and inflammation, especially in alcoholic liver disease patients [6]. However, alcoholic beverages consumed in moderation should be examined as a complex intake, since the ethanol delivery would be within the low range, and its detrimental effects, if existent at all, might be overruled by the beneficial effects of the bioactive compounds present, for example, in fermented drinks, such as wine or beer. In this sense, the fermentation process in the gut yields energy for microbiotic proliferation and metabolite production, e.g., short chain fatty acids (SCFA) [7] for regulation of inflammatory responses [8] and gut hormone secretion [9] in the host [5]. Regarding gut microbiota composition, the dietary administration of red wine polyphenols in a mouse model of carcinogenesis changed the fecal microbiota composition to a predominance of *Bacteroides, Lactobacillus,* and *Bifidobacterium* spp., compared to the most abundant *Bacteroides, Clostridium,* and *Propionibacterium* spp. in the control animals [10]. Despite this finding, the number of published studies on fermented alcoholic beverages and gut microbiota is scarce, both in animals and humans, especially considering moderate alcohol consumption. However, some studies have proven the interactions between beer polyphenols and gut microbiota, and have been recently summarized in a review [11]. Thus, results on the most widely studied flavonoids, such as quercetin or catechin, which are present in beer, have shown their capacity to modulate gut microbiota. Tzounis et al. showed in an in vitro model that (−)-epicatechin and (+)-catechin promote the growth of the *Clostridium coccoides-Eubacterium rectale* group, and inhibit the growth of the *Clostridium histolyticum* group [12]. In addition, isoxanthohumol present in hops, together with other prenylated flavonoids, can be metabolized to render 8-prenylnaringenin, through an O-demethylation that is carried out by *Eubacterium limosum* [13]. Moreover, metabolization by *Eubacterium ramulus* transforms prenylated flavonoids into its chalcones, and likely affects both the activity and toxicity of ingested molecules [14].

Regarding dietary fiber in beer, an interesting study fed rats with different diets containing fiber from barley malts, brewer's spent grain, and barley extracts, resulting in varying amounts of β-glucan, soluble arabinoxylan, and insoluble arabinoxylan in the diets [15]. The results showed that there is a potential to stimulate butyrate- and propionate-producing bacteria in the cecum of rats with malt products of specific fiber properties, compared with a fiber-free diet. The authors also pointed out that a complex mixture of fiber, as in the malts, is of greater importance for microbiota diversity than purer fiber extracts. A different study in high-fat fed rats showed that barley and barley malt differed in the microbiota modulatory effects, compared to values in the cecum of control rats. Those fed malt showed higher *Roseburia, Coprococcus,* and *Lactobacillus,* which was related to the changing characteristics of β-glucans during malting, while barley was associated with higher *Blautia* and *Akkermansia* [16].

SCFAs are formed during the fermentation of undigested carbohydrates by the lower gut microbiota [17]. The main SCFAs produced (acetate, propionate, and butyrate) are rapidly absorbed by the colonocytes, and used as an energy source and precursors in anti-inflammatory mechanisms [17].

In addition, SCFAs are key substrates in the cross-feeding web, which comprises the intestinal microbiota. As an example, mutual cross-feeding interactions occur between *Bifidobacterium longum* and *Eubacterium rectale*. Both strains consume AXOS, but the bifidobacterial strain is additionally stimulated by consuming the monosaccharides released by the extracellular degradation of AXOS by the *E. rectale* strain, leading to cross-feeding interactions that are mutually beneficial [18]. Moreover, *Faecalibacterium prausnitzii* uses lactate produced by certain *Bifidobacterium* spp. to produce butyrate [19]. In the presence of AXOS, Bifidobacteria and butyrate-producing colon bacteria (*F. prausnitzii, Eubacterium,* and *Roseburia* spp.) are stimulated simultaneously, with a significant increase of butyrate production as a result [18].

Despite the above evidence that beer components could potentially have an influence on gut microbiota composition, there is a gap in research regarding beer consumption and microbiota modulation in humans. Only one intervention study has been published so far aimed at studying the effects of both, non-alcoholic and alcoholic beer on the microbiota of healthy adults, which were 89% between 21 and 35 years old [20]. The results showed that non-alcoholic beer increased the diversity of the microbiota after 30 days of intervention, while alcoholic beer did not, but both favored the proliferation of the *Bacteroidetes* phylum in relation to the *Firmicutes* phylum. Significant changes were observed in the relative abundance of a number of taxa with both types of beer, which in the case of non-alcoholic beer the authors propose as enrichment with beneficial bacteria [20]. Thus, in order to reduce the shortage of information published in this field, the aim of this work was to study the associations between beer consumption and gut microbiota composition in healthy adults as well as the concentration of SCFA. To this end, men and women selected from a larger cohort were studied in two groups, differing in their alcohol consumption pattern. The first group included abstainers, and the second group subjects consuming beer as their fundamental alcoholic drink choice.

2. Results

2.1. Anthropometric, Lifestyle, and Dietary Profile Characteristics of the Beer Consumption Groups

As observed in Table 1, the men in the abstainers (ABS) group showed a higher body mass index (BMI) than that of the subjects in the beer consumption (BEER) group ($p = 0.038$). Furthermore, they also showed a tendency to have higher body fat and visceral fat (both $p = 0.057$). On the other hand, the women in the BEER group tended to show higher total dietary energy consumption (kcal/d) than women in the ABS group ($p = 0.054$; Table S1). Although most of these results are trends, these variables were taken into account as indicated in the statistical methodology when analyzing the significant differences in the bacterial taxa studied. Regarding alcohol consumption by type of drink, the median of beer consumption, in men of the BEER group was 11.6 alcohol g/day and in women was 13.5 alcohol g/day, which is 232 and 270 mL of beer in men and women, respectively. In addition, the median for alcohol consumption from wine and spirits in men was 0 and 1 alcohol g/day, equivalent to 2 mL spirits per day. In women of the BEER group, these were 0.39 and 0.36 alcohol g/day respectively, equivalent to 3 mL and 0.9 mL of wine and spirits per day, respectively (Table S2).

Table 1. Anthropometric characteristics and lifestyle in moderate beer consumers and abstainers (by sex).

| | Beer Consumption Group | | | | |
Men	ABS (n = 18)	BEER (n = 15)	p^Y	p^*	$p^¥$
Age (years)	37.23 (5.99)	34.05 (6.39)	0.151	–	–
BMI (kg/m^2)	26.47 (3.20)	24.23 (2.64)	0.038	–	–
BMI-Fat (%)					
Normal weight	44.4	73.3	–	–	0.095
Overweight	55.6	26.7			
Body fat (%)	20.47 (6.32)	16.48 (5.06)	0.057	–	–
Visceral fat index	7.00 (4.00–9.00)	4.00 (3.00–6.00)	–	0.057	–
MEDAS total score	6.778 (2.264)	6.867 (1.767)	0.902	–	–
Capital (%)					
Low (<50,000 €)	33.3	40.0			
Medium (50,000–200,000 €)	50.0	40.0	–	–	0.848
High (>200,000 €)	16.7	20.0			
Smoking habits (%)					
Non-smokers	5.6	13.3			
Current smokers	11.1	20.0	–	–	0.530
Former smokers	83.3	66.7			
Physical activity (kcal/wk) [†φ]	5588 (3793)	7280 (4497)	0.152	–	–
Sleep (h/d) [†]	7.45 (1.05)	7.64 (0.87)	0.535	–	–
Women	ABS (n = 26)	BEER (n = 19)	p^Y	p^*	$p^¥$
Age (years)	36.78 (7.18)	34.70 (6.48)	0.323	–	–
BMI (kg/m^2)	22.80 (3.01)	23.75 (2.25)	0.253	–	–
BMI-Fat (%)					
Normal weight	69.2	63.2	–	–	0.670
Overweight	30.8	36.8			
Body fat (%)	27.60 (8.12)	29.11 (5.07)	0.480	–	–
Visceral fat index	3.00 (2.00–4.13)	4.00 (3.00–4.00)	–	0.476	–
MEDAS total score	7.269 (1.756)	7.526 (1.679)	0.624	–	–
Capital (%)					
Low (<50,000 €)	57.7	47.4			
Medium (50,000–200,000 €)	30.8	47.4	–	–	0.471
High (>200,000 €)	11.5	5.3			
Smoking habits (%)					
Non-smokers	19.2	36.8			
Current smokers	30.8	31.6	–	–	0.339
Former smokers	50.0	31.6			
Physical activity (kcal/wk) [†φ]	3156 (2242)	2617 (2359)	0.433	–	–
Sleep (h/d) [†]	7.75 (0.88)	7.75 (0.78)	0.943	–	–

MEDAS: Mediterranean diet adherence score. [Y] Student's t test for parametric variables. [*] Mann–Whitney U (MW-U) test for non-parametric variables. [¥] Chi-square test for categorical variables. [φ] Physical activity corresponds to regular (daily/weekly/monthly) activities and excludes occasional activities. [†] Variables transformed to logarithmic scale (Ln). h/d, hours/day. Statistical significance was set at $p < 0.05$.

2.2. Gut Microbiota Diversity and Relative Abundance

The analysis of α diversity showed that the Chao1 index (richness) and the Shannon index were similar between subjects consuming beer and abstainers ($p = 0.330$ and $p = 0.871$, respectively) (Figure 1). No differences were found, either, in β diversity ($p = 0.332$) (Figure S1) between beer consumption groups in a principal coordinates analysis (PCoA), based in the Bray-Curtis distance matrix with permutational multivariate analysis of variance (PERMANOVA) test.

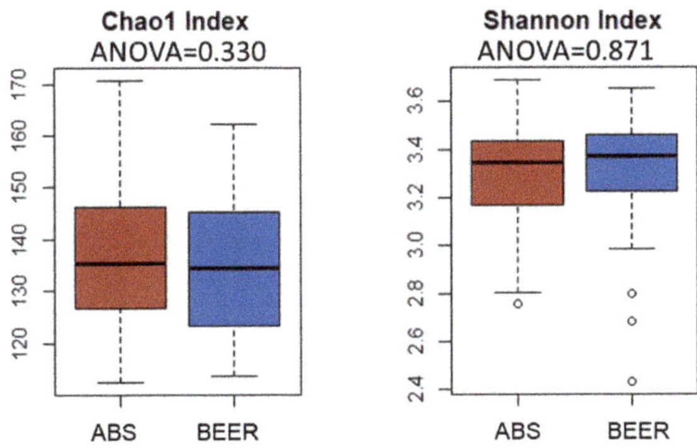

Figure 1. Differences in α diversity indices (Chao1 and Shannon) between beer consumption groups. ANOVA test.

The relative abundances of phyla, families, genera, and species that can metabolize beer components were selected from the metagenomic sequence analysis aggregate tables and compared between alcohol consumption groups. No differences in phyla abundances were observed between the groups (Table S3). Moreover, out of the eight bacterial families studied, differences were only observed in the *Clostridiaceae* family ($p = 0.009$), which presented a lower relative abundance in the BEER group compared to the ABS group (Table 2, Figure S2).

Table 2. Relative abundances [%] of bacterial families in moderate beer consumers and abstainers.

	Beer Consumption Groups				
	ABS (n = 44)	**BEER (n = 34)**	p [#]	p [λ]	p [*]
Lachnospiraceae	16.16 (5.61)	17.68 (4.01)	–	0.189	–
Ruminococcaceae	15.16 (4.77)	15.28 (4.30)	0.958	–	–
Clostridiaceae [†]	11.31 (5.93)	8.20 (3.61)	–	0.009	–
Bacteroidaceae	10.62 (7.89–17.71)	13.89 (7.75–20.34)	–	–	0.228
Bifidobacteriaceae	2.342 (0.935–5.441)	1.348 (0.516–3.260)	–	–	0.303
Peptococcaceae	0.268 (0.156–0.410)	0.237 (0.166–0.353)	–	–	0.323
Eubacteriaceae	0.146 (0.118–0.187)	0.141 (0.121–0.161)	–	–	0.465
Lactobacillaceae	0.134 (0.054–0.249)	0.095 (0.059–0.147)	–	–	0.181

Data shown as mean (SD) or median (interquartile range, IQR) for normally and non-normally distributed variables, respectively. [#] Group effect in a general linear model with fixed factors "Group" and "BMI-fat". [λ] ANOVA test. [*] MW-U test. [†] Variable transformed to logarithmic scale (Ln + 1).

Regarding genera (Table 3), *Blautia* and *Pseudobutyrivibrio* showed higher levels in the BEER group compared to the ABS group ($p = 0.044$ and $p = 0.037$, respectively) (Figure 2A,B). Furthermore, *Alkaliphilus* showed lower levels in the BEER group compared to the ABS group, but only in men ($p = 0.025$) (Table 3; Figure 2C.1). Finally, non-significant trends were found for lower levels of *Clostridium* genus ($p = 0.056$), and higher levels of *Johnsonella* ($p = 0.051$) and *Butyrivibrio* ($p = 0.056$) in the BEER group compared to the ABS group.

Table 3. Relative abundances [%] of bacterial genera in moderate beer consumers and abstainers.

	Beer Consumption Groups		
	ABS (n = 44)	**BEER (n = 34)**	***p*** *
	Bacteroidaceae		
Bacteroides	10.62 (7.89–17.71)	13.60 (7.78–19.90)	0.268
	Lachnospiraceae		
Blautia	6.419 (5.041–8.822)	8.098 (6.801–9.043)	0.044
Lachnospira			
Normal weight	2.268 (1.082–3.408)	1.951 (1.428–4.011)	0.795
Overweight	1.110 (0.653–1.662)	1.297 (0.992–1.822)	0.387
Coprococcus			
Normal weight	1.919 (0.957–2.580)	1.321 (0.970–1.822)	0.065
Overweight	1.908 (1.762–3.091)	3.223 (1.119–4.252)	0.188
Roseburia	1.424 (0.675–2.639)	2.196 (1.278–2.978)	0.118
Dorea	0.442 (0.236–0.617)	0.332 (0.233–0.570)	0.438
Pseudobutyrivibrio	0.224 (0.104–0.364)	0.323 (0.159–0.591)	0.037
Butyrivibrio	0.080 (0.034–0.179)	0.166 (0.065–0.296)	0.056
Anaerostipes	0.072 (0.028–0.116)	0.052 (0.007–0.121)	0.276
Johnsonella	0.035 (0.022–0.073)	0.055 (0.036–0.071)	0.051
Oribacterium			
Normal weight	0.019 (0.011–0.033)	0.028 (0.015–0.035)	0.200
Overweight	0.014 (0.009–0.023)	0.010 (0.007–0.018)	0.438
Lachnobacterium	0.009 (0.002–0.027)	0.014 (0.006–0.078)	0.107
Shuttleworthia	0.008 (0.003–0.016)	0.009 (0.002–0.017)	0.525
Catonella	0.001 (0.000–0.001)	0.001 (0.000–0.001)	0.674
	Ruminococcaceae		
Faecalibacterium			
Normal weight	8.007 (4.794–9.872)	7.530 (6.330–9.767)	0.857
Overweight	6.067 (3.362–9.512)	6.663 (4.237–8.366)	0.842
Ruminococcus	4.796 (3.410–6.352)	4.019 (2.998–5.724)	0.144
Oscillospira	3.776 (2.436–4.839)	3.755 (2.721–4.344)	0.856
Anaerofilum	0.082 (0.048–0.149)	0.072 (0.049–0.131)	0.896
Anaerotruncus	0.059 (0.034–0.094)	0.063 (0.044–0.095)	0.643
Ethanoligenens	0.000 (0.000–0.001)	0.000 (0.000–0.001)	0.938
	Clostridiaceae		
Clostridium	5.192 (3.514–6.378)	4.013 (3.161–5.072)	0.056
Alkaliphilus			
Men	2.914 (1.570–5.391)	1.535 (1.318–2.590)	0.025
Women	2.100 (0.951–3.672)	1.200 (0.463–2.608)	0.198
Caloramator	0.450 (0.157–1.226)	0.218 (0.080–1.144)	0.204
	Eubacteriaceae		
Acetobacterium	0.144 (0.108–0.181)	0.134 (0.122–0.156)	0.724
Eubacterium	0.104 (0.038–0.503)	0.217 (0.017–0.981)	0.526
Anaerofustis	0.001 (0.000–0.003)	0.000 (0.000–0.001)	0.068
	Bifidobacteriaceae		
Bifidobacterium	2.333 (0.930–5.425)	1.500 (0.509–3.804)	0.426
	Lactobacillaceae		
Lactobacillus	0.131 (0.046–0.245)	0.083 (0.058–0.142)	0.201
	Peptococcaceae		
Desulfotomaculum	0.070 (0.048–0.108)	0.064 (0.043–0.092)	0.450
Peptococcus	0.042 (0.020–0.095)	0.029 (0.018–0.073)	0.403
Sporotomaculum	0.024 (0.017–0.034)	0.024 (0.019–0.035)	0.896
Desulfosporosinus	0.023 (0.008–0.034)	0.013 (0.004–0.025)	0.090
Dehalobacterium	0.012 (0.001–0.028)	0.007 (0.003–0.014)	0.343
Desulfurispora	0.005 (0.003–0.016)	0.005 (0.003–0.008)	0.632
Pelotomaculum	0.003 (0.002–0.005)	0.003 (0.002–0.005)	0.545
Desulfitobacterium	0.001 (0.000–0.005)	0.001 (0.000–0.003)	0.402

Data shown as median (IQR). * Mann Whitney-U test. Statistical significance was set at $p < 0.05$.

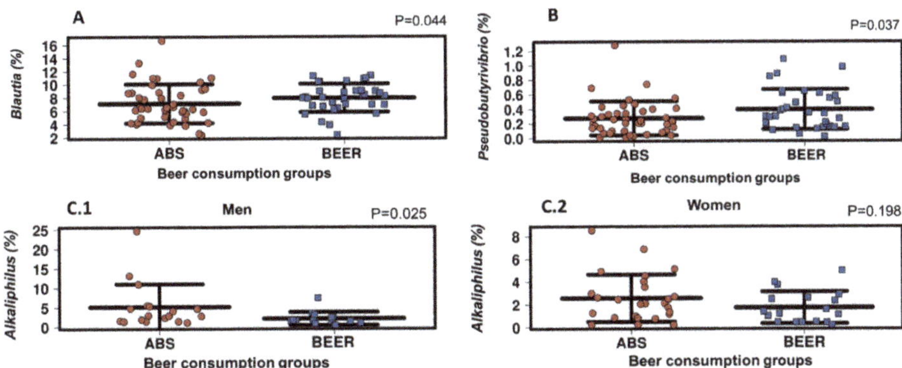

Figure 2. Differences between beer consumption groups in the abundance of the genus Blautia (**A**), Pseudobutyrivibrio (**B**), and Alkaliphilus (**C**). The Alkaliphilus genus is stratified in men (**C.1**) and women (**C.2**). MW-U test. Significance at $p < 0.05$.

Regarding species analysis within the genera that differed between consumption groups, the results showed that *Alkaliphilus peptidifermentans* ($p = 0.028$) and *Clostridium hiranonis* ($p = 0.006$) were less abundant in the BEER group than in the ABS group (Table 4; Figure S3A,B, respectively). In contrast, *Blautia coccoides* ($p = 0.027$), *Pseudobutyrivibrio xylanivorans* ($p = 0.037$), *Johnsonella ignava* ($p = 0.046$), and *Blautia producta* ($p = 0.039$) showed higher levels in the BEER group compared to the ABS group (Table 4; Figure S3C–F, respectively).

Table 4. Relative abundances [%] of bacterial species in moderate beer consumers and abstainers.

	Beer Consumption Groups		
	ABS (n = 44)	**BEER (n = 34)**	**p ***
Blautia coccoides	1.752 (1.142–2.212)	2.254 (1.724–2.912)	0.027
Alkaliphilus peptidifermentans	1.022 (0.270–2.407)	0.574 (0.211–1.008)	0.028
Alkaliphilus crotonatoxidans	0.920 (0.295–1.906)	0.376 (0.178–0.934)	0.054
Clostridium alkalicellulosi	0.629 (0.393–0.932)	0.668 (0.444–0.886)	0.747
Blautia hanseii	0.400 (0.218–0.526)	0.343 (0.285–0.492)	0.928
Blautia wexlerae	0.384 (0.236–0.658)	0.522 (0.341–0.819)	0.133
Pseudobutyrivibrio xylanivorans	0.224 (0.104–0.364)	0.323 (0.159–0.591)	0.037
Clostridium cadaveris	0.170 (0.072–0.577)	0.124 (0.070–0.310)	0.190
Clostridium histolyticum	0.144 (0.093–0.245)	0.155 (0.108–0.209)	0.904
Clostridium frigoris	0.102 (0.020–0.514)	0.167 (0.020–0.565)	0.657
Butyrivibrio proteoclasticus	0.080 (0.034–0.179)	0.166 (0.065–0.276)	0.058
Blautia obeum	0.079 (0.031–0.160)	0.084 (0.055–0.198)	0.328
Clostridium caenicola	0.056 (0.029–0.080)	0.053 (0.039–0.077)	0.832
Clostridium hiranonis	0.046 (0.012–0.110)	0.013 (0.002–0.038)	0.006
Clostridium fallax	0.038 (0.010–0.100)	0.040 (0.006–0.116)	0.687
Johnsonella ignava	0.034 (0.020–0.072)	0.055 (0.035–0.069)	0.046
Clostridium thermosuccinogenes	0.030 (0.021–0.048)	0.039 (0.020–0.116)	0.230
Clostridium taeniosporum	0.020 (0.013–0.049)	0.040 (0.015–0.078)	0.131
Clostridium thermoalcaliphilum			
Men	0.029 (0.011–0.038)	0.017 (0.011–0.044)	0.442
Women	0.016 (0.008–0.037)	0.010 (0.007–0.017)	0.103
Clostridium termitidis			
Men	0.011 (0.005–0.041)	0.008 (0.006–0.019)	0.442
Women	0.036 (0.013–0.066)	0.015 (0.011–0.050)	0.301
Blautia schinkii	0.009 (0.004–0.017)	0.010 (0.004–0.015)	0.856
Blautia glucerasea	0.008 (0.003–0.016)	0.010 (0.004–0.017)	0.665
Clostridium hveragerdense	0.008 (0.003–0.022)	0.009 (0.003–0.020)	0.956

Table 4. *Cont.*

Beer Consumption Groups		
Clostridium cavendishii	0.006 (0.002–0.012)	0.004 (0.003–0.008)	0.420
Clostridium malenominatum	0.005 (0.002–0.014)	0.004 (0.001–0.016)	0.519
Clostridium straminisolvens	0.003 (0.001–0.009)	0.002 (0.000–0.005)	0.263
Clostridium proteolyticus	0.002 (0.001–0.005)	0.002 (0.001–0.004)	0.532
Alkaliphilus metalliredigens	0.001 (0.000–0.002)	0.001 (0.000–0.001)	0.267
Blautia hydrogenotrophica	0.001 (0.000–0.006)	0.001 (0.000–0.004)	0.830
Clostridium tepidiprofundi	0.001 (0.000–0.002)	0.001 (0.001–0.002)	0.944
Clostridium chartatabidum	0.001 (0.000–0.003)	0.001 (0.000–0.002)	0.235
Clostridium aestuarii	0.001 (0.000–0.002)	0.000 (0.000–0.001)	0.128
Blautia producta	0.000 (0.000–0.003)	0.003 (0.000–0.028)	0.039

Data shown as median (IQR). * Mann Whitney-U test. Statistical significance was set at $p < 0.05$.

2.3. SCFA Concentration in the Gut

Since SCFAs could be produced following the fermentation of beer components by the microbial taxa analyzed, their concentration was measured in feces. A higher concentration of butyric acid was observed in the BEER group compared to the ABS group ($p = 0.032$; Figure S4), while the rest of the SCFA analyzed did not show significant differences (Table 5).

Table 5. SCFA concentration in moderate beer consumers and abstainers.

Beer Consumption Groups		
	ABS (n = 44)	BEER (n = 34)	p^{λ}
Acetic acid (µM/g) [†]	33.24 (16.29)	37.79 (15.51)	0.158
Propionic acid (µM/g) [†]	11.66 (7.08)	13.32 (6.55)	0.133
Butyric acid (µM/g) [†]	8.831 (5.383)	11.35 (6.538)	0.032
Isobutyric acid (µM/g) [†]	1.857 (0.942)	1.639 (0.710)	0.351
Valeric acid (µM/g) [†]	1.854 (1.436)	1.923 (0.972)	0.376
Isovaleric acid (µM/g) [†]	2.679 (1.691)	2.371 (1.259)	0.599

Data are shown as mean (SD). [λ] ANOVA test. [†] Variables transformed to logarithmic scale (Ln). Statistical significance was set at $p < 0.05$.

The correlations between the abundance of those taxa showing different levels in beer consumers and abstainers and the SCFA concentrations were analyzed. As observed in Figure 3, the three main SCFAs, acetic, propionic, and butyric acids, were positively correlated to *Pseudobutyrivibrio* and *Pseudobutyrivibrio xylanivorans* (all $p < 0.015$). On the other hand, acetic and propionic acids were negatively correlated, although with marginal significance, to *Johnsonella* and *Johnsonella ignava* (all $p < 0.040$). Furthermore, propionic acid was also negatively correlated with *Clostridiaceae* and *Alkaliphilus* (both $p < 0.010$).

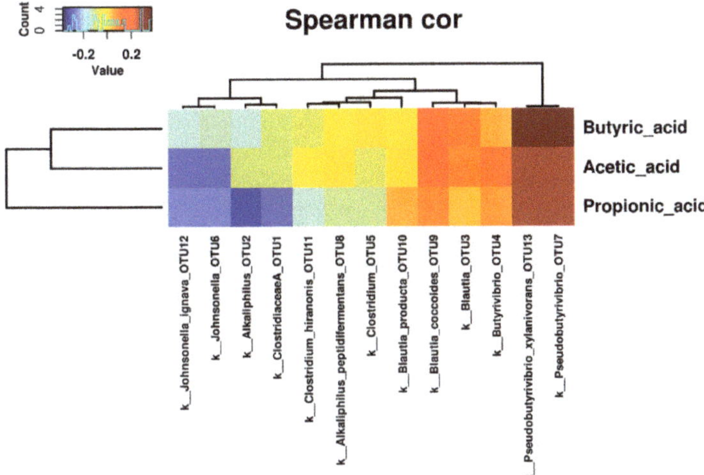

Figure 3. Heatmap representation of correlation coefficients between SCFA concentrations and gut microbiota composition. Only taxa with significantly different abundance in both beer consumption groups were included.

3. Discussion

Beer is the most abundantly consumed drink among alcoholic beverages, but its effects on health are still controversial. Given the high proportion of alcohol abuse in many populations and the high prevalence of consumption patterns that fall far from the "moderate and regular", stronger evidence on the healthy properties of beer or its components is warranted, in order to balance recommendations for this ancient drink. This work analyzed the changes in the intestinal microbial groups that could be associated with regular and moderate consumption of beer as compared to abstinence. The study was directed towards those taxa that could use some of the components of beer as a metabolic substrate, such as fiber and polyphenols. A few taxa showed changes between consumption groups, which together with the higher concentration of butyric acid in consumers versus non-consumers of beer, suggest that the effect of beer consumption might be beneficial for intestinal health, and deserves further investigation.

When studying possible demographic, anthropometric, or lifestyle differences of the volunteers, a high similarity was found in most of the variables between the BEER and ABS groups. It should be noted, as an exception, that abstemious men showed a trend towards a higher percentage of body fat, specifically visceral fat. For this reason, the composite variable "BMI-fat" was considered as a potential confusion factor in the subsequent analyses of the effect of beer consumption on the intestinal microbiota and SCFA. According to the results regarding the amount of alcohol intake from each drink shown in Table S2, these are consistent with a moderate consumption, this being considered up to one drink (typically a can of beer, 330 mL, containing about 4% *w/v* alcohol) per day in women, and up to two in men [1]. It is worth pinpointing, however, that the amount of beer consumed was very similar in men and women in this study, probably because among Spanish men there is a higher prevalence of mixed drinking, including beer, wine, and spirits, and this particular pattern was excluded when selecting the current study population from the ALMICROBHOL study cohort.

Regarding the microbiota, the results showed no between-group differences in α and β diversity. The same was observed in an intervention study in which 355 mL of non-alcoholic or alcoholic beer were consumed daily by the participants. This moderate alcoholic beer consumption did not affect α or β diversity in comparison with baseline values [20]. On the other hand, in the current study, other potential confounding factors were included in the analysis of diversity, such as sex, age, or the composite BMI-fat, and no significant effect was observed for any of these in this population.

Focusing on the relative abundance of the studied taxa, a lower relative abundance of the *Clostridiaceae* family, and specifically the *Clostridium* and *Alkaliphilus* genera, was observed in individuals who consumed beer than in abstainers, while *Blautia* genus abundance was higher in beer consumers. As mentioned above, scientific evidence on the effect of beer on gut microbiota is very scarce, although there are some animal and in vitro studies on the effect of some of the main beer compounds. Thus, beer has proven to be a relatively rich source of AXOS and β-glucans. Two published works carried out in rats reported increased *Blautia* levels associated with malt, barley, or beer derived compounds [15,16]. First, in rats fed with different types of dietary fiber from barley malts, brewer's spent grain, and barley extracts, an increase in *Blautia* was observed with the diet containing β-glucan extract [15]. This finding is coincident with the observations in the BEER group of the current study. Similarly, an increase in *Blautia* was observed in a group of high-fat diet fed rats, when barley or barley malt were incorporated to the diet, as compared to control rats on a fiber-free diet [16].

According to scientific evidence, the prebiotic intake tends to change the composition of the intestinal microbiota towards a relative increase in species belonging to *Bifidobacterium* and/or *Lactobacillus* genera. First, the endoxylanase enzymatic activity that is carried out by *Roseburia* and *Bacteroides* species degrades cereal arabinoxylans to AXOS, and these same species, together with specialized *Bifidobacterium* possessing arabinofuranosidase and xylosidase enzymes, further degrade AXOS, such as those in beer, to the monosaccharides arabinose and xylose; these leading finally to SCFA as the main fermentation output [21]. In a study based on continuous in vitro fermentation, simulating the human colon, Vardakou et al. found that the AXOS produced from the treatment of wheat with arabinoxylan endoxylanase significantly increased the levels of *Bifidobacterium* spp., while reducing the levels of *Clostridium* and *Bacteroides* [22]. In the current work, no significant changes were found in *Bifidobacterium* or *Bacteroides*, probably because the free-living human source of the samples is not directly comparable to the in vitro fermentation simulator, and the amount of AXOS would be relatively lower in real life conditions; however, a decrease in *Clostridium* was observed, in agreement with the mentioned work [21]. On the other hand, an increase in *Lactobacillus* was found with several of the malt-based ingredients or its extracts tested in rats [15]. The current results did not show differences in *Lactobacillus* levels, perhaps due to a lower amount of usable substrates in the case of moderate beer intake in the study subjects, compared to the supplemented diet in the animal model. In this sense, regarding the bifidogenic power of AXOS, supplementation studies have shown that 1.4 g AXOS for 28 days was the lowest daily dose that showed bifidogenic effects in humans [23]. Considering a moderate consumption of beer (500 mL/day) with an average amount of AXOS (range 0.49 to 1.90 g/L), it can be estimated that beer could contribute 18–68% to that minimum daily amount of 1.4 g, which might be considered a relevant contribution to the bifidogenic effect.

The *Lachnospiraceae* family is relevant among SCFA producers. While this family did not show differences between the BEER and ABS groups in the current study, several genera within the family showed significant, or almost significant, differences between the groups. The highly abundant genus *Blautia*, which has already been mentioned, and the minor genera *Pseudobutyrivibrio*, *Butyrivibrio*, and *Johnsonella*, presented higher levels in beer consumers than in abstainers. The last three genera are butyric acid producers [24–26]; thus, their increased abundance may be related to the increased fecal concentration of butyric acid found in the BEER group. In contrast, *Blautia* is mainly an acetate-producing bacteria [26]; however, the current study results did not show significant differences in acetate levels between groups, and no significant associations were observed between *Blautia* and acetic acid. In the only human intervention study with beer supplementation, and including microbiota and SCFA analysis, an increase in the *Bacillus* genus was found, and also in Proteobacteria such as *Pseudomonas*, *Succinivibrio*, and *Aeromonadales* [20]. Unfortunately, these observations are not in agreement with this study's results. No differences in SCFA were found either, after the 30-day beer supplementation [20], as opposed to the butyric acid increase with moderate beer consumption reported here. The different consumption habits in individuals in the intervention study [20] and in the present observational study, involving ad-libitum consumption and a more prolonged habit, might perhaps account for the discrepancies regarding microbiota composition and SCFA production.

According to the dietary assessment results, moderate beer consumption was not associated with changes in nutrient intakes. ABS and BEER subjects showed a similar dietary profile and, specifically, the total fiber consumption, as well as the different food group contributions to fiber intake (Table S1), were similar between groups. For example, fiber from fruits and vegetables and fiber from cereals, which include abundant β-glucans and AXOS, respectively, did not differ between BEER and ABS groups, suggesting that the changes observed in the composition of the intestinal microbiota and the production of SCFA are related with the different beer intake habit.

In vitro studies performed with beer polyphenols have also been published. For example, one study suggested that (−)-epicatechin and (+)-catechin may be able to influence the gut microbiota, even in the presence of other nutrients such as carbohydrates and proteins [12]. Specifically, (+)-catechin significantly inhibited the growth of *Clostridium histolyticum*, while it promoted that of the *Clostridium coccoides-Eubacterium rectale* group, of which *Blautia coccoides* (formerly *Clostridium coccoides*) is a member. These results are in accordance with findings in the current study relative to the *Clostridium* and *Blautia* genera. In this in vitro study, *Bifidobacterium* and *Lactobacillus* spp. remained relatively unaffected, which is also in agreement with the present results [12]. Furthermore, red wine polyphenols, were also associated with significantly lower levels of *Clostridium* spp. in a murine model [10]; however higher levels of *Bacteroides*, *Bifidobacterium*, and *Lactobacillus* spp. were also reported in this study, which highlights the need to differentiate the effects of different polyphenols and alcoholic drinks, and also the exposure dose.

4. Materials and Methods

4.1. Experimental Design

This is an observational study based on a convenience sub-sample from the ALMICROBHOL study population [27]. Two hundred and sixty-one adults between 25 and 50 years, and with a BMI between 18.5 and 35 kg/m^2, participated in the later project. The exclusion criteria in the mother study were: (1) pathological conditions such as type 1 diabetes, cancer, chronic liver, heart, kidney or lung disease, brain disorders, congenital metabolic diseases, autoimmune diseases (including thyroid disease), inflammatory bowel disease, human immunodeficiency virus (HIV), Cushing's syndrome, or diagnosed food intolerances; (2) prescription of chronic medication; (3) antibiotics use in the last two months; (4) to be on any type of special diet; (5) having undergone a surgical procedure in the last month.

Criteria based on the alcohol consumption profiles detected in the ALMICROBHOL population were used to select the study sub-sample as follows: (1) abstainers or subjects consuming <1.5 g alcohol/day with nil beer consumption (ABS; n = 44); (2) beer consumers: ≥70% beer contribution to total alcohol consumption and 10–30 alcohol g from beer/day, while wine consumption <2 alcohol g/day and spirits <6 alcohol g/day (BEER; n = 34). These amounts of alcohol in the BEER group are equivalent to between 200 and 600 mL of beer (5% vol.) per day, while consuming less than 15 mL of wine (13% vol.) and less than 15 mL of spirits (40% vol.) per day. Thus, a total of 78 individuals were included in this study, and all of them had maintained a stable behavior regarding alcohol consumption at least for the last year.

During this observational study, subjects attended the research center twice to participate in individual interviews with trained nutritionists who collected data on their lifestyle habits. In the first visit, the nutritionist administered an ad hoc frequency recall questionnaire on alcoholic beverage consumption [28]. The questionnaire recorded the intake of wine, beer, champagne, cider, liquors, spirits, and all the mixtures by estimations over the last year. Reference drink sizes were considered. Frequency of intake was registered using a continuous scale as follows: never or almost never (0 to once every 2 months), 1 to 3 times per month, times per week, or number of times per day. Habitual intake was recorded for working days and separately for weekends. Total alcohol intake (g/d) was calculated using average grams of alcohol content per 100 mL of each alcoholic beverage. In this first visit, the overall health status, diagnosed diseases, symptoms, drug prescriptions, and sleep quality were assessed through the Spanish National Health Survey. The self-estimation of the total capital, assisted by the interviewer, was used for the socioeconomic status (SES) classification as follows: (1) "low-intermediate": 10,000 € to 50,000 €; (2) "intermediate-high": 50,000 € to 200,000 € and (3) "high": above 200,000 €. Height (Soehnle), body weight, and bioimpedance analysis without shoes, and with light clothing (Tanita BC 601) were also measured in this first visit. Since the body mass index (BMI = weight (kg)/height (m^2)) does not represent an accurate measurement of body fat, the optimal body fat percentages were considered separately for men and women; and the subjects were divided into two groups: (1) normal weight (BMI < 25 kg/m^2) or overweight (BMI = 25–30 kg/m^2) plus normal body fat percentages (21–32% for women; 10–20% for men); and (2) overweight (BMI = 25–30 kg/m^2) or obese (BMI > 30 kg/m^2) plus high body fat percentages (>32% for women; >20% for men). Cut-off criteria for body fat percentages were taken from the Tanita guidelines. In addition, subjects were instructed to collect a stool sample, in sterile conditions, and bring it frozen with the aid of cold bricks to the study center on a second visit.

During the second visit, participants completed the Minnesota Leisure-Time Physical Activity Questionnaire (MLTPAQ, Spanish version) and went through a dietary assessment. They were asked to complete a validated food frequency questionnaire, which estimates the amounts and frequency of consumption of 104 items over the past one-year period [29]. The interviewer asked for both the quantity (referred to a standard size) and the frequency of consumption of each item, which was registered as never or almost never/number of times per month (1 to 3)/number of times per week (1 to 6)/number of times a day. Consumption variability, according to the seasonal availability, especially for vegetables and fruits, was also considered. Food and beverage intake were converted into energy and nutrients using the food composition tables by Mataix et al. [30].

4.2. Gut Microbiota Analysis

The fecal samples from the ALMICROBHOL project were collected in sterile containers at home, immediately frozen at −20 °C and transported on the next day, in refrigerated conditions, to the study center, where they were stored at −80 °C until analyses. Starting from 180–220 mg of each fecal sample, bacterial DNA was extracted using an optimized protocol [31]. After recovery of the supernatant by centrifugation, ammonium acetate was added for protein precipitation and the resulting supernatant was treated with isopropanol during 30 min for DNA precipitation. After centrifugation, the resulting pellet was washed with 70% ethanol, dried at 37 °C until ethanol evaporation, and then washed with Tris-Ethylenediaminetetraacetic acid (EDTA) buffer. RNase was added for RNA removal, and DNA was finally recovered with the commercial QIAamp DNA Stool Mini Kit (QIAGEN GmbH), following the manufacturer's instructions. DNA was quantified in a NanoDrop ND-1000 spectrophotometer (NanoDrop Technologies, Wilmington, DE, USA) and diluted (0.5 ng/µL) for library preparation before 16S rRNA sequencing. which involved several steps: a first polymerase chain reaction (PCR) to amplify the V3-V4 region of the 16S rRNA gene, using the primers 341F (5′-CCTACGGGNNGGCWGCAG-3′) and 785R (5′-GACTACHVGGTATCTAATCC-3′). Then, 1.5% agarose gel electrophoresis (EX 2% agarose, Invitrogen, Life Technologies, Grand Island, NY, USA) was performed to check the integrity of the amplicons and to estimate the dilution necessary for a second PCR, with adapter and barcode sequences, to facilitate sequence allocation, before loading libraries into the sequencer. After this second PCR, all the samples were run in a bioanalyzer, subsequently an equimolar pool was made, taking into account the data obtained in the bioanalyzer. This pool was cleaned with Ampure beads and run again in the bioanalyzer to check that there were no impurities. Samples sequencing was performed with a MiSeq Illumina system, using the V3 kit (Illumina, San Diego, CA, USA), and generating 2 × 270 bp reads. The analysis of microbial communities was done with the Metagenomics workflow in MiSeq Reporter (v2.3) software (San Diego, CA, USA), including demultiplexing and FASTq (text files containing sequence data with a quality score for each base) generation, obtaining 37,793,518 high-quality reads (144,803 mean reads/sample). Sequences were then clustered into operational taxonomic units (OTU) with Classify Reads, a high-performance implementation of the Ribosomal Database Project (RDP), based on the Greengenes database, obtaining 20 phyla, 243 families, 651 genera, and 1492 species (Classify Reads accuracy was 100%, 99.97%, 99.65%, and 98.65%, respectively) [32]. Taxa with relative abundance <0.001% of the total readings, and also those with a prevalence of <10 subjects, were removed, leaving a total of 20 phyla, 151 families, 364 genera, and 511 species. However, from these taxa only the beer-fermenting bacteria or their components were selected, such as the taxa belonging to *Lachnospiraceae*, *Ruminococcaceae*, *Clostridiaceae*, *Eubacteriaceae*, *Peptococcaceae* families, *Bifidobacteria* spp., *Lactobacilli* spp., and *Bacteroides* genus, for statistical analysis, obtaining 3 phyla, 8 families, 36 genera, and 41 species.

Next, the raw sequences (FASTq files) related to the 78 volunteers were selected to estimate α and β diversity using mothur. From the FASTq files, the direct and indirect readings (R1 and R2) of each sample were combined. Sequences with more than 3 ambiguous bases and more than 465 bp were removed. Once this first cleaning was done, the *unique.seqs* command was executed, which groups together identical sequences and accounts for their abundance. The sequences alignment was performed with the *align.seqs* command that matches the samples' sequences to the Greengenes reference database (May 2013 version). Non-aligned sequences were removed, and afterwards, filtering, clustering, and chimeric sequence cleaning were performed. The *classify.seqs* command was executed, which assigns the sequences a taxonomy, using the reference database; and the *remove.lineage* command, with which all those sequences that do not correspond to bacteria or archaea were eliminated, according to the taxonomic assignment. Therefore, at the end of the process, a total of 7,791,070 sequences were obtained, of which 3,222,457 were unique sequences. To start the analysis of diversity, the phylotype command was used, which clusters the sequences into phylotypes according to the taxonomic classification, and generates an OTUs abundance ".shared" file. At this point, all those OTUs that did not have more than 0.001% relative abundance were removed. Finally, the *classify.otu* command was executed, detecting a total of 189 OTUs. Rarefaction curves were generated and the α diversity indexes Chao and Shannon, were calculated using the ".shared" file. Regarding β diversity, the *dist.shared* command was executed to generate Bray-Curtis distance matrixes; as well as the *pcoa* command to generate a file that allows viewing a principal coordinates analysis (PCoA) graph in R, using the *plotPCOA* command.

4.3. Short Chain Fatty Acids Analysis

An aliquot of 100 mg of frozen fecal sample was diluted in 1 mL of 5% phosphoric acid, followed by homogenization and freezing of the fecal homogenates for dry matter precipitation. Stool samples were then thawed and centrifuged for 5 min at 112× *g* (Jouan Centrifuge A14, Saint Herblain, France). The SCFA acetic, propionic, butyric, isobutyric, valeric, and isovaleric acids were quantified in the supernatants by gas chromatography and flame ionization detection (GC-FID, Agilent 6890A, Agilent Technologies, Waldbronn, Germany). The capillary chromatographic column used was a DB-WAXtr column (100% polyethylene glycol, 60 m, 0.325 × 0.25) and helium was used as the carrier gas at 1.5 mL/min. Injection was made in splitless mode, with an injection volume of 1 µL and a temperature of 260 °C. Methyl valeric was used as an internal standard, and the standard curve was prepared in a similar way to the samples. The detector temperature was 260 °C. The column was heated at 50 °C for 2 min, followed by an increase of 15 °C every min to 150 °C, 5 °C every min to 200 °C, and finally 15 °C every min to 240 °C. The different SCFAs were identified by the retention time of the standard compounds.

4.4. Statistical Analysis

The normal distribution of the data was checked prior to analysis, and logarithmic transformation was applied for data normalization of some variables (physical activity, hours of sleep, *Clostridiaceae* family, and SCFAs). Descriptive measures used were mean ± standard deviation (SD), and median and interquartile range (IQR) for normal and non-normally distributed variables, respectively. Regarding the demographic characteristics, parametric (T test) and non-parametric (Mann–Whitney U, MW-U) tests were applied for between-group comparisons and the Chi-square test for categorical variables. General linear models, adjusted sequentially for gender, age, BMI-fat, total energy, fiber (g/1000 kcal), and MEDAS score, were used to assess the associations of beer consumption with parametric variables of the gut microbiota. Only significantly contributing factors were retained in the model. When no factor influenced the model, a one-way ANOVA test with the factor "consumption group" was used. For variables not fitting a normal distribution the MW-U test was used to compare groups defined by sex, BMI-fat status, and beer consumption, and when necessary, the population was split by sex or BMI-fat (or both) prior to beer consumption group comparison. The gut microbiota analysis was restricted to those taxa that could potentially metabolize beer compounds. Since this was a targeted analysis, no adjustment of multiple comparisons was deemed necessary. The analysis of α diversity (Chao and Shannon indexes) and SCFAs by consumption group was performed by one-way ANOVA, since other factors such as sex, age, BMI-fat, and total alcohol showed not to have an effect in the general linear model analyses. For β diversity, a Bray-Curtis distance matrix was used, and a PERMANOVA test was performed, introducing one by one, along with the "group" factor, the variables gender, age, BMI-fat, and total energy, using the adonis command from the vegan package of R. Values of $p < 0.05$ were considered significant. Data analysis was performed with SPSS software (v.23) (Chicago, IL, USA), while mothur (v.1.35.1) and R (v.3.5.3) were used to analyze α and β diversity.

5. Conclusions

In conclusion, this observational study on the intestinal microbiota composition in a healthy adult population revealed certain differences between regular consumers of beer in moderate amounts and non-consumers. In the absence of other dietary differences between the groups studied, moderate beer consumption was associated with higher levels of *Blautia*, *Pseudobutyrivibrio*, *Butyrivibrio*, and *Johnsonella*, and increased butyric acid, which can be attributed to the higher levels of the last three genera. In contrast, beer consumption was not associated with changes in the diversity of the gut microbiota. Intervention studies with different types of beer are needed to help provide more evidence to support the present results.

Supplementary Materials: The following are available online. Table S1: Dietary profile in moderate beer consumers and abstainers (by sex), Table S2. Total alcohol intake and amount of alcohol intake by type of drink in moderate beer consumers and abstainers (by sex), Table S3: Relative abundances [%] of bacterial phylum in moderate beer consumers and abstainers, Figure S1: Two-dimensional scatter plot of ABS and BEER groups generated using PCoA from the Bray-Curtis distance matrix, Figure S2: Differences in *Clostridiaceae* levels between beer consumption groups, Figure S3: Differences between beer consumption groups in the levels of (A) *Alkaliphilus peptidifermentans*; (B) *Clostridium hiranonis*; (C) *Blautia coccoides*; (D) *Pseudobutyrivibrio xilanivorans*; (E) *Johnsonella ignava*; and (F) *Blautia producta*, Figure S4: Differences in the production of butyric acid between beer consumption groups.

Author Contributions: E.N., A.M., and N.R.-U., study design and methodology set-up. N.R.-U., L.E.D. and S.G.-M., subject interviews and laboratory procedures; N.G.-Z., E.N., and N.R.-U., formal analysis, resources and data curation; N.G.-Z., E.N. and N.R.-U., writing—original draft preparation; E.N. and A.M. writing—review and editing; A.M. and E.N., funding acquisition. All authors have read and agreed to the published version of the manuscript.

Funding: The present work received funding from ERAB (European Foundation for Alcohol Research), Ref: EA 14. 44., and N.G.-Z received a grant from "Manuel de Oya–Beer, Health and Nutrition (2018)" from the organization Beer and Health Information Center (Spain). We acknowledge support of the publication fee by the CSIC Open Access Publication Support Initiative through its Unit of Information Resources for Research (URICI).

Conflicts of Interest: The authors declare no conflict of interest.

References

1. de Gaetano, G.; Costanzo, S.; Di Castelnuovo, A.; Badimon, L.; Bejko, D.; Alkerwi, A.; Chiva-Blanch, G.; Estruch, R.; La Vecchia, C.; Panico, S.; et al. Effects of moderate beer consumption on health and disease: A consensus document. *Nutr. Metab. Cardiovasc. Dis.* **2016**, *26*, 443–467. [CrossRef] [PubMed]

2. Goñi, I.; Díaz-Rubio, M.E.; Saura-Calixto, F. Dietary fiber in beer: Content, composition, colonic fermentability, and contribution to the diet. In *Beer in Health and Disease Prevention*; Elsevier: Amsterdam, The Netherlands, 2008; pp. 299–307, ISBN 9780123738912.

3. Pérez-Jiménez, J.; Díaz-Rubio, M.E.; Saura-Calixto, F. Non-extractable polyphenols, a major dietary antioxidant: Occurrence, metabolic fate and health effects. *Nutr. Res. Rev.* **2013**, *26*, 118–129. [CrossRef] [PubMed]

4. Neish, A.S. Microbes in gastrointestinal health and disease. *Gastroenterology* **2009**, *136*, 65–80. [CrossRef]

5. Hansen, N.W.; Sams, A. The microbiotic highway to health—New perspective on food structure, gut microbiota, and host inflammation. *Nutrients* **2018**, *10*, 1590. [CrossRef] [PubMed]

6. Engen, P.A.; Green, S.J.; Voigt, R.M.; Forsyth, C.B.; Keshavarzian, A. The gastrointestinal microbiome: Alcohol effects on the composition of intestinal microbiota. *Alcohol Res. Curr. Rev.* **2015**, *37*, 223–236.

7. Bach Knudsen, K.E. Microbial degradation of whole-grain complex carbohydrates and impact on short-chain fatty acids and health. *Adv. Nutr.* **2015**, *6*, 206–213. [CrossRef]

8. Maslowski, K.M.; Vieira, A.T.; Ng, A.; Kranich, J.; Sierro, F.; Yu, D.; Schilter, H.C.; Rolph, M.S.; MacKay, F.; Artis, D.; et al. Regulation of inflammatory responses by gut microbiota and chemoattractant receptor GPR43. *Nature* **2009**, *461*, 1282–1286. [CrossRef] [PubMed]

9. Tolhurst, G.; Heffron, H.; Lam, Y.S.; Parker, H.E.; Habib, A.M.; Diakogiannaki, E.; Cameron, J.; Grosse, J.; Reimann, F.; Gribble, F.M. Short-chain fatty acids stimulate glucagon-like peptide-1 secretion via the G-protein-coupled receptor FFAR2. *Diabetes* **2012**, *61*, 364–371. [CrossRef]

10. Dolara, P.; Luceri, C.; De Filippo, C.; Femia, A.P.; Giovannelli, L.; Caderni, G.; Cecchini, C.; Silvi, S.; Orpianesi, C.; Cresci, A. Red wine polyphenols influence carcinogenesis, intestinal microflora, oxidative damage and gene expression profiles of colonic mucosa in F344 rats. *Mutat. Res. Fundam. Mol. Mech. Mutagen.* **2005**, *591*, 237–246. [CrossRef]

11. Quesada-Molina, M.; Muñoz-Garach, A.; Tinahones, F.J.; Moreno-Indias, I. A new perspective on the health benefits of moderate beer consumption: Involvement of the gut microbiota. *Metabolites* **2019**, *9*, 272. [CrossRef]

12. Tzounis, X.; Vulevic, J.; Kuhnle, G.G.C.; George, T.; Leonczak, J.; Gibson, G.R.; Kwik-Uribe, C.; Spencer, J.P.E. Flavanol monomer-induced changes to the human faecal microflora. *Br. J. Nutr.* **2008**, *99*, 782–792. [CrossRef]

13. Possemiers, S.; Bolca, S.; Grootaert, C.; Heyerick, A.; Decroos, K.; Dhooge, W.; De Keukeleire, D.; Rabot, S.; Verstraete, W.; Van de Wiele, T. The Prenylflavonoid Isoxanthohumol from Hops (*Humulus lupulus* L.) Is Activated into the Potent Phytoestrogen 8-Prenylnaringenin In Vitro and in the Human Intestine. *J. Nutr.* **2006**, *136*, 1862–1867. [CrossRef] [PubMed]

14. Paraiso, I.L.; Plagmann, L.S.; Yang, L.; Zielke, R.; Gombart, A.F.; Maier, C.S.; Sikora, A.E.; Blakemore, P.R.; Stevens, J.F. Reductive Metabolism of Xanthohumol and 8-Prenylnaringenin by the Intestinal Bacterium *Eubacterium ramulus*. *Mol. Nutr. Food Res.* **2019**, *63*, e1800623. [CrossRef]

15. Teixeira, C.; Prykhodko, O.; Alminger, M.; Fåk Hållenius, F.; Nyman, M. Barley Products of Different Fiber Composition Selectively Change Microbiota Composition in Rats. *Mol. Nutr. Food Res.* **2018**, *62*, 1701023. [CrossRef]

16. Zhong, Y.; Nyman, M.; Fåk, F. Modulation of gut microbiota in rats fed high-fat diets by processing whole-grain barley to barley malt. *Mol. Nutr. Food Res.* **2015**, *59*, 2066–2076. [CrossRef] [PubMed]

17. Cummings, J.H. Short-chain fatty acid enemas in the treatment of distal ulcerative colitis. *Eur. J. Gastroenterol. Hepatol.* **1997**, *9*, 149–153. [CrossRef]

18. Rivière, A.; Selak, M.; Lantin, D.; Leroy, F.; De Vuyst, L. Bifidobacteria and Butyrate-Producing Colon Bacteria: Importance and Strategies for Their Stimulation in the Human Gut. *Front. Microbiol.* **2016**, *7*, 979. [CrossRef]

19. Ramirez-Farias, C.; Slezak, K.; Fuller, Z.; Duncan, A.; Holtrop, G.; Louis, P. Effect of inulin on the human gut microbiota: Stimulation of *Bifidobacterium adolescentis* and *Faecalibacterium prausnitzii*. *Br. J. Nutr.* **2009**, *101*, 541–550. [CrossRef]

20. Hernández-Quiroz, F.; Nirmalkar, K.; Villalobos-Flores, L.E.; Murugesan, S.; Cruz-Narváez, Y.; Rico-Arzate, E.; Hoyo-Vadillo, C.; Chavez-Carbajal, A.; Pizano-Zárate, M.L.; García-Mena, J. Influence of moderate beer consumption on human gut microbiota and its impact on fasting glucose and β-cell function. *Alcohol* **2020**, *85*, 77–94. [CrossRef]

21. Broekaert, W.F.; Courtin, C.M.; Verbeke, K.; van de Wiele, T.; Verstraete, W.; Delcour, J.A. Prebiotic and other health-related effects of cereal-derived arabinoxylans, arabinoxylan-oligosaccharides, and xylooligosaccharides. *Crit. Rev. Food Sci. Nutr.* **2011**, *51*, 178–194. [CrossRef]

22. Vardakou, M.; Nueno Palop, C.; Gasson, M.; Narbad, A.; Christakopoulos, P. In vitro three-stage continuous fermentation of wheat arabinoxylan fractions and induction of hydrolase activity by the gut microflora. *Int. J. Biol. Macromol.* **2007**, *41*, 584–589. [CrossRef] [PubMed]

23. Na, M.H.; Kim, W.K. Effects of Xylooligosaccharide Intake on Fecal Bifidobacteria, Lactic acid and Lipid Metabolism in Korean Young Women. *Korean J. Nutr.* **2007**, *40*, 154–161.

24. Moore, L.V.H.; Moore, W.E.C. *Oribaculum catoniae* gen. nov., sp. nov.; *Catonella morbi* gen. nov., sp. nov.; *Hallella seregens* gen. nov., sp. nov.; *Johnsonella ignava* gen. nov., sp. nov.; and *Dialister pneumosintes* gen. nov., comb. nov., nom. rev., anaerobic gram-negative bacilli from the human gingival crevice. *Int. J. Syst. Bacteriol.* **1994**, *44*, 187–192. [CrossRef] [PubMed]

25. Kopečný, J.; Zorec, M.; Mrázek, J.; Kobayashi, Y.; Marinšek-Logar, R. *Butyrivibrio hungatei* sp. nov. and *Pseudobutyrivibrio xylanivorans* sp. nov., butyrate-producing bacteria from the rumen. *Int. J. Syst. Evol. Microbiol.* **2003**, *53*, 201–209. [CrossRef]

26. Rajilić-Stojanović, M.; de Vos, W.M. The first 1000 cultured species of the human gastrointestinal microbiota. *FEMS Microbiol. Rev.* **2014**, *38*, 996–1047. [CrossRef] [PubMed]

27. Redondo-Useros, N.; Gheorghe, A.; Díaz-Prieto, L.E.; Villavisencio, B.; Marcos, A.; Nova, E. Associations of Probiotic Fermented Milk (PFM) and Yogurt Consumption with *Bifidobacterium* and *Lactobacillus* Components of the Gut Microbiota in Healthy Adults. *Nutrients* **2019**, *11*, 651. [CrossRef] [PubMed]

28. Nova, E.; San Mauro-Martín, I.; Díaz-Prieto, L.E.; Marcos, A. Wine and beer within a moderate alcohol intake is associated with higher levels of HDL-c and adiponectin. *Nutr. Res.* **2019**, *63*, 42–50. [CrossRef] [PubMed]

29. Martin-Moreno, J.M.; Boyle, P.; Gorgojo, L.; Maisonneuve, P.; Fernandez-Rodriguez, J.C.; Salvini, S.; Willett, W.C. Development and validation of a food frequency questionnaire in Spain. *Int. J. Epidemiol.* **1993**, *22*, 512–519. [CrossRef] [PubMed]

30. Mataix, J. Tablas de Composición de Alimentos [Food Composition Tables]. Available online: http://www.sennutricion.org/es/2013/05/11/tablas-de-composicin-de-alimentos-mataix-et-al (accessed on 27 May 2019).

31. Dore, J.; Ehrlich, S.D.; Levenez, F.; Pelletier, E.; Alberti, A.; Bertrand, L.; Bork, P.; Costea, P.I.; Sunagawa, S.; Guarner, F.; et al. IHMS_SOP 06 V1: Standard Operating Procedure for Faecal Samples DNA Extraction, Protocol, Q. In *International Human Microbiome Standards*; INRA: Paris, France, 2015.

32. 16S Metagenomics App. Illumina. Available online: https://support.illumina.com/help/BaseSpace_App_16S_Metagenomics_help/16S_Metagenomics_App_Help.htm (accessed on 27 May 2019).

Sample Availability: Not available.

Publisher's Note: MDPI stays neutral with regard to jurisdictional claims in published maps and institutional affiliations.

Article

Ethylchloroformate Derivatization for GC–MS Analysis of Resveratrol Isomers in Red Wine

Elisa Di Fabio, Alessio Incocciati, Federica Palombarini, Alberto Boffi, Alessandra Bonamore *,† and Alberto Macone *,†

Department of Biochemical Sciences, "Sapienza" University of Rome, p.le A.Moro 5, 00185 Rome, Italy; elisa.difabio@uniroma1.it (E.D.F.); incocciati.1750499@studenti.uniroma1.it (A.I.); federica.palombarini@uniroma1.it (F.P.); alberto.boffi@uniroma1.it (A.B.)
* Correspondence: alessandra.bonamore@uniroma1.it (A.B.); alberto.macone@uniroma1.it (A.M.)
† These authors contributed equally to this work.

Received: 17 September 2020; Accepted: 9 October 2020; Published: 9 October 2020

Abstract: Resveratrol (3,5,4′-trihydroxystilbene) is a natural compound that can be found in high concentrations in red wine and in many typical foods found in human diet. Over the past decades, resveratrol has been widely investigated for its potential beneficial effects on human health. At the same time, numerous analytical methods have been developed for the quantitative determination of resveratrol isomers in oenological and food matrices. In the present work, we developed a very fast and sensitive GC–MS method for the determination of resveratrol in red wine based on ethylchloroformate derivatization. Since this reaction occurs directly in the water phase during the extraction process itself, it has the advantage of significantly reducing the overall processing time for the sample. This method presents low limits of quantification (LOQ) (25 ng/mL and 50 ng/mL for *cis*- and *trans*-resveratrol, respectively) and excellent accuracy and precision. Ethylchloroformate derivatization was successfully applied to the analysis of resveratrol isomers in a selection of 15 commercial Italian red wines, providing concentration values comparable to those reported in other studies. As this method can be easily extended to other classes of molecules present in red wine, it allows further development of new GC–MS methods for the molecular profiling of oenological matrices.

Keywords: resveratrol; red wine; ethylchloroformate; gas chromatography–mass spectrometry

---

1. Introduction

Resveratrol (RSV) is a natural compound that can be found in high concentrations in red wine and in many foods found in the human diet [1–4]. For many years, this molecule has been the subject of many studies concerning its potential benefits for human health. It was shown that RSV is an excellent antioxidant [5], and it can have effects on many cellular processes, from aging and inflammation to stress resistance and cell survival [6–9]. At the same time, more and more analytical methods have been developed for the quantitative analysis of RSV isomers in oenological and food matrices as well as in biological fluids [10]. Most of these methods are based on HPLC or UPLC coupled with electrochemical, mass spectrometric, or photometric detectors; GC–MS; and capillary electrophoresis [11–14]. Developing an analytical method for both RSV isomers may be challenging because only *trans*-RSV is commercially available. Typically, *cis*-RSV standard is produced by exposing *trans*-RSV to UV rays (Figure 1).

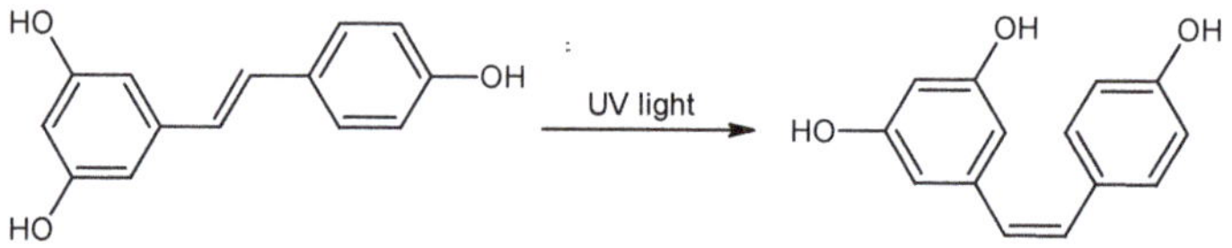

Figure 1. *trans*-RSV to *cis*-RSV isomerization induced by UV light.

Our research team has recently developed an analytical method for the simultaneous analysis in red wine of TBDMS derivatives of both RSV isomers and 2,4,6-trihydroxyphenanthrene, a RSV derivative which can be detected in red wine following exposure to UV rays [13,15]. In the present work, we developed a particularly fast and highly sensitive GC–MS method for the analysis of RSV in the form of ethoxycarbonyl derivative. Ethylchloroformate (ECF) derivatization has been known for a long time [16] and has been applied to different classes of molecules including amino and fatty acids, polyamines, and phenolic acids [17–19]. Surprisingly, to the best of our knowledge, it has never been applied to the qualitative and quantitative analysis of RSV isomers. Considering that the derivatization reaction with chloroformates occurs in the presence of water during the extraction process itself, this technique could be advantageous for the analysis of RSV because it significantly reduces the processing times of the sample, thus limiting the formation of artifacts. Unlike the classic GC–MS derivatization techniques, which very often require incubation at high temperatures for variable times, ECF can react almost immediately at room temperature with the molecules to be derivatized.

In the present work, we report for the first time the typical mass spectra of the isomers of RSV in the form of ethoxycarbonyl derivatives. We also show that this type of derivatization/extraction is suitable for the development of an analytical method with excellent characteristics of accuracy, precision, linearity, and sensitivity. Once validated, the analytical method will be used for the analysis of RSV in a selection of Italian red wines.

2. Results and Discussion

2.1. Derivatization Process

In the present paper, we set up a fast and practical analytical method for the determination of RSV isomers in red wine using ECF as a derivatizing agent (Figure 2).

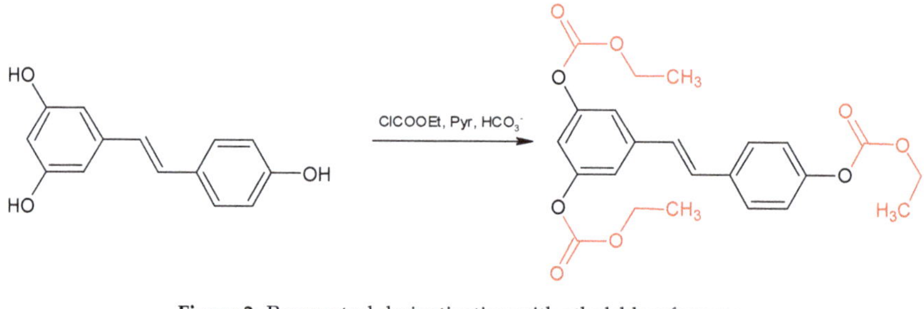

Figure 2. Resveratrol derivatization with ethylchloroformate.

Chloroformates are well known as efficient derivatizing reagents that are able to react in aqueous media, shortening the time required for sample processing. This derivatization methodology requires the use of pyridine as a catalyst, as well as the use of an alkaline environment that allows the ethoxycarbonylation of the phenolic hydroxyl groups. The development of the derivatization conditions was done only with *trans*-RSV. This allows to evaluate any *cis* isomerization inherent in the procedure itself, thus excluding the occurrence of artifacts in the analysis of wines. The first step in developing the method was the optimization of the extraction and of ECF concentration. The solvents generally used in this procedure are either hexane or more polar solvents such as chloroform and ethylacetate [17–19]. Therefore, we decided to evaluate the derivatization/extraction efficiency *trans*-RSV with these three solvents using methyl heptadecanoate. This molecule is completely soluble in these three solvents used to test the derivatization/extraction efficiency and it does not have functional groups that can react with ECF, so its concentration does not vary in all phases of the derivatization/extraction process. For this purpose, we extracted 0.5 mL of alkalinized wine with 2 mL of solvent containing methyl heptadecanoate in the presence of a fixed quantity of ECF (20 µL) and pyridine (10 µL) as a catalyst. The results obtained show that extraction with hexane provides the best recovery yields (Figure 3).

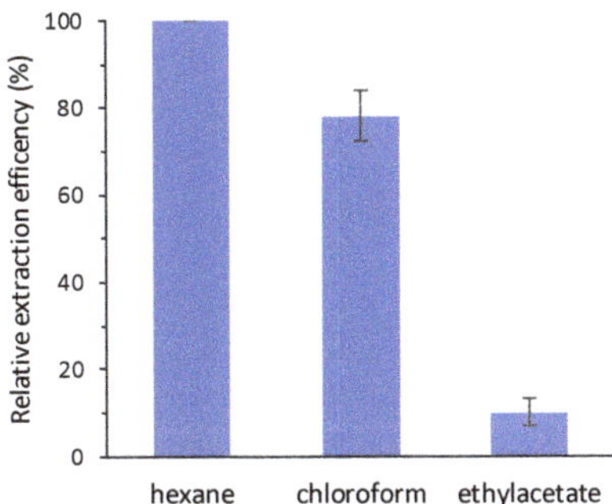

Figure 3. Relative extraction efficiency of derivatized *trans*-RSV with different organic solvents. Relative extraction efficiency of the organic solvents was obtained by setting the highest *trans*-RSV: methyl heptadecanoate peak area ratio equal to 100% (*n* = 3; mean ± SD).

By setting hexane as the solvent of choice, we then assessed the amount of ECF to be used. The derivatization efficiency was the same for quantities tested (10, 20, 30, 40, and 50 µL). Eventually, 30 µL was chosen as model concentration, considering that there may be wines richer in RSV (or in other molecules with similar derivatization potential) that could require a higher ECF concentration.

Given the complexity of the oenological matrix, we assessed whether introducing a second extraction step with chloroform could improve the efficiency of the process. At the same time, we decided to perform this second extraction by adding an additional 20 µL of ECF. This further step improves the overall extraction process (+15.33% ± 3.59%; *n* = 3; mean ± SD). However, chloroform extraction has two main drawbacks: the organic phase, representing the bottom layer, is more difficult to recover while an insoluble material is deposited at the interface, which makes quantitative recovery of the organic phase challenging. For this reason, it was decided to reduce the effect of the matrix by extracting half of the starting volume of wine (0.25 mL instead of 0.5 mL). Surprisingly, the yields have doubled, probably due to the extremely favorable organic solvent/aqueous phase ratio (8:1 vs. 4:1).

Since red wine contains on average 13% ethanol, we evaluated whether the extraction/derivatization process was affected by an increase in its concentration. It is known that ethanol is used to promote the formation of ethyl esters of carboxylic acids when ECF is used. Indeed, we have observed that the extraction/derivatization process of molecules bearing carboxylic groups (e.g., gallic acid) is greatly influenced by the concentration of ethanol (data not shown). This is different than in the case of RSV that lacks carboxyl groups. To confirm this assumption, we tested the extraction/derivatization of wine samples in the presence of a higher concentration of ethanol and we observed that, as expected, the derivatization yield remained identical.

2.2. GC–MS Characterization of ECF Derivatives

Derivatization with ECF makes the molecules particularly suitable for gas chromatographic analysis. RSV isomers and pinostilbene (in the form of ethoxycarbonyl derivatives) are well separated on the HP5–MS chromatographic column with the following retention times: 17.6 min for *cis*-RSV, 18.5 min for pinostilbene (internal standard), and 21.7 min for *trans*-RSV. Furthermore, ethoxycarbonylation of hydroxyl groups is quantitative as no peaks related to partially derivatized species can be detected. In Figure 4, the mass spectra of the ethoxycarbonyl derivatives obtained are reported. The molecular ion is present in the mass spectra of all derivatized species. In addition, a prominent peak corresponding to [M-73]+ ion is detected, which corresponds to the ion formed from the loss of ethoxycarbonyl radical ($CO_2C_2H_5$) form the molecular ion (M^+). In addition, ions corresponding to the molecular weight of underivatized molecules are always present (m/z 228 for *cis*- and *trans*-RSV which show the same fragmentation pattern; m/z 242 for pinostilbene). Given their abundance, those ions were chosen for the validation of the analytical method.

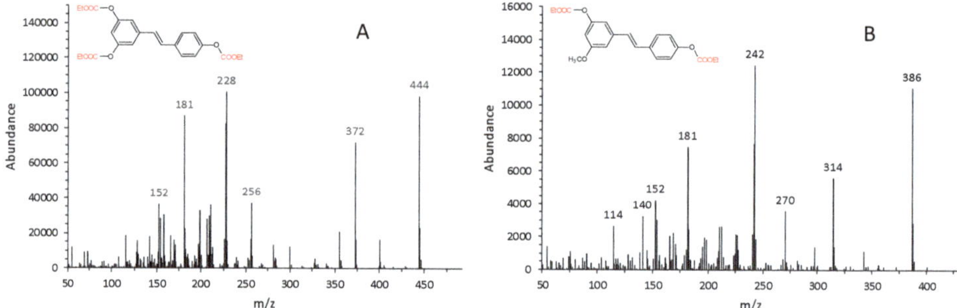

Figure 4. Electron ionization (EI) mass spectra of (**A**) *trans/cis*-resveratrol as tri-ethoxycarbonyl derivatives (the isomers show an identical fragmentation pattern) and (**B**) pinostilbene as di-ethoxycarbonyl derivative (internal standard).

2.3. Method Validation

Given the complexity of red wine and the variability of its composition, in our previous work, we developed a wine-like matrix that contains its main constituents [13]. The same matrix was used in the present work as well. In particular, the validation of the analytical method was performed using a pH 3.3 solution containing 13% ethanol and 0.3% *v/v* tartaric acid. We opted for this value, as it is reported that in musts from grapes produced from vineyards located in northern regions, the concentration of tartaric acid is higher than 6 g/L, while in musts from southern regions, that concentration does not exceed 2–3 g/L [20]. Therefore, we chose an average value of 4 g/L.

The matrix effect was evaluated for both RSV isomers by comparing the slopes of regression lines in wine-like matrix with the slopes calculated for each isomer in the control wine sample. The experiments were performed in triplicate and the slopes obtained for *trans*-RSV were $0.0033 \pm 5.55 \times 10^{-5}$ in wine-like matrix and $0.0034 \pm 1 \times 10^{-4}$ in red wine. These values do not significantly differ as assessed by Student's *t*-test ($p = 0.649$). Similar results were obtained for *cis*-RSV ($0.0126 \pm 3.06 \times 10^{-4}$ in wine-like matrix vs $0.0126 \pm 1.53 \times 10^{-4}$ in red wine; $p = 1.00$). Based on these results, the calibration obtained with wine-like matrix can be used for quantification purposes. In addition, the lack of detectable matrix effect can be explained by the lack of interfering peaks at the retention times of derivatized *trans*- and *cis*-RSV.

The linearity of the method was tested separately on *trans*- and *cis*-RSV. For both of the analytes, a good linearity was achieved (Table 1) with an R^2 coefficient always ≥ 0.999.

Limit of quantification (LOQ) corresponds to the lowest concentration value used in the calibration plot, i.e., 50 ng/mL for *trans*-RSV and 25 ng/mL for *cis*-RSV. At lower concentrations, at S/N = 3, it was not possible to identify RSV isomers in a reliable way. Thus, in this specific case, LOQ and limit of detection (LOD) values are the same. The same experimental observation was reported also by Paulo et al. [21].

Table 1. Validation parameters.

Compound	Range (ng/mL)	Slope	Intercept	R^2	LOQ (LOD) * (ng/mL)	Concentration (ng/mL)	Accuracy (recovery %)	Precision (RSD %)
trans-RSV	50–3000	0.0033	0.0907	0.9992	50	200	99.02	5.46
						2000	99.20	3.28
cis-RSV	25–1000	0.0126	0.08456	0.9991	25	100	103.11	4.19
						1000	99.88	1.58

* LOQ and LOD values are the same for both RSV isomers.

In comparison to all protocols that require one or more extraction steps with organic solvents followed by a derivatization step, the use of ECF has the advantage of directly derivatizing the molecules in the presence of the aqueous phase while the extraction process is taking place. This allows a considerable reduction of the sample's processing times while ensuring almost complete substrate recovery. At least two consecutive extraction/derivatization steps were needed to fully recover the analytes from wine-like matrix. As reported in Table 1, the recovery of each RSV isomer at two different concentrations (*trans*-RSV: 200 ng/mL and 2000 ng/mL; *cis*-RSV: 100 ng/mL and 1000 ng/mL) was >99%. Concerning precision, the % RSD values obtained both for *trans*- and for *cis*-RSV fell well within the criteria normally accepted in bioanalytical method validation being lower than 10% [22].

2.4. Red Wine Analysis

Red wine has been consumed by humans for hundreds of years and its beneficial effects on human health are well described [23,24]. The antioxidant activity of red wine is due to the synergy of *cis*- and *trans*-RSV with other molecules such as catechins, anthocyanins, polyphenols, and flavanols, which are particularly abundant in this specific oenological matrix [25]. In the early 1990s, RSV became popular as it was recognized as one of the main components of red wine responsible for the so-called French paradox, according to which the French have a low incidence of coronary heart disease despite consuming a diet rich in saturated fats [24]. Since then, several GC–MS analytical methods have been developed for the quantitative analysis of RSV isomers in red wine [14]. Since both RSV isomers show remarkable antioxidant properties [26], it is essential to determine also the *cis* isomer, which is present in non-negligible quantities in wine. The method validated in this work, unlike the others, has the advantage of being particularly fast as the derivatization with the ECF proceeds directly in the aqueous phase at room temperature. This method was applied to the quantitative analysis of the RSV isomers in 15 wines from different Italian regions that differ in vintage and grape variety.

In Figure 5, a typical GC–MS chromatographic profile of a wine sample submitted to ECF derivatization is reported.

RSV isomers and the internal standard are well resolved and elute in a part of the chromatogram free of interfering peaks. In addition, in the first 15 min of elution, it is possible to observe the presence of numerous peaks, among which there are molecules with acidic functional groups that are derivatized as ethyl esters (e.g., gallic acid which elutes at 14.4 min). This reaction is possible as about 13% ethanol is normally present in red wine. As expected, the quantitative analysis shows that RSV content can vary significantly between wines (Table 2).

This parameter is influenced mainly, but not exclusively, by the grapes that are used for red wine production, as RSV is found in widely varying amounts among grape varieties. For example, it is known that the grape variety known as "pinot noir" is particularly rich in RSV as well as the wine derived from it [27]. In the present work, we obtained a similar result in that, among the wines tested, pinot noir (#3) shows the highest concentration of both *trans*- and *cis*-RSV.

Overall, the total RSV content in the 15 wines tested in this paper ranges from a minimum of 336.42 ng/mL to a maximum of 3095.70 ng/mL. These data are comparable with those reported in other papers, where RSV was determined with different analytical methods on red wines from different geographical origin [13,21,26].

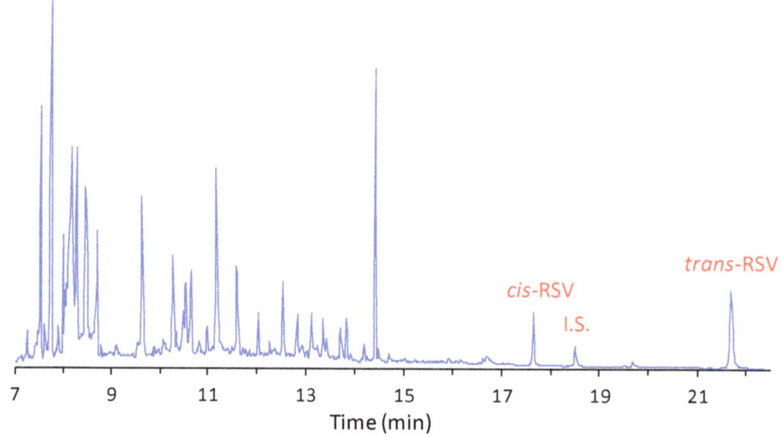

Figure 5. Typical GC–MS chromatogram of a red wine sample derivatized with ethylchloroformate.

Table 2. *trans-* and *cis*-resveratrol content in a selection of Italian red wines (values are the mean of two measurements).

Wine	Vintage	Italian Region	Varieties	*trans*-RSV (ng/mL)	*cis*-RSV (ng/mL)	Total RSV (ng/mL)
#1	2018	Piemonte	100% Barbera	1185.06	343.27	1528.33
#2	2019	Alto Adige	100% Lagrain	475.97	170.77	646.74
#3	2017	Alto Adige	100% Pinot Noir	1772.94	1322.76	3095.70
#4	2017	Veneto	70% Corvina, 30% Rondinella	275.97	67.84	343.81
#5	2015	Friuli Venezia Giulia	100% Cabernet Franc	766.88	302.48	1069.36
#6	2016	Toscana	90% Sangiovese, 10% Merlot	885.06	290.44	1175.50
#7	2016	Toscana	100% Sangiovese	1339.61	688.73	2028.34
#8	2018	Umbria	70% Sangiovese, 15% Merlot, 15% Sagrantino	688.09	145.98	834.07
#9	2018	Lazio	100% Cesanese	385.06	156.88	541.94
#10	2019	Lazio	100% Cabernet Sauvignon	594.15	154.67	748.82
#11	2016	Campania	100% Aglianico	254.76	81.66	336.42
#12	2016	Puglia	100% Primitivo	891.12	243.44	1134.56
#13	2018	Puglia	100% Negramaro	945.67	408.77	1354.44
#14	2018	Sicilia	60% Merlot, 40% Cabernet Sauvignon	485.06	123.66	608.72
#15	2018	Sicilia	100% Syrah	530.52	213.86	744.38

In addition, two of the wines analyzed in this study (#6 and #11) had already been analyzed with a different analytical method developed by our research group [13]. This method, which involved extraction with organic solvents and derivatization with TBDMS, had provided *trans-* and *cis*-RSV values equal to 266.4 ng/mL and 77.6 ng/mL, respectively, for wine #11, and 848.2 ng/mL and 283.7 ng/mL, respectively, for wine #6. These data are comparable to those reported in Table 2 with a variation in total RSV content lower than 4%. These data further confirm the accuracy of the analytical method here developed.

In conclusion, we have set up an analytical method for the analysis of RSV in red wines based on ECF derivatization. This method is fast, sensitive, and specific, providing low LOD and LOQ. Precision and accuracy are in conformity with the criteria normally accepted in methods validation, with practically total recovery and a percentage RSD lower than 5%.

Finally, this work demonstrates that derivatization with ECF can also be extended to other classes of molecules present in wine such as polyphenols. We have, in fact, observed that by varying the concentration of ethanol during the extraction step, it is possible to optimize the analysis of phenolic acids such as gallic acid as well. This provides a future blueprint for the development of new analytical methods in GC–MS aimed at the molecular characterization of oenological matrices.

3. Materials and Methods

3.1. Reagents and Standards

Hexane, chloroform, ethylchloroformate, *trans*-RSV, and pinostilbene (internal standard) were purchased from Sigma-Aldrich (Germany). Standard stock solutions were prepared by dissolving *trans*-RSV and pinostilbene in ethanol. All calibrations were performed by diluting the stock solutions in wine-like matrix (50 mL final volume containing: 150 mg disodium tartrate, 6.5 mL ethanol (13% final concentration), 43.5 mL H_2O adjusted to pH 3.3 with an aqueous tartaric acid solution). *cis*-RSV standard solution was prepared according to Francioso et al. [13]. Briefly, 30 µg/mL *trans*-RSV in wine-like matrix were exposed for 2 min to UV light (20 cm from the irradiation source, with a 14.7 W UV-B fluorescent tube emitting at wavelengths of 270–320 nm with a peak at 312 nm). Conversion rate was estimated using GC–MS by comparing *trans*-RSV peak areas before and after the treatment with UV light.

3.2. Extraction/Derivatization Procedure

RSV isomers were subjected to derivatization with ECF according to the following protocol: 0.25 mL of wine sample or standards dissolved in wine-like matrix (containing 500 ng of pinostilbene) were put in a 10 mL glass tube. This tube was shielded from light by wrapping it with aluminum foil to minimize light-induced isomerization of *trans*-RSV. The solution was made alkaline (pH > 9) by adding 65 µL of 0.6 M $NaHCO_3$. Hexane (2 mL) and ECF (30 µL) were added to this solution followed by the slow addition of 10 µL of pyridine as catalyst. The tube was left uncapped for a few seconds to allow the releasing of CO_2. After 2 min shaking, the organic layer was removed, and a second extraction step was performed with chloroform (2 mL) containing further 20 µL of ECF. The lower organic layer was removed, combined with the hexane extract, and dried using a nitrogen stream. The sample was resuspended in 75 µL of chloroform and subjected to GC–MS analysis.

Extraction efficiency of derivatized *trans*-RSV was tested with three different organic solvents (hexane, chloroform, or ethylacetate) was tested in the same wine sample using methyl heptadecanoate as an internal standard. Methyl heptadecanoate was dissolved in the solvents used for the extraction at a final concentration of 500 ng/mL. Relative derivatization efficiency of the organic solvents was obtained by setting the highest *trans*-RSV: methyl heptadecanoate peak area ratio equal to 100%.

3.3. Gas Chromatography–Mass Spectrometry

GC–MS analyses were performed with an Agilent 7890B gas chromatograph coupled to a 5977B quadrupole mass selective detector (Agilent Technologies, Palo Alto, CA, USA). Chromatographic separations were carried out with an Agilent HP5ms fused-silica capillary column (30 m × 0.25 mm i.d.) coated with 5% phenyl 95%-dimethylpolysiloxane (film thickness 0.25 µm) as stationary phase. Injection mode: splitless at a temperature of 260 °C. Column temperature program: 70 °C (1 min) then to 300 °C at a rate of 15 °C/min and held for 5 min, solvent delay: 7 min. The carrier gas was helium at a constant flow of 1.0 mL/min. The spectra were obtained in the electron impact mode at 70 eV ionization energy; ion source 280 °C; ion source vacuum 10^{-5} Torr. Mass spectrometric analysis was performed simultaneously in TIC (mass range scan from m/z 50 to 650 at a rate of 0.42 scans s^{-1}) and SIM mode (selected ions: m/z 242 for internal standard and m/z 444 for *cis/trans*-RSV).

3.4. Method Validation

Calibrations were performed adding increasing amounts of RSV isomers to 0.25 mL of wine-like matrix containing 500 ng of pinostilbene as internal standard. The calibration samples were subjected to the ECF derivatization as described above.

Calibration plot for *trans*-RSV was performed in the range of 50–3000 ng/mL (seven calibration points). Calibration curve of *cis*-RSV standard was performed in a separate experiment in the range of 25–1000 ng/mL (six calibration points). *cis*-RSV standard was obtained at a final concentration of 10 μg/mL by exposing *trans*-RSV (30 μg/mL in wine-like matrix) to UV light for 2 min. Three replicate analyses were performed at each concentration in wine-like matrix. The calibration curves were obtained by plotting the peak area ratio between each analyte and the internal standard versus analyte concentration.

Accuracy and precision were determined in a wine sample spiked with *trans*-RSV and *cis*-RSV at two different final concentrations (*trans*-RSV: 200 ng/mL and 2000 ng/mL; *cis*-RSV: 100 ng/mL and 1000 ng/mL) analyzing five replicates for each concentration in the same day. Spiked and unspiked wine samples were subjected to ECF derivatization and analyzed by GC–MS.

Accuracy was evaluated through standard recovery experiments. A comparison of the amount found versus the amount added provides the recovery of the method (%) which is an estimate of the accuracy of the method itself. The same samples reported above were also used to determine method precision expressed as % relative standard deviation (% RSD).

LOD and LOQ were determined by the analysis of wine-like matrix with decreasing concentrations of *trans*-RSV and *cis*-RSV. The limit of detection (LOD) of the target compounds is taken at S/N = 3, whereas the limit of quantification (LOQ) was set to S/N = 10.

The matrix effect was evaluated by analyzing *trans*-RSV and *cis*-RSV both in wine-like matrix and in red wine, and by comparing the slopes of the regression plots by Student's *t*-test.

3.5. Red Wine Analysis

trans-RSV and *cis*-RSV were measured in fifteen different red wines from different Italian regions, grape varieties, and vintage. 0.25 mL of each red wine were spiked with 500 ng of pinostilbene as internal standard and submitted to derivatization/extraction procedure as described above.

Author Contributions: Conceptualization, A.B. (Alessandra Bonamore) and A.M.; methodology, A.M.; validation, A.I., E.D.F., and F.P.; formal analysis, A.I., E.D.F., and F.P.; data curation, A.B. (Alessandra Bonamore) and A.M.; writing—original draft preparation, A.B. (Alberto Boffi), A.B. (Alessandra Bonamore), and A.M.; writing—review and editing, A.B. (Alberto Boffi), A.B. (Alessandra Bonamore), and A.M. All authors have read and agreed to the published version of the manuscript.

Funding: This research was funded by Sapienza University of Rome, "Progetto Ateneo 2019" to A.B. The APC was funded by Sapienza University of Rome, "Progetto Ateneo 2019".

Conflicts of Interest: The authors declare no conflict of interest.

References

1. Castaldo, L.; Narváez, A.; Izzo, L.; Graziani, G.; Gaspari, A.; Di Minno, G.; Ritieni, A. Red Wine Consumption and Cardiovascular Health. *Molecules* **2019**, *24*, 3626. [CrossRef] [PubMed]

2. Bavaresco, L.; Lucini, L.; Busconi, M.; Flamini, R.; De Rosso, M. Wine Resveratrol: From the Ground Up. *Nutrients* **2016**, *8*, 222. [CrossRef] [PubMed]

3. Wang, J.F.; Ma, L.; Xi, H.F.; Wang, L.J.; Li, S.H. Resveratrol synthesis under natural conditions and after UV-C irradiation in berry skin is associated with berry development stages in 'Beihong' (V. vinifera × V. amurensis). *Food Chem.* **2015**, *168*, 430–438. [CrossRef]

4. Sobolev, V.S.; Cole, R.J. Trans-resveratrol content in commercial peanuts and peanut products. *J. Agric. Food Chem.* **1999**, *47*, 1435–1439. [CrossRef] [PubMed]

5. Nardini, M.; Foddai, M.S. Phenolics Profile and Antioxidant Activity of Special Beers. *Molecules* **2020**, *25*, 2466. [CrossRef]

6. Silva, P.; Sureda, A.; Tur, J.A.; Andreoletti, P.; Cherkaoui-Malki, M.; Latruffe, N. How efficient is resveratrol as an antioxidant of the Mediterranean diet, towards alterations during the aging process? *Free Radic. Res.* **2019**, *53*, 1101–1112. [CrossRef]

7. Vervandier-Fasseur, D.; Latruffe, N. The Potential Use of Resveratrol for Cancer Prevention. *Molecules* **2019**, *24*, 4506. [CrossRef]

8. Velmurugan, B.K.; Rathinasamy, B.; Lohanathan, B.P.; Thiyagarajan, V.; Weng, C.F. Neuroprotective Role of Phytochemicals. *Molecules* **2018**, *23*, 2485. [CrossRef]

9. Leri, M.; Scuto, M.; Ontario, M.L.; Calabrese, V.; Calabrese, E.J.; Bucciantini, M.; Stefani, M. Healthy Effects of Plant Polyphenols: Molecular Mechanisms. *Int. J. Mol. Sci.* **2020**, *21*, 1250. [CrossRef]

10. Fiod Riccio, B.V.; Fonseca-Santos, B.; Colerato Ferrari, P.; Chorilli, M. Characteristics, Biological Properties and Analytical Methods of Trans-Resveratrol: A Review. *Crit. Rev. Anal. Chem.* **2020**, *50*, 339–358. [CrossRef]

11. Svilar, L.; Martin, J.; Defoort, C.; Paut, C.; Tourniaire, F.; Brochot, A. Quantification of trans-resveratrol and its metabolites in human plasma using ultra-high performance liquid chromatography tandem quadrupole-orbitrap mass spectrometry. *J. Chromatogr. B Analyt. Technol. Biomed. Life Sci.* **2019**, *1104*, 119–129. [CrossRef] [PubMed]

12. Ji, M.; Li, Q.; Ji, H.; Lou, H. Investigation of the distribution and season regularity of resveratrol in Vitis amurensis via HPLC-DAD-MS/MS. *Food Chem.* **2014**, *142*, 61–65. [CrossRef] [PubMed]

13. Francioso, A.; Laštovičková, L.; Mosca, L.; Boffi, A.; Bonamore, A.; Macone, A. Gas Chromatographic–Mass Spectrometric Method for the Simultaneous Determination of Resveratrol Isomers and 2,4,6-Trihydroxyphenanthrene in Red Wines Exposed to UV-Light. *J. Agric. Food Chem.* **2019**, *67*, 11752–11757. [CrossRef] [PubMed]

14. Fan, E.; Lin, S.; Du, D.; Jia, Y.; Kang, L.; Zhang, K. Currents separative strategies used for resveratrol determination from natural sources. *Anal. Methods* **2011**, *3*, 2454–2462. [CrossRef]

15. Francioso, A.; Boffi, A.; Villani, C.; Manzi, L.; D'Erme, M.; Macone, A.; Mosca, L. Isolation and identification of 2,4,6-trihydroxyphenanthrene as a byproduct of trans-resveratrol photochemical isomerization and electrocyclization. *J. Org. Chem.* **2014**, *79*, 9381–9384. [CrossRef]

16. Husek, P. Chloroformates in gas chromatography as general purpose derivatizing agents. *J. Chromatogr. B Biomed. Sci. Appl.* **1998**, *717*, 57–91. [CrossRef]

17. Reddy, B.S.; Chary, V.N.; Pavankumar, P.; Prabhakar, S. Characterization of N-methylated amino acids by GC-MS after ethyl chloroformate derivatization. *J. Mass Spectrom.* **2016**, *51*, 638–650. [CrossRef]

18. Boffi, A.; Favero, G.; Federico, R.; Macone, A.; Antiochia, R.; Tortolini, C.; Sanzò, G.; Mazzei, F. Amine oxidase-based biosensors for spermine and spermidine determination. *Anal. Bioanal. Chem.* **2015**, *407*, 1131–1137. [CrossRef]

19. Citová, I.; Sladkovský, R.; Solich, P. Analysis of phenolic acids as chloroformate derivatives using solid phase microextraction-gas chromatography. *Anal. Chim. Acta* **2006**, *573*, 231–241. [CrossRef]

20. Vera, L.; Mestres, M.; Boqué, R.; Busto, O.; Guash, J. Use of synthetic wine for models transfer in wine analysis by HS-MS e-nose. *Sens. Actuat. B Chem.* **2010**, *143*, 689–695. [CrossRef]

21. Paulo, L.; Domingues, F.; Queiroz, J.A.; Gallardo, E. Development and validation of an analytical method for the determination of trans- and cis-resveratrol in wine: Analysis of its contents in 186 Portuguese red wines. *J. Agric. Food Chem.* **2011**, *59*, 2157–2168. [CrossRef] [PubMed]

22. Tiwari, G.; Tiwari, R. Bioanalytical method validation: An updated review. *Pharm. Methods* **2010**, *1*, 25–38. [CrossRef] [PubMed]

23. Natella, F.; Macone, A.; Ramberti, A.; Forte, M.; Mattivi, F.; Matarese, R.M.; Scaccini, C. Red wine prevents the postprandial increase in plasma cholesterol oxidation products: A pilot study. *Br. J. Nutr.* **2011**, *105*, 1718–1723. [CrossRef] [PubMed]

24. Haseeb, S.; Alexander, B.; Baranchuk, A. Wine and cardiovascular health: A comprehensive review. *Circulation* **2017**, *136*, 1434–1448. [CrossRef] [PubMed]

25. Di Majo, D.; La Guardia, M.; Giammanco, S.; La Neve, L.; Giammanco, M. The antioxidant capacity of red wine in relationship with its polyphenolic constituents. *Food Chem.* **2008**, *111*, 45–49. [CrossRef]

26. Orallo, F. Comparative studies of the antioxidant effects of cis- and trans-resveratrol. *Curr. Med. Chem.* **2006**, *13*, 87–98. [CrossRef] [PubMed]

27. Weiskirchen, S.; Weiskirchen, R. Resveratrol: How Much Wine Do You Have to Drink to Stay Healthy? *Adv. Nutr.* **2016**, *7*, 706–718. [CrossRef]

Review

Effects of the Non-Alcoholic Fraction of Beer on Abdominal Fat, Osteoporosis, and Body Hydration in Women

Marta Trius-Soler [1,2,3,†], Arnau Vilas-Franquesa [4,†], Anna Tresserra-Rimbau [1,2,3], Gemma Sasot [1,2], Carolina E. Storniolo [1,2,3], Ramon Estruch [3,5] and Rosa M. Lamuela-Raventós [1,2,3,*]

[1] Department of Nutrition, Food Sciences and Gastronomy, XaRTA, School of Pharmacy and Food Sciences, University of Barcelona, 08028 Barcelona, Spain; mtrius@ub.edu (M.T.-S.); annatresserra@ub.edu (A.T.-R.); gemmamsf@gmail.com (G.S.); carolina.e.storniolo@gmail.com (C.E.S.)

[2] INSA-UB, Nutrition and Food Safety Research Institute, University of Barcelona, 08921 Santa Coloma de Gramanet, Spain

[3] CIBER Fisiopatología de la Obesidad y Nutrición (CIBEROBN), Instituto de Salud Carlos III, 28029 Madrid, Spain; restruch@clinic.cat

[4] Centre d'Innovació, Recerca i Transferència en Tecnologia dels Aliments (CIRTTA), XaRTA, TECNIO, MALTA-Consolider, Departament de Ciència Animal i dels Aliments, Facultat de Veterinària, Universitat Autònoma de Barcelona (Cerdanyola del Vallès), 08193 Bellaterra, Spain; arnauvilas.nl@gmail.com

[5] Department of Internal Medicine, Hospital Clínic, Institut d'Investigacions Biomèdiques August Pi i Sunyer (IDIBAPS), University of Barcelona, 08036 Barcelona, Spain

[*] Correspondence: lamuela@ub.edu; Tel.: +34-934-034-843

[†] These authors contributed equally to this work.

Academic Editor: Mirella Nardini

Received: 22 July 2020; Accepted: 25 August 2020; Published: 27 August 2020

Abstract: Several studies have shown that binge drinking of alcoholic beverages leads to non-desirable outcomes, which have become a serious threat to public health. However, the bioactive compounds in some alcohol-containing beverages might mitigate the negative effects of alcohol. In beer, the variety and concentration of bioactive compounds in the non-alcoholic fraction suggests that its consumption at moderate levels may not only be harmless but could also positively contribute to an improvement of certain physiological states and be also useful in the prevention of different chronic diseases. The present review focuses on the effects of non-alcoholic components of beer on abdominal fat, osteoporosis, and body hydration in women, conditions selected for their relevance to health and aging. Although beer drinking is commonly believed to cause abdominal fat deposition, the available literature indicates this outcome is inconsistent in women. Additionally, the non-alcoholic beer fraction might improve bone health in postmenopausal women, and the effects of beer on body hydration, although still unconfirmed seem promising. Most of the health benefits of beer are due to its bioactive compounds, mainly polyphenols, which are the most studied. As alcohol-free beer also contains these compounds, it may well offer a healthy alternative to beer consumers.

Keywords: hops; malt; health; menopause; polyphenol; phytoestrogen; prenylnarigenin; humulones; ethanol; bioactives

---

1. Introduction

Beer, an alcoholic drink composed of four main ingredients (water, malt, hops, and yeast) [1], is one of the most consumed beverages in the world [2]. From a nutritional point of view, its main components are water (around 90%), followed by carbohydrates, ethanol, minerals, vitamins, and bioactive compounds such as polyphenols and organic acids (iso-α-humulones). Beer composition, as well as its flavor, taste, and texture, differs considerably according to the ingredients and processing techniques [3]. Besides their health benefits, the bioactive compounds are also linked to the sensory characteristics of beer [4].

In view of the worldwide growth in beer consumption, studies investigating possible links between beer and different health outcomes are of utmost importance. Among others (i.e., liver disease), recently, one of the most important consequences of a high beer consumption is a greater risk of developing different site-specific cancers (e.g., colorectal [5], lung [6,7], prostate [8], and oral cavity, esophagus, and larynx cancer [9]). It is also known that high alcohol intake help to develop a dilated cardiomyopathy and also may trigger certain cardiovascular events [10,11]. Nevertheless, a moderate consumption of beer may also help to prevent these type of events [12,13].

Clinical evidence about beer consumption effects needs to be more specific on sex-related differences and health outcomes. Postmenopausal women due to the estrogen depletion suffer body changes [14] and there is an accumulation of abdominal fat [15], an increasing risk of osteoporosis [16] and a loss of body hydration [14] among other health issues. Interestingly, some studies have pointed out that bioactive compounds of beer may help to mitigate some of these adverse effects.

In a unit of beer the main bioactive compounds with health benefits described in several studies [9,17,18] are depicted in Table 1. Particular attention has been given to the polyphenols found in malt (75%) and hops (25%), due to their antioxidant and anti-inflammatory properties [19,20]. Polyphenols are also critical to the flavor, astringency, bitterness, haze, and body of beer [21,22], and their concentration varies according to the ingredients and processing [23,24]. Regular beer, both ale and lager beers, is richer in polyphenol content compared to alcohol-free beers [25].

Table 1. Mean content of selected bioactive compounds in a standard drink of regular beer.

Bioactive Compound	Avarege Level (mg/330 mL)
Phytoestrogens	
Xanthohumol	4.653×10^{-3}
6-Prenylnaringenin	8.547×10^{-3}
8-Prenylnaringenin	3.432×10^{-3}
Isoxanthohumol	0.132
Bitter acids	
$\alpha+\beta$ acids	0.891 [a]
Iso-α-humulones	9.207 [a]
Minerals	
Silicon	6.336
Sodium	14.883
Potassium	116.589

[a] mean value from three beer samples. Content of phytoestrogens from Rothwell et al. (2013) [26], bitter acids from Česlová et al. (2009) [27], silicon from Jugdaohsingh (2007) [28] and sodium and potassium derived from *the Food composition data of 16 European countries* via www.EuroFIR.org.

Among polyphenols, a particular group has attracted special interest for their estrogen-like properties [29]. Hops (*Humulus lupulus* L.) are a source of prenylflavonoids, a class of phytoestrogens, predominantly xanthohumol (XN), that during the brewing process isomerizes into isoxanthohumol (IX), 6-prenylnaringenine (6-PN), and 8-prenylnaringenine (8-PN) [30]. These compounds can mimic and modulate the action of estrogenic hormones by epigenetic mechanisms, via binding with cell surface receptors or by interacting with estrogen receptors (ERs). In particular, 8-PN has been described as the most estrogenic phytoestrogen, surpassing those typically found in soya products [31].

The aim of the present review is to summarize the available literature on the health outcomes of beer consumption in women, focusing on three specific health-related conditions: increased abdominal fat, osteoporosis, and overall body hydration. In particular, findings related to the beer bioactive compounds are discussed.

2. Beer Consumption Related to Health and Disease in Women

2.1. Beer, Abdominal Fat, and Weight Gain

A widely held belief is that beer consumption directly contributes to an increase in abdominal fat and ultimately leads to overweight and obesity. This assumption might be due to the nutritional value of beer, since it contains not only alcohol but also more carbohydrates than other alcoholic drinks [32]. In this section, we assess whether or not beer consumption can increase abdominal fat and site-specific adiposity in women, central obesity being the most relevant sign of metabolic syndrome (MetS) [33].

The type of alcoholic drink, as well as dose, frequency and time of consumption play a role in how alcohol drinking may change fat distribution [34,35]. Additional factors such as genetics, gender, and age may also be important determinants of central body fat [34]. Thus, for instance, drinking alcoholic beverages during meals was significantly more prevalent in females than in males in one study population [35]. In addition, it has been suggested that enlarged waist circumference (WC), known as "beer belly", commonly observed in regular beer consumers might be more due to unhealthy lifestyle factors and drinking patterns (e.g., physical inactivity and smoking) rather than to beer consumption alone [36].

Women seem to be more prone to fat deposition than men upon the consumption of high doses of alcohol [37]. In general, postmenopausal women have a higher total body fat mass and more abdominal fat than premenopausal women. More specifically, despite exhibiting a similar mean body mass index (BMI), postmenopausal women have a larger WC [15]. While both genders experience somatic changes with aging, in women they particularly affect the WC and waist-to-hip ratio (WHR) [33,38]. Interestingly, both visceral and subcutaneous adipocytes express estrogen and androgen receptors such as ER-α, a regulator of adipocyte activity and fat distribution responsible for these gender differences and hyperandrogenism in postmenopausal women [15,39]. As increased visceral abdominal fat deposition causes metabolic changes in fatty acid metabolism, it would be useful to know which foods and ingredients may be more effective for counteracting this fat accumulation in postmenopausal women [15].

Several studies have investigated the effects of gender in the relationship between beer consumption and abdominal adiposity [32,40]. A systematic review of observational studies published before November 2010 indicates that there is an inverse or no association between general obesity and moderate beer consumption in women, while findings referring to abdominal obesity seem to be inconsistent [40]. The authors pointed out that these conflicting observational data may be explained by the small proportion of women beer drinkers and their relatively low beer intake in the studies analyzed [40].

Alcohol or beer consumption and abdominal fat or weight gain have been described as having a U-shaped relationship, with the lowest BMI values observed in women who consumed an average of 6–24 g/day of alcohol [41]. In another study, women with a low beer consumption (maximum 1.32 L/week) also had the lowest WHR values, whereas non-consumers had the highest WC [33]. In the Third National Health and Nutrition Examination Survey (NHANES III), the lowest MetS and WC values were observed in the mild to moderate beer and wine drinkers [42]. Consequently, it can be stated that excessive beer intake may contribute to a higher WC and WHR, and even a higher overall BMI, yet the regular consumption of less than 0.5 L/day of beer (4% alcohol) seems unlikely to have this effect, according to the data available in cross-sectional and prospective observational studies [40]. Women studies evaluating the relationship between beer consumption and abdominal fat increase has been summarized in Table 2 [33,35–37,41,43–55].

In a study focused on the effects of a moderate beer intake on the body composition of healthy adults undergoing a high-intensity interval training, the group consuming alcohol-free beer experienced a significant decrease in visceral adipose tissue and WC, and a clear decreasing trend in the WHR. The other groups (consuming beer or water supplemented with vodka ethanol) did not show any changes in these variables [56].

Now, we should look for the compounds of regular and non-alcoholic beer responsible of these effects. The main bitter compounds of beer are iso-α-acids or iso-α-humulones, derived from the isomerization of α-acids in hops during brewing [57,58]. A study of mice fed with a high-fat diet (HFD) supplemented with iso-α-acids reported significantly reduced body weight, epididymal fat weight, and plasma triglyceride levels after the intervention, whereas in the control group the values increased [59]. As in other studies, it was concluded that iso-humulones might have a protective effect on internal organs damaged by obesity, making this a promising line of future research [59,60]. Iso-α-acids bind and activate both peroxisome proliferator-activated receptors α (PPARα) and γ (PPARγ), which exhibit anti-obesity and anti-inflammatory activities in vivo [59–61]. Regular beers contain 20–40 mg/L of iso-α-acids [27,62,63], and some bitter beers up to 50–80 mg/L [62].

A clinical trial with prediabetes subjects found that 32–48 mg/day of iso-humulones lowered the fasting blood glucose and hemoglobin A1c after 8 weeks, while the total fat and BMI in participants receiving 48 mg/day decreased at 12 weeks [62]. However, some effective concentrations of iso-humulones reported in the literature, such as 500 mg/kg body weight in mice, would be impossible to ingest through moderate or even high beer consumption [60]. Additionally, it would be difficult to formulate a food other than beer with 10–100 mg/L of iso-humulones and an effective dose of iso-α-acids because of their strong bitterness [57].

Matured hop bitter acids (MHBA) are components derived from α-acid oxidation and bear a β-tricarbonyl moiety in their structure such as α-, β-, and iso-α-acids. The bitterness of α-acid oxidation products is described as being more acceptable for the consumer compared to iso-α-acids, and some studies of the bioactive properties of MHBA have been carried out [57]. Weight gain in six-week-old male C57BL/6J mice, a model of MetS, was significantly suppressed when their high fat diet was supplemented with MHBA [64]. Additionally, MHBA administration induced cholecystokinin secretion and signal transduction in the rat gastrointestinal tract, resulting in an increase in the brown adipose tissue temperature. Moreover, MHBA may target TAS2 receptors (TAS2Rs) because they share a similar structure with iso-α-acid [57]. Although 25 TAS2 bitter taste receptors have been determined in humans, only TAS2R1, TAS2R14, and TAS2R40 have been reported to mediate psychophysical responses to bitter hop-derived compounds [65]. Specifically, TAS2R1 and TAS2R40 are expressed in enteroendocrine cells, responsible for incretin hormone secretion [66–68]. There is also interesting evidence that the consumption of mature hop extract significantly reduces abdominal visceral fat of healthy overweight subjects [58].

On the other hand, it has been found that a XN-rich hop extract (17.8% XN and 12.4% IX) prevents fat gain due to overnutrition by modulating preadipocyte differentiation in a 3T3-L1 mouse fibroblast cell line [69]. Furthermore, oral administration of 30 and 60 mg/kg/day of XN during 12-weeks in a C57BL/6J mice model improved markers of inflammation and MetS and decreased BMI in a dose-dependent manner. Nevertheless, the authors concluded that because XN concentrations found in beer are only about 0.2 mg/L, XN taken in the form of beer would be unlikely to have a protective effect against MetS [70]. Two other studies performed in the same C57BL/6J mice model demonstrated that XN derivatives [71] and IX [72] significantly changed the gut microbiota profile, constituting a potential mechanism against obesity and MetS [71,72].

Table 2. Women studies evaluating the relationship between beer consumption and abdominal fat increase.

Authors Year [Ref]	Type of Study	Study Population	Key Finding
Lapidus et al., 1989 [43]	Cross-sectional	1462 women 38–60 years-old	No correlation was found between WHR and beer consumption.
Slattery et al., 1992 [44]	Cross-sectional	1447 black women 1284 white women 18–30 years-old	Higher beer consumption was associated with a higher WHR among white and black women.
Kahn et al., 1997 [45]	Prospective observational	44080 women 40–54 years-old	OR of abdominal weight gain was positively associated in women drinking >0 to <5 days per week and no associated in women drinking <5 days per week versus non-drinkers
Dallongeville et al., 1998 [37]	Cross-sectional	11730 women 35–64 years-old	Beer & cider consumption was associated with a higher WHR.
Rosmond & Bjorntorp 1999 [46]	Cross-sectional	1137 women 40 years-old	Beer consumption was negatively correlated to WHR.
Machado & Sichieri 2002	Cross-sectional	1396 women 20–60 years-old	No trend association for OR for WHR >0.80 across beer consumption categories was found.
Vadstrup et al., 2003 [48]	Prospective observational	3970 women 20–83 years-old	Positive trend association was found for WC at follow-up across beer intake categories.
Bobak et al., 2003 [49]	Cross-sectional	1098 women 25–64 years-old	Beer intake was not associated with an increase in WHR.
Dorn et al., 2003 [35]	Cross-sectional	1322 women 53.3 ± 9.4 years-old	No trend association was found between sagittal abdominal diameter and beer consumption.
Halkjaer et al., 2004 [50]	Prospective observational	1131 women 30–60 years-old	Women consuming >4 drinks of beer per week have higher WC, while no significance increase in WC was found in the group drinking 1–3 drinks of beer per week compared to non-drinkers.
Deschamps et al., 2004 [52]	Cross-sectional	284 women 42.4 ± 4.6 years-old	Women drinking >1 glass of beer per day have a higher WRC than abstainers and those who drink <1 glass of beer per day. No trend association was found for WC.
Lukasiewicz et al., 2005 [53]	Cross-sectional	1268 women 47.7 ± 6.6 years-old	No trend association was found between beer consumption and WHC.
Halkjaer et al., 2006	Prospective observational	22570 women 55 (50–64) years-old	No trend association was found between ΔWC and beer consumption.
Krachler et al., 2006 [54]	Cross-sectional	3087 women 25–64 years-old	Increased beer consumption was not significantly associated to WC.
Tolstrup et al., 2008 [55]	Prospective observational	1610 women 50–65 years-old	Negative association was found for OR of WC across beer intake frequency categories among women who preferred beer.
Schütze et al. [36] 2009	Cross-sectional	2749 women 35–65 years-old	Positive trend association for ΔWC and ΔWHR was found across beer consumption categories.
Schütze et al., 2009 [36]	Prospective observational	12749 women 35–65 years-old	No trend association for WC was found across beer consumption categories.
Bergmann et al., 2011 [41]	Cross-sectional	158796 women 52.9 ± 9 years-old	Positive association was found for OR of WC and WHR for women drinking <6 versus ≤ 6 g per day of alcohol from beer.
Zugravu et al., 2019 [33]	Cross-sectional	784 women >18 years-old	No linear trend association was found between beer consumption and WC or WHR.

WC: waist circumference; WHR: waist-hip ratio.

2.2. Beer and Osteoporosis

Known as one of the most important health-related conditions of aging, osteoporosis is attributed to a decrease of bone mineral density (BMD), which ultimately leads to increased bone fragility [73]. Although common, the condition is underdiagnosed and undertreated, and clinical trials and public health strategies are needed to improve screening and management [74]. Nutrition, exercise and lifestyle are recognized as important aspects in osteoporosis prognosis [75], so modifiable environmental factors such as diet should be considered in its management [76].

Postmenopausal status has been described as a risk factor of BMD loss [16]. As a long-term consequence of the lack of estrogenic stimulation, menopausal bone loss has been linked to an accelerated bone turnover combined with an imbalance that favors bone resorption rather than formation [29,77]. The risk of osteoporosis is six times higher in postmenopausal versus premenopausal women [74]. One of the main mechanisms underlying the protective effect of estrogen against osteoporosis could be an enhanced expression of the vitamin D receptor in the duodenal mucosa and responsiveness to endogenous 1,25-dihydroxycolecalciferol [78].

Certain dietary factors, such as moderate alcohol consumption, have been positively associated with BMD values in postmenopausal women and in the general population [16,79,80]. A study found that women who consumed more than 1 drink of alcohol/day (i.e., 270 mL of beer, 100 mL of wine, or 27 mL of liquor) had a significantly higher femoral neck and lumbar spine BMD than non-alcohol consumers, in a lifestyle adjusted model [81]. Among alcoholic drink subtypes, only beer and low-alcohol beer (but not wine or liquors) seemed to have a significantly positive effect on lumbar spine BMD in older women [81,82]. Similarly, in a cohort of elderly men and women, the lowest hazard ratios for hip fracture tended to be among beer consumers [83]. Also, quantitative bone ultrasound values were higher in women who consumed beer compared to the non-beer or wine drinkers, independently of their gonadal status. This result could be explained by the phytoestrogen content and low grade of alcohol in beer [84]. In contrast, other studies have found positive associations between wine or wine preference and spine BMD in a postmenopausal population group, but not for beer or spirits [76,85]. Women studies evaluating the relationship between beer consumption and osteoporosis has been summarized in Table 3 [76,81,82,84].

In 2008, a systematic review and meta-analysis concluded that subjects consuming 0.5–1 drink/day, equivalent to 7–14 g alcohol/day, had a lower hip fracture risk than abstainers, whereas those consuming more than 2 drinks/day had a greater risk [86]. Thus, abstainers and heavy drinkers have a higher risk of hip fractures than light-moderate drinkers, with a U-shaped relationship between the variables [83,86]. Supporting these results, abnormal bone histology and decreasing bone formation and mineralization have been described in alcoholics [87]. The tendency of a higher association between BMD and beer or wine consumption compared to liquor suggests that other compounds besides ethanol may contribute to bone health [4].

Most of the positive effects of beer on osteoporosis in postmenopausal women have been attributed to the non-alcoholic fraction, specifically to polyphenols, silicon and α-acids. Among phenolic compounds, flavonoids have been inversely linked to bone resorption biomarkers in Scottish women aged 45–54 years. The flavonoids most consumed by the participants were catechins, demonstrating the significant contribution of these compounds to improving BMD [88,89]. The bioactive compounds in hops have been proposed as an alternative to conventional hormone replacement therapy. In particular, the phenolic phytoestrogens from hop extract seem to exhibit estrogen-like effects on bone metabolism [90]. A recent study in animals found that hop extract containing phytoestrogens and iso-α-acids attenuated bone loss and reversed high bone turnover in ovariectomy mice [91]. Furthermore, in vitro experiments demonstrate that hop phytoestrogens (XN, IX, 6-PN, and 8-PN) regulate both osteoblast and osteoclast activities, while α-acids exert a strong bone resorption inhibitory activity, however, the recommended dosage is still unclear [90–92].

The phytoestrogen XN inhibits the receptor activator for the nuclear factor κ B ligand (RANKL) signaling pathway, which has been identified as critical to osteoclast formation and bone resorption [93,94]. XN has also been reported to promote osteoblast differentiation, up-regulate alkaline phosphatase activity, and increase the expression of osteogenic marker genes in osteoblastic cell lines [95]. Interestingly, Prouillet et al. (2004) had previously suggested that one of the consequences of increased alkaline phosphatase activity could be an activation of the ER [94], and another study described an inhibitory resorption effect of XN in a dose-dependent manner [92]. Regarding 8-PN, a recent review of its therapeutic perspectives discusses plausible mechanisms for the anti-osteoporotic properties of this intestinal metabolite. 8-PN has preferential binding to ER-α, which is the prevailing

ER in bone tissue, and its prenyl group seems to be essential for the anti-osteoporotic mechanism [29]. In summary, the beneficial effects of 8-PN, promoting bone formation and inhibiting bone resorption, are mediated by ER-α instead of ER-β, and it is more potent than the isoflavones genistein and daidzein [96].

Silicon from malt has been reported to facilitate bone mineralization and regeneration [75,97], which are essential for bone formation [97]. Some alcoholic beverages such as beer or wine contain significant amounts of silicon [98], although due to the processing of barley and hops, beer is a better source than wine or other alcoholic beverages, with an average content of 19.2 mg/L and non-significant differences among different types of beer [28,75]. Moreover, silicon in beer has a high bioavailability [98,99]. Tucker et al. (2009) showed that adjustment for silicon intake mitigates the positive effect of beer consumption on BMD in older men and women [4].

To sum up, bone remodeling is a slow process and aging affects bone turnover [100]. The phenolic fraction of beer, including phytoestrogens and iso-α-acids from hops, and the silicon from malt seem to play a role in osteoporosis prevention. However, long-term clinical trials are needed to better predict the impact of beer consumption on bone mass, a major concern for postmenopausal women suffering from bone loss.

Table 3. Women studies evaluating the relationship between beer consumption and osteoporosis.

Authors Year [Ref]	Type of Study	Study Population	Key Finding
Pedrera-Zamorano et al., 2009 [86]	Cross-sectional	1697 women (710 premenopausal; 176 perimenopausal and 811 postmenopausal) 48.8 ± 12.59 years-old	Light or moderate consumption of beer was associated to higher bone mass in women independently on their gonadal status.
Fairweather-Tait et al., 2011 [76]	Cross-sectional	2464 postmenopausal women twins 56.3 ± 11.9 years-old	Beer consumption was not associated with higher BMD.
Yin et al., 2011 [82]	Cross-sectional	428 women 62.6 ± 7.2 years-old	Low alcohol beer consumption frequency was positively associated with BMD at lumbar spine.
Yin et al., 2011 [82]	Prospective observational	428 women 62.6 ± 7.2 years-old	No association between beer consumption frequency and BMD at hip was found.
McLenon et al., 2012 [81]	Prospective observational	3173 women 50–62 years-old	Moderate beer consumption had a positive significant effect on lumbar spine BMD after adjustment for lifestyle.
Kubo et al., 2013 [85]	Prospective observational	115,655 postmenopausal women 50–79 years-old	No association was observed between ≥ 1 servings of beer per week and risk of hip fracture.

BMD: bone mineral density.

2.3. Beer and Body Hydration

Hydration has a crucial impact on a variety of factors related to the correct functioning of the body and specific recommendations are needed for each population group. Female sex hormones affect the body water balance, although it is still unclear how the regulation of hydration in women may enhance wellness, safety, and mental and physical performance [101]. Estrogen and progesterone levels have been correlated with body fluid regulation and thermoregulation changes [101]. As more water is retained in the body when estrogen levels are high [102], hormonal depletion in menopause results in a loss of hydration, which should be carefully monitored. Current literature reports that estrogen therapy increases osmotic sensitivity and water retention, helping menopausal women to control diuresis and prevent dehydration [14]. The effect of estrogen on fluid regulation in older women seems to be related to sodium retention [102,103]. Not only the menopause but aging itself affects the fluid balance [14].

An estimated intake of 2.5 L of water/day is considered necessary under normal conditions or 3.5 L of water/day in hot weather or when exercising [104]. Perspiration while exercising may cause an important depletion of water and electrolytes [105], as well as part of the body's stored glycogen. Most recommendations for sustaining the nutritional state and optimizing water absorption during exercise include the intake of beverages containing carbohydrates and electrolytes, in particular glucose–fructose and sodium [106]. Besides the main components of water and carbohydrates, beer also contains electrolytes, which may play a role in maintaining water and electrolyte balance, although the ethanol content may counteract these positive effects.

The effect of beer consumption on the overall hydration status has been studied among men. Unfortunately, no studies on this issue have been performed in women. Hobson and Maughan (2010) investigated the effect of low-alcohol doses on induced euhydration or hypohydration [107], administering alcohol-free or alcoholic beer in each case to create four experimental conditions. In the euhydrated group, those consuming alcoholic beer produced more total urine in the 4 h after intake and for 3 h also exhibited considerably higher serum osmolality, a parameter associated with fluid balance, although the difference had disappeared at 4 h, the end of the monitoring period. The authors also mentioned that sodium excretion was notably lower in the alcohol consumers [108]. In an elderly population with more hydration problems, Polhuis et al. (2017) observed a temporary diuretic effect only after moderate consumption of stronger alcoholic beverages (wine, spirits), but not beer. This demonstrates that: (i) moderate consumption of beer and other weak alcoholic beverages may be safe in terms of hydration for the elderly and (ii) the diuretic effect was plainly triggered by the amount of alcohol in the beverage [108].

Several studies have investigated the effect of beer or its components in those practicing sports, monitoring hydration status, muscle performance, environmental conditions, and duration of exercise in male athletes [105,109,110]. The most controversial component of beer is ethanol. An early study from 1997 reported that the retention volume of the total fluid ingested was about 20% lower in those who consumed an alcohol-free beer supplemented with 4% alcohol compared to those who drank non-supplemented alcohol-free beer, following intermittent cycle ergometer exercises in the heat that induced dehydration of up to 2% of body mass [111]. Alcohol itself undoubtedly has a negative effect on exercise performance, although its extent may also depend on other factors, such as the mode and duration of exercise [109]. In extreme conditions, when the body requires greater hydration, any diuretic or anti-hydration effect of the ethanol in beer is more easily noted. Jiménez-Pavón et al. (2015) observed that consumption of 660 mL of regular beer (4% alcohol) after 1 h of running in hot conditions had no deleterious effect on any hydration marker [106]. Two other studies evaluated the effect of water, beer or alcohol-free beer on fluid and electrolyte homeostasis in male athletes or physically active men [112,113]. Castro-Sepulveda et al. (2016) reported that an intake of 700 mL of alcoholic beer before aerobic exercising increased plasma K^+ and decreased plasma Na^+ during the exercise activity, with a negative impact on athletic performance. Notably, this effect was not observed when alcohol-free beer was administered, to the extent that the decrease in plasma Na^+ during exercise was lower than after the ingestion of water. Accordingly, alcohol-free beer might be an effective sports drink for maintaining electrolyte homeostasis in males when taken before exercise [113]. In contrast, another study found that rehydration of young, healthy, and physically active males with non-alcoholic beer was not advantageous with regard to water [112]. A more recent study evaluated the effects of ingesting isotonic drinks or beer with different alcohol concentrations after mild dehydration or exercise among males. The net fluid balance was measured after a 5-hour observation period and the lowest rate of fluid retention (21%) was obtained for beer with 5% alcohol, whereas the highest (42%) was recorded for an isotonic sports drink [114]. Interestingly, the effects of modifying the sodium and alcohol content of beer have also been studied [115,116]. Participants consumed low-alcohol beer (2% alcohol + 25 or 50 mM/L of sodium) or normal beer (3.5% alcohol + 25 mM/L of sodium) and after exercise, the greatest fluid retention was observed in consumers of beer with the highest electrolyte content and the lowest concentration of alcohol (2% alcohol + 50 mM/L of sodium) [116].

While non-alcoholic beer has promising effects in terms of fluid homeostasis in the context of aerobic exercise, a low dose of alcohol (0.5 g/kg of body weight) consumed before muscle damage-inducing anaerobic exercise had no impact on the posterior muscle performance or related water loss in ten healthy young males [110].

Notably, all the aforementioned studies were performed in men. More research is needed to understand the effects of different types of drinks on the hydration state of female athletes, in order to improve performance and provide personalized supplementation recommendations [101].

3. Implications and Future Research

Most of the health benefits of beer are thought to be originated by its non-alcoholic components, mainly polyphenols. Although found in small quantities in the final product, the flavonoid XN (whose only source is hops) is of particular interest. Intestinal metabolites of related flavonoids, notably 8-PN, could also have an important role in human health. Other components, such as silicon or bitter acids, may help to explain other health effects of beer consumption, such as improvement in bone density. Nevertheless, the beneficial properties of beer components outlined in this review have not been extensively studied because of the adverse effects of ethanol. Human interventional trials are required to elucidate the real association between beer intake and health benefits in women, but the consumption of ethanol is an important obstacle for their development. We, therefore, suggest a directional change towards the non-alcoholic fraction of beer and its effect on the female population as an interesting target for future studies. With some authors already using this strategy, a greater focus on alcohol-free beer will lead to the emergence of more human trials and new evidence in this field. Finally, new long-term randomized trials on the effects of moderate alcoholic and non-alcoholic beer consumption (and other alcoholic beverages) on health and diseases, including cardiovascular disease, obesity, diabetes, cancer, cognitive decline, osteoporosis, and others in women (and also in women) are needed to better define the protective role (or not) of beer consumption, independent of other lifestyle factors, on the aforementioned conditions.

4. Conclusions

Although the results of studies on abdominal fat deposition in female beer consumers are inconsistent, moderate consumption appears not to have a significant effect on adiposity. Moderate beer intake has also been associated with improved bone health in elderly women in observational studies. Moreover, the non-alcoholic fraction of beer is of potential interest as a counteracting agent for bone mass loss after menopause.

In the elderly, beer intake does not seem to pose a risk for hydration. When ingested before exercise, beer with lower alcohol content has a better rehydration effect, and the consumption of alcohol-free beer may even have a positive impact on electrolyte homeostasis. However, the effects of beer on hydration in women still need to be investigated.

Author Contributions: Conceptualization, A.T.-R, R.E., and R.M.L.-R.; Acquisition and interpretation of data, M.T.-S., A.V.-F., G.S., and C.E.S.; writing—original draft preparation, M.T.-S. and A.V.-F.; writing—review and editing, A.T.-R, G.S., C.E.S, R.E., and R.M.L.-R. All authors have read and agreed to the published version of the manuscript.

Funding: This work was supported by the European Foundation for Alcohol Research (ERAB) (EA 1514, EA 1515, and EA 1517), the CICYT (AGL2016-79113-R), the Instituto de Salud Carlos III (ISCIII) (CIBEROBN) from the Ministerio de Economía, Industria y Competitividad (MEIC) (AEI/FEDER, UE) and Generalitat de Catalunya (GC) (2017 SGR196). Marta Trius-Soler is thankful for the APIF 2018-2019 fellowship from the University of Barcelona.

Conflicts of Interest: Anna Tresserra-Rimbau and Rosa M. Lamuela-Raventós have received funding from The European Foundation for Alcohol Research (ERAB). Rosa M. Lamuela-Raventós has received lecture fees and travel support from Cerveceros de España. Ramon Estruch is a Board Membership of Cerveza y Salud, Madrid (Spain) and of Fundación Dieta Mediterránea, Barcelona (Spain), and has received lecture fees and travel support from Brewers of Europe, Brussels (Belgium) and Organización Interprofesional del Aceite de Oliva, Madrid (Spain).

References

1. Buiatti, S. Beer Composition: An Overview. In *Beer in Health and Disease Prevention*; Elsevier: London, UK, 2009; pp. 213–225, ISBN 9780123738912.
2. Colen, L.; Swinnen, J. Economic growth, globalisation and beer consumption. *J. Agric. Econ.* **2016**, *67*, 186–207. [CrossRef]
3. *Handbook of Brewing*, 2nd ed.; Stewart, G.G.; Priest, F.G. (Eds.) CRC Press: Boca Raton, FL, USA, 2006; ISBN 9780429116179.
4. Tucker, K.L.; Jugdaohsingh, R.; Powell, J.J.; Qiao, N.; Hannan, M.T.; Sripanyakorn, S.; Cupples, L.A.; Kiel, D.P. Effects of beer, wine, and liquor intakes on bone mineral density in older men and women. *Am. J. Clin. Nutr.* **2009**, *89*, 1188–1196. [CrossRef] [PubMed]
5. Zhang, C.; Zhong, M. Consumption of beer and colorectal cancer incidence: A meta-analysis of observational studies. *Cancer Causes Control* **2015**, *26*, 549–560. [CrossRef] [PubMed]
6. Benedetti, A.; Parent, M.E.; Siemiatycki, J. Consumption of alcoholic beverages and risk of lung cancer: Results from two case-control studies in Montreal, Canada. *Cancer Causes Control* **2006**, *17*, 469–480. [CrossRef]
7. Chao, C. Associations between beer, wine, and liquor consumption and lung cancer risk: A meta-analysis. *Cancer Epidemiol. Biomark. Prev.* **2007**, *16*, 2436–2447. [CrossRef]
8. Demoury, C.; Karakiewicz, P.; Parent, M.-E. Association between lifetime alcohol consumption and prostate cancer risk: A case-control study in Montreal, Canada. *Cancer Epidemiol.* **2016**, *45*, 11–17. [CrossRef]
9. de Gaetano, G.; Costanzo, S.; Di Castelnuovo, A.; Badimon, L.; Bejko, D.; Alkerwi, A.; Chiva-Blanch, G.; Estruch, R.; La Vecchia, C.; Panico, S.; et al. Effects of moderate beer consumption on health and disease: A consensus document. *Nutr. Metab. Cardiovasc. Dis.* **2016**, *26*, 443–467. [CrossRef]
10. Reynolds, K.; Lewis, L.B.; Nolen, J.D.L.; Kinney, G.L.; Sathya, B.; He, J. Alcohol consumption and risk of stroke: A meta-analysis. *J. Am. Med. Assoc.* **2003**, *289*, 579–588. [CrossRef]
11. Mukamal, K.J. Alcohol, beer, and ischemic stroke. In *Beer in Health and Disease Prevention*; Elsevier Inc.: Amsterdam, The Netherlands, 2008; pp. 623–634, ISBN 9780123738912.
12. Costanzo, S.; Di Castelnuovo, A.; Donati, M.B.; Iacoviello, L.; de Gaetano, G. Wine, beer or spirit drinking in relation to fatal and non-fatal cardiovascular events: A meta-analysis. *Eur. J. Epidemiol.* **2011**, *26*, 833–850. [CrossRef]
13. Di Castelnuovo, A.; Rotondo, S.; Iacoviello, L.; Donati, M.B.; De Gaetano, G. Meta-analysis of wine and beer consumption in relation to vascular risk. *Circulation* **2002**, *105*, 2836–2844. [CrossRef]
14. Stachenfeld, N.S. Hormonal changes during menopause and the impact on fluid regulation. *Reprod. Sci.* **2014**, *21*, 555–561. [CrossRef] [PubMed]
15. Ko, S.H.; Kim, H.S. Menopause-associated lipid metabolic disorders and foods beneficial for postmenopausal women. *Nutrients* **2020**, *12*, 202. [CrossRef] [PubMed]
16. Bainbridge, K.; Sowers, M.; Lin, X.; Harlow, S. Risk factors for low bone mineral density and the 6-year rate of bone loss among premenopausal and perimenopausal women. *Osteoporos. Int.* **2004**, *15*, 439–446. [CrossRef]
17. Sripanyakorn, S.; Jugdaohsingh, R.; Mander, A.; Davidson, S.L.; Thompson, R.P.H.; Powell, J.J. Moderate ingestion of alcohol is associated with acute ethanol-induced suppression of circulating CTX in a PTH-independent fashion. *J. Bone Miner. Res.* **2009**, *24*, 1380–1388. [CrossRef] [PubMed]
18. Arranz, S.; Chiva-Blanch, G.; Valderas-Martínez, P.; Medina-Remón, A.; Lamuela-Raventós, R.M.; Estruch, R. Wine, beer, alcohol and polyphenols on cardiovascular disease and cancer. *Nutrients* **2012**, *4*, 759–781. [CrossRef]
19. Proestos, C.; Komaitis, M. *Antioxidant Capacity of Hops*; Elsevier Inc.: Amsterdam, The Netherlands, 2008; ISBN 9780123738912.
20. Feick, P.; Gerloff, A.; Singer, M.V. The effect of beer and its non-alcoholic constituents on the exocrine and endocrine pancreas as well as on gastrointestinal hormones. In *Beer in Health and Disease Prevention*; Elsevier: Amsterdam, The Netherlands, 2009; pp. 587–601, ISBN 9780123738912.
21. Callemien, D.; Collin, S. Structure, organoleptic properties, quantification methods, and stability of phenolic compounds in beer—A review. *Food Rev. Int.* **2010**, *26*, 1–84. [CrossRef]
22. Intelmann, D.; Haseleu, G.; Dunkel, A.; Lagemann, A.; Stephan, A.; Hofmann, T. Comprehensive sensomics analysis of hop-derived bitter compounds during storage of beer. *J. Agric. Food Chem.* **2011**, *59*, 1939–1953. [CrossRef]

23. Rivero, D.; Pérez-Magariño, S.; González-Sanjosé, M.L.; Valls-Belles, V.; Codoñer, P.; Muñiz, P. Inhibition of induced DNA oxidative damage by beers: Correlation with the content of polyphenols and melanoidins. *J. Agric. Food Chem.* **2005**, *53*, 3637–3642. [CrossRef]

24. Venturelli, S.; Burkard, M.; Biendl, M.; Lauer, U.M.; Frank, J.; Busch, C. Prenylated chalcones and flavonoids for the prevention and treatment of cancer. *Nutrition* **2016**, *32*, 1171–1178. [CrossRef]

25. Boronat, A.; Soldevila-Domenech, N.; Rodríguez-Morató, J.; Martínez-Huélamo, M.; Lamuela-Raventós, R.M.; de la Torre, R. Beer phenolic composition of simple phenols, prenylated flavonoids and alkylresorcinols. *Molecules* **2020**, *25*, 2582. [CrossRef]

26. Rothwell, J.A.; Perez-Jimenez, J.; Neveu, V.; Medina-Remon, A.; M'Hiri, N.; Garcia-Lobato, P.; Manach, C.; Knox, C.; Eisner, R.; Wishart, D.S.; et al. Phenol-Explorer 3.0: A major update of the Phenol-Explorer database to incorporate data on the effects of food processing on polyphenol content. *Database* **2013**, *2013*, bat070. [CrossRef] [PubMed]

27. Česlová, L.; Holčapek, M.; Fidler, M.; Drštičková, J.; Lísa, M. Characterization of prenylflavonoids and hop bitter acids in various classes of Czech beers and hop extracts using high-performance liquid chromatography-mass spectrometry. *J. Chromatogr. A* **2009**, *1216*, 7249–7257. [CrossRef] [PubMed]

28. Jugdaohsingh, R. Silicon and bone health. *J. Nutr. Health Aging* **2007**, *11*, 99–110. [PubMed]

29. Štulíková, K.; Karabín, M.; Nešpor, J.; Dostálek, P. Therapeutic perspectives of 8-prenylnaringenin, a potent phytoestrogen from hops. *Molecules* **2018**, *23*, 660. [CrossRef]

30. Quifer-Rada, P.; Vallverdú-Queralt, A.; Martínez-Huélamo, M.; Chiva-Blanch, G.; Jáuregui, O.; Estruch, R.; Lamuela-Raventós, R. A comprehensive characterisation of beer polyphenols by high resolution mass spectrometry (LC–ESI-LTQ-Orbitrap-MS). *Food Chem.* **2015**, *169*, 336–343. [CrossRef] [PubMed]

31. Omoruyi, I.M.; Pohjanvirta, R. Estrogenic activities of food supplements and beers as assessed by a yeast bioreporter assay. *J. Diet. Suppl.* **2018**, *15*, 665–672. [CrossRef] [PubMed]

32. Wannamethee, S.G. *Beer and Adiposity*; Elsevier Inc.: Amsterdam, The Netherlands, 2009; ISBN 9780123738912.

33. Zugravu, C.-A.; Pătrașcu, D.; Otelea, M. Central obesity and beer consumption. *Ann. Univ. Dunarea Jos Galati Fascicle VI Food Technol.* **2019**, *43*, 110–124. [CrossRef]

34. Ferreira, M.G.; Valente, J.G.; Gonçalves-Silva, R.M.V.; Sichieri, R. Alcohol consumption and abdominal fat in blood donors. *Rev. Saude Publica* **2008**, *42*, 1067–1073. [CrossRef] [PubMed]

35. Dorn, J.M.; Hovey, K.; Muti, P.; Freudenheim, J.L.; Russell, M.; Nochajski, T.H.; Trevisan, M. Alcohol drinking patterns differentially affect central adiposity as measured by abdominal height in women and men. *J. Nutr.* **2003**, *133*, 2655–2662. [CrossRef]

36. Schütze, M.; Schulz, M.; Steffen, A.; Bergmann, M.M.; Kroke, A.; Lissner, L.; Boeing, H. Beer consumption and the "beer belly": Scientific basis or common belief? *Eur. J. Clin. Nutr.* **2009**, *63*, 1143–1149. [CrossRef]

37. Dallongeville, J.; Marécaux, N.; Ducimetière, P.; Ferrières, J.; Arveiler, D.; Bingham, A.; Ruidavets, J.; Simon, C.; Amouyel, P. Influence of alcohol consumption and various beverages on waist girth and waist-to-hip ratio in a sample of French men and women. *Int. J. Obes.* **1998**, *22*, 1178–1183. [CrossRef] [PubMed]

38. Wong, M.C.S.; Huang, J.; Wang, J.; Chan, P.S.F.; Lok, V.; Chen, X.; Leung, C.; Wang, H.H.X.; Lao, X.Q.; Zheng, Z.-J. Global, regional and time-trend prevalence of central obesity: A systematic review and meta-analysis of 13.2 million subjects. *Eur. J. Epidemiol.* **2020**. [CrossRef] [PubMed]

39. Marchand, G.B.; Carreau, A.-M.; Weisnagel, S.J.; Bergeron, J.; Labrie, F.; Lemieux, S.; Tchernof, A. Increased body fat mass explains the positive association between circulating estradiol and insulin resistance in postmenopausal women. *Am. J. Physiol. Metab.* **2018**, *314*, E448–E456. [CrossRef] [PubMed]

40. Bendsen, N.T.; Christensen, R.; Bartels, E.M.; Kok, F.J.; Sierksma, A.; Raben, A.; Astrup, A. Is beer consumption related to measures of abdominal and general obesity? A systematic review and meta-analysis. *Nutr. Rev.* **2013**, *71*, 67–87. [CrossRef]

41. Bergmann, M.M.; Schütze, M.; Steffen, A.; Boeing, H.; Halkjaer, J.; Tjonneland, A.; Travier, N.; Agudo, A.; Slimani, N.; Rinaldi, S.; et al. The association of lifetime alcohol use with measures of abdominal and general adiposity in a large-scale European cohort. *Eur. J. Clin. Nutr.* **2011**, *65*, 1079–1087. [CrossRef]

42. Freiberg, M.S.; Cabral, H.J.; Heeren, T.C.; Vasan, R.S.; Curtis Ellison, R. Alcohol consumption and the prevalence of the metabolic syndrome in the U.S.: A cross-sectional analysis of data from the Third National Health and Nutrition Examination Survey. *Diabetes Care* **2004**, *27*, 2954–2959. [CrossRef]

43. Lapidus, L.; Bengtsson, C.; Hällström, T.; Björntorp, P. Obesity, adipose tissue distribution and health in women-Results from a population study in Gothenburg, Sweden. *Appetite* **1989**, *13*, 25–35. [CrossRef]

44. Slattery, M.L.; McDonald, A.; Bild, D.E.; Caan, B.J.; Hilner, J.E.; Jacobs, D.R.; Liu, K. Associations of body fat and its distribution with dietary intake, physical activity, alcohol, and smoking in blacks and whites. *Am. J. Clin. Nutr.* **1992**, *55*, 943–949. [CrossRef]

45. Kahn, H.S.; Tatham, L.M.; Heath, C.W. Contrasting factors associated with abdominal and peripheral weight gain among adult women. *Int. J. Obes.* **1997**, *21*, 903–911. [CrossRef]

46. Rosmond, R.; Björntorp, P. Psychosocial and socio-economic factors in women and their relationship to obesity and regional body fat distribution. *Int. J. Obes.* **1999**, *23*, 138–145. [CrossRef]

47. Machado, P.A.N.; Sichieri, R. Relação cintura-quadril e fatores de dieta em adultos. *Rev. Saude Publica* **2002**, *36*, 198–204. [CrossRef] [PubMed]

48. Vadstrup, E.; Petersen, L.; Sørensen, T.; Grønbaek, M. Waist circumference in relation to history of amount and type of alcohol: Results from the Copenhagen City Heart Study. *Int. J. Obes. Relat. Metab. Disord.* **2003**, *27*, 238–246. [CrossRef] [PubMed]

49. Bobak, M.; Skodova, Z.; Marmot, M. Beer and obesity: A cross-sectional study. *Eur. J. Clin. Nutr.* **2003**, *57*, 1250–1253. [CrossRef]

50. Halkjær, J.; Sørensen, T.I.; Tjønneland, A.; Togo, P.; Holst, C.; Heitmann, B.L. Food and drinking patterns as predictors of 6-year BMI-adjusted changes in waist circumference. *Br. J. Nutr.* **2004**, *92*, 735–748. [CrossRef] [PubMed]

51. Halkjær, J.; Tjønneland, A.; Thomsen, B.L.; Overvad, K.; Sørensen, T.I.A. Intake of macronutrients as predictors of 5-y changes in waist circumference. *Am. J. Clin. Nutr.* **2006**, *84*, 789–797. [CrossRef]

52. Deschamps, V.; Alamowitch, C.; Borys, J. Boissons alcooliques, poids et paramètres d'adiposité chez 520 adultes issus de l'étude Fleurbaix Laventie Ville Santé. *Cah. Nutr. Diététique* **2004**, *39*, 262–268. [CrossRef]

53. Lukasiewicz, E.; Mennen, L.I.; Bertrais, S.; Arnault, N.; Preziosi, P.; Galan, P.; Hercberg, S. Alcohol intake in relation to body mass index and waist-to-hip ratio: The importance of type of alcoholic beverage. *Public Health Nutr.* **2005**, *8*, 315–320. [CrossRef]

54. Krachler, B.; Eliasson, M.; Stenlund, H.; Johansson, I.; Hallmans, G.; Lindahl, B. Reported food intake and distribution of body fat: A repeated cross-sectional study. *Nutr. J.* **2006**, *5*, 1–11. [CrossRef]

55. Tolstrup, J.S.; Halkjær, J.; Heitmann, B.L.; Tjønneland, A.M.; Overvad, K.; Sørensen, T.I.A.; Grønbæk, M.N. Alcohol drinking frequency in relation to subsequent changes in waist circumference. *Am. J. Clin. Nutr.* **2008**, *87*, 957–963. [CrossRef]

56. Molina-Hidalgo, C.; De-Lao, A.; Jurado-Fasoli, L.; Amaro-Gahete, F.J.; Castillo, M.J. Beer or ethanol effects on the body composition response to high-intensity interval training. The BEER-HIIT study. *Nutrients* **2019**, *11*, 909. [CrossRef]

57. Yamazaki, T.; Morimoto-Kobayashi, Y.; Koizumi, K.; Takahashi, C.; Nakajima, S.; Kitao, S.; Taniguchi, Y.; Katayama, M.; Ogawa, Y. Secretion of a gastrointestinal hormone, cholecystokinin, by hop-derived bitter components activates sympathetic nerves in brown adipose tissue. *J. Nutr. Biochem.* **2019**, *64*, 80–87. [CrossRef] [PubMed]

58. Morimoto-Kobayashi, Y.; Ohara, K.; Ashigai, H.; Kanaya, T.; Koizumi, K.; Manabe, F.; Kaneko, Y.; Taniguchi, Y.; Katayama, M.; Kowatari, Y.; et al. Matured hop extract reduces body fat in healthy overweight humans: A randomized, double-blind, placebo-controlled parallel group study. *Nutr. J.* **2015**, *15*, 25. [CrossRef] [PubMed]

59. Ayabe, T.; Ohya, R.; Kondo, K.; Ano, Y. Iso-α-acids, bitter components of beer, prevent obesity-induced cognitive decline. *Sci. Rep.* **2018**, *8*, 4760. [CrossRef] [PubMed]

60. Miura, Y.; Hosono, M.; Oyamada, C.; Odai, H.; Oikawa, S.; Kondo, K. Dietary isohumulones, the bitter components of beer, raise plasma HDL-cholesterol levels and reduce liver cholesterol and triacylglycerol contents similar to PPARα activations in C57BL/6 mice. *Br. J. Nutr.* **2005**, *93*, 559–567. [CrossRef]

61. Dostálek, P.; Karabín, M.; Jelínek, L. Hop phytochemicals and their potential role in metabolic syndrome prevention and therapy. *Molecules* **2017**, *22*, 1761. [CrossRef]

62. Obara, K.; Mizutani, M.; Hitomi, Y.; Yajima, H.; Kondo, K. Isohumulones, the bitter component of beer, improve hyperglycemia and decrease body fat in Japanese subjects with prediabetes. *Clin. Nutr.* **2009**, *28*, 278–284. [CrossRef]

63. Vanhoenacker, G.; De Keukeleire, D.; Sandra, P. Analysis of iso-α-acids and reduced iso-α-acids in beer by direct injection and liquid chromatography with ultraviolet absorbance detection or with mass spectrometry. *J. Chromatogr. A* **2004**, *1035*, 53–61. [CrossRef]

64. Morimoto-Kobayashi, Y.; Ohara, K.; Takahashi, C.; Kitao, S.; Wang, G.; Taniguchi, Y.; Katayama, M.; Nagai, K. Matured hop bittering components induce thermogenesis in brown adipose tissue via sympathetic nerve activity. *PLoS ONE* **2015**, *10*, e131042. [CrossRef]

65. Intelmann, D.; Batram, C.; Kuhn, C.; Haseleu, G.; Meyerhof, W.; Hofmann, T. Three TAS2R bitter taste receptors mediate the psychophysical responses to bitter compounds of hops (*Humulus lupulus* L.) and beer. *Chemosens. Percept.* **2009**, *2*, 118–132. [CrossRef]

66. Kok, B.P.; Galmozzi, A.; Littlejohn, N.K.; Albert, V.; Godio, C.; Kim, W.; Kim, S.M.; Bland, J.S.; Grayson, N.; Fang, M.; et al. Intestinal bitter taste receptor activation alters hormone secretion and imparts metabolic benefits. *Mol. Metab.* **2018**, *16*, 76–87. [CrossRef]

67. Kidd, M.; Modlin, I.M.; Gustafsson, B.I.; Drozdov, I.; Hauso, O.; Pfragner, R. Luminal regulation of normal and neoplastic human EC cell serotonin release is mediated by bile salts, amines, tastants, and olfactants. *Am. J. Physiol. Gastrointest. Liver Physiol.* **2008**, *295*, 260–272. [CrossRef] [PubMed]

68. Wu, S.V.; Rozengurt, N.; Yang, M.; Young, S.H.; Sinnett-Smith, J.; Rozengurt, E. Expression of bitter taste receptors of the T2R family in the gastrointestinal tract and enteroendocrine STC-1 cells. *Proc. Natl. Acad. Sci. USA* **2002**, *99*, 2392–2397. [CrossRef] [PubMed]

69. Kiyofuji, A.; Yui, K.; Takahashi, K.; Osada, K. Effects of xanthohumol-rich hop extract on the differentiation of preadipocytes. *J. Oleo Sci.* **2014**, *63*, 593–597. [CrossRef] [PubMed]

70. Miranda, C.L.; Elias, V.D.; Hay, J.J.; Choi, J.; Reed, R.L.; Stevens, J.F. Xanthohumol improves dysfunctional glucose and lipid metabolism in diet-induced obese C57BL/6J mice. *Arch. Biochem. Biophys.* **2016**, *599*, 22–30. [CrossRef] [PubMed]

71. Zhang, Y.; Bobe, G.; Revel, J.S.; Rodrigues, R.R.; Sharpton, T.J.; Fantacone, M.L.; Raslan, K.; Miranda, C.L.; Lowry, M.B.; Blakemore, P.R.; et al. Improvements in metabolic syndrome by xanthohumol derivatives are linked to altered gut microbiota and bile acid metabolism. *Mol. Nutr. Food Res.* **2020**, *64*, 1900789. [CrossRef] [PubMed]

72. Yamashita, M.; Fukizawa, S.; Nonaka, Y. Hop-derived prenylflavonoid isoxanthohumol suppresses insulin resistance by changing the intestinal microbiota and suppressing chronic inflammation in high fat diet-fed mice. *Eur. Rev. Med. Pharmacol. Sci.* **2020**, *24*, 1537–1547. [CrossRef] [PubMed]

73. Pietschmann, P.; Rauner, M.; Sipos, W.; Kerschan-Schindl, K. Osteoporosis: An age-related and gender-specific disease—A mini-review. *Gerontology* **2009**, *55*, 3–12. [CrossRef]

74. Biino, G.; Casula, L.; De Terlizzi, F.; Adamo, M.; Vaccargiu, S.; Francavilla, M.; Loi, D.; Casti, A.; Atzori, M.; Pirastu, M. Epidemiology of osteoporosis in an isolated sardinian population by using quantitative ultrasound. *Am. J. Epidemiol.* **2011**, *174*, 432–439. [CrossRef]

75. Price, C.T.; Koval, K.J.; Langford, J.R. Silicon: A review of its potential role in the prevention and treatment of postmenopausal osteoporosis. *Int. J. Endocrinol.* **2013**, *2013*, 316783. [CrossRef]

76. Fairweather-Tait, S.J.; Skinner, J.; Guile, G.R.; Cassidy, A.; Spector, T.D.; MacGregor, A.J. Diet and bone mineral density study in postmenopausal women from the twinsUK registry shows a negative association with a traditional english dietary pattern and a positive association with wine. *Am. J. Clin. Nutr.* **2011**, *94*, 1371–1375. [CrossRef]

77. Marrone, J.A.; Maddalozzo, G.F.; Branscum, A.J.; Hardin, K.; Cialdella-Kam, L.; Philbrick, K.A.; Breggia, A.C.; Rosen, C.J.; Turner, R.T.; Iwaniec, U.T. Moderate alcohol intake lowers biochemical markers of bone turnover in postmenopausal women. *Menopause* **2012**, *19*, 974–979. [CrossRef] [PubMed]

78. Liel, Y.; Shany, S.; Smirnoff, P.; Schwartz, B. Estrogen increases 1,25-dihydroxyvitamin D receptors expression bioresponse in the rat duodenal mucosa. *Endocrinology* **1999**, *140*, 280–285. [CrossRef] [PubMed]

79. Tucker, K.L. Osteoporosis prevention and nutrition. *Curr. Osteoporos. Rep.* **2009**, *7*, 111. [CrossRef]

80. Sommer, I.; Erkkilä, A.T.; Järvinen, R.; Mursu, J.; Sirola, J.; Jurvelin, J.S.; Kröger, H.; Tuppurainen, M. Alcohol consumption and bone mineral density in elderly women. *Public Health Nutr.* **2013**, *16*, 704–712. [CrossRef] [PubMed]

81. McLernon, D.J.; Powell, J.J.; Jugdaohsingh, R.; Macdonald, H.M. Do lifestyle choices explain the effect of alcohol on bone mineral density in women around menopause? *Am. J. Clin. Nutr.* **2012**, *95*, 1261–1269. [CrossRef] [PubMed]

82. Yin, J.; Winzenberg, T.; Quinn, S.; Giles, G.; Jones, G. Beverage-specific alcohol intake and bone loss in older men and women: A longitudinal study. *Eur. J. Clin. Nutr.* **2011**, *65*, 526–532. [CrossRef]

83. Mukamal, K.J.; Robbins, J.A.; Cauley, J.A.; Kern, L.M.; Siscovick, D.S. Alcohol consumption, bone density, and hip fracture among older adults: The cardiovascular health study. *Osteoporos. Int.* **2007**, *18*, 593–602. [CrossRef]

84. Pedrera-Zamorano, J.D.; Lavado-Garcia, J.M.; Roncero-Martin, R.; Calderon-Garcia, J.F.; Rodriguez-Dominguez, T.; Canal-Macias, M.L. Effect of beer drinking on ultrasound bone mass in women. *Nutrition* **2009**, *25*, 1057–1063. [CrossRef]

85. Kubo, J.T.; Stefanick, M.L.; Robbins, J.; Wactawski-Wende, J.; Cullen, M.R.; Freiberg, M.; Desai, M. Preference for wine is associated with lower hip fracture incidence in post-menopausal women. *BMC Womens Health* **2013**, *13*, 36. [CrossRef]

86. Berg, K.M.; Kunins, H.V.; Jackson, J.L.; Nahvi, S.; Chaudhry, A.; Harris, K.A.; Malik, R.; Arnsten, J.H. Association between alcohol consumption and both osteoporotic fracture and bone density. *Am. J. Med.* **2008**, *121*, 406–418. [CrossRef]

87. Hansen, S.A.; Folsom, A.R.; Kushi, L.H.; Sellers, T.A. Association of fractures with caffeine and alcohol in postmenopausal women: The Iowa Women's Health Study. *Public Health Nutr.* **2000**, *3*, 253–261. [CrossRef] [PubMed]

88. Hardcastle, A.C.; Aucott, L.; Reid, D.M.; MacDonald, H.M. Associations between dietary flavonoid intakes and bone health in a scottish population. *J. Bone Miner. Res.* **2011**, *26*, 941–947. [CrossRef] [PubMed]

89. Welch, A.; MacGregor, A.; Jennings, A.; Fairweather-Tait, S.; Spector, T.; Cassidy, A. Habitual flavonoid intakes are positively associated with bone mineral density in women. *J. Bone Miner. Res.* **2012**, *27*, 1872–1878. [CrossRef] [PubMed]

90. Effenberger, K.E.; Johnsen, S.A.; Monroe, D.G.; Spelsberg, T.C.; Westendorf, J.J. Regulation of osteoblastic phenotype and gene expression by hop-derived phytoestrogens. *J. Steroid Biochem. Mol. Biol.* **2005**, *96*, 387–399. [CrossRef]

91. Xia, T.S.; Lin, L.Y.; Zhang, Q.Y.; Jiang, Y.P.; Li, C.H.; Liu, X.Y.; Qin, L.P.; Xin, H.L. Humulus lupulus, L. extract prevents ovariectomy-induced osteoporosis in mice and regulates activities of osteoblasts and osteoclasts. *Chin. J. Integr. Med.* **2019**, 1–8. [CrossRef]

92. Tobe, H.; Muraki, Y.; Kitamura, K.; Komiyama, O.; Sato, Y.; Sugioka, T.; Maruyama, H.B.; Matsuda, E.; Nagai, M. Bone resorption inhibitors from hop extract. *Biosci. Biotechnol. Biochem.* **1997**, *61*, 158–159. [CrossRef]

93. Lambert, M.N.T.; Hu, L.M.; Jeppesen, P.B. A systematic review and meta-analysis of the effects of isoflavone formulations against estrogen-deficient bone resorption in peri- and postmenopausal women. *Am. J. Clin. Nutr.* **2017**, *106*, ajcn151464. [CrossRef]

94. Prouillet, C.; Mazière, J.-C.; Mazière, C.; Wattel, A.; Brazier, M.; Kamel, S. Stimulatory effect of naturally occurring flavonols quercetin and kaempferol on alkaline phosphatase activity in MG-63 human osteoblasts through ERK and estrogen receptor pathway. *Biochem. Pharmacol.* **2004**, *67*, 1307–1313. [CrossRef]

95. Jeong, H.M.; Han, E.H.; Jin, Y.H.; Choi, Y.H.; Lee, K.Y.; Jeong, H.G. Xanthohumol from the hop plant stimulates osteoblast differentiation by RUNX2 activation. *Biochem. Biophys. Res. Commun.* **2011**, *409*, 82–89. [CrossRef]

96. Luo, D.; Kang, L.; Ma, Y.; Chen, H.; Kuang, H.; Huang, Q.; He, M.; Peng, W. Effects and mechanisms of 8-prenylnaringenin on osteoblast MC3T3-E1 and osteoclast-like cells RAW264.7. *Food Sci. Nutr.* **2014**, *2*, 341–350. [CrossRef]

97. Dong, M.; Jiao, G.; Liu, H.; Wu, W.; Li, S.; Wang, Q.; Xu, D.; Li, X.; Liu, H.; Chen, Y. Biological silicon stimulates collagen type 1 and osteocalcin synthesis in human osteoblast-like cells through the BMP-2/Smad/RUNX2 signaling pathway. *Biol. Trace Elem. Res.* **2016**, *173*, 306–315. [CrossRef] [PubMed]

98. Boguszewska-Czubara, A.; Pasternak, K. Silicon in medicine and therapy. *J. Elem.* **2011**, *16*, 489–497. [CrossRef]

99. Suh, K.S.; Rhee, S.Y.; Kim, Y.S.; Lee, Y.S.; Choi, E.M. Xanthohumol modulates the expression of osteoclast-specific genes during osteoclastogenesis in RAW264.7 cells. *Food Chem. Toxicol.* **2013**, *62*, 99–106. [CrossRef] [PubMed]

100. Chen, Y.-M.; Ho, S.C.; Lam, S.S.H.; Ho, S.S.S.; Woo, J.L.F. Soy isoflavones have a favorable effect on bone loss in chinese postmenopausal women with lower bone mass: A double-blind, randomized, controlled trial. *J. Clin. Endocrinol. Metab.* **2003**, *88*, 4740–4747. [CrossRef] [PubMed]

101. Giersch, G.E.W.; Charkoudian, N.; Stearns, R.L.; Casa, D.J. Fluid balance and hydration considerations for women: Review and ruture directions. *Sport. Med.* **2020**, *50*, 253–261. [CrossRef]
102. Stachenfeld, N.S.; DiPietro, L.; Palter, S.F.; Nadel, E.R. Estrogen influences osmotic secretion of AVP and body water balance in postmenopausal women. *Am. J. Physiol. Regul. Integr. Comp. Physiol.* **1998**, *274*, 187–195. [CrossRef]
103. Stachenfeld, N.S.; Splenser, A.E.; Calzone, W.L.; Taylor, M.P.; Keefe, D.L. Selected contribution: Sex differences in osmotic regulation of AVP and renal sodium handling. *J. Appl. Physiol.* **2001**, *91*, 1893–1901. [CrossRef]
104. González-SanJosé, M.L.; Rodríguez, P.M.; Valls-Bellés, V. Beer and its role in human health. In *Fermented Foods in Health and Disease Prevention*; Elsevie: Amsterdam, The Netherlands, 2017; pp. 365–384, ISBN 9780128023099.
105. Jiménez-Pavón, D.; Cervantes-Borunda, M.S.; Díaz, L.E.; Marcos, A.; Castillo, M.J. Effects of a moderate intake of beer on markers of hydration after exercise in the heat: A crossover study. *J. Int. Soc. Sports Nutr.* **2015**, *12*, 26. [CrossRef]
106. Orrù, S.; Imperlini, E.; Nigro, E.; Alfieri, A.; Cevenini, A.; Polito, R.; Daniele, A.; Buono, P.; Mancini, A. Role of functional beverages on sport performance and recovery. *Nutrients* **2018**, *10*, 1470. [CrossRef]
107. Hobson, R.M.; Maughan, R.J. Hydration status and the diuretic action of a small dose of alcohol. *Alcohol Alcohol.* **2010**, *45*, 366–373. [CrossRef]
108. Polhuis, K.C.M.M.; Wijnen, A.H.C.; Sierksma, A.; Calame, W.; Tieland, M. The diuretic action of weak and strong alcoholic beverages in elderly men: A randomized diet-controlled crossover trial. *Nutrients* **2017**, *9*, 660. [CrossRef] [PubMed]
109. Shirreffs, S.M.; Maughan, R.J. The effect of alcohol on athletic performance. *Curr. Sports Med. Rep.* **2006**, *5*, 192–196. [CrossRef] [PubMed]
110. Barnes, M.J.; Mündel, T.; Stannard, S.R. A low dose of alcohol does not impact skeletal muscle performance after exercise-induced muscle damage. *Eur. J. Appl. Physiol.* **2011**, *111*, 725–729. [CrossRef] [PubMed]
111. Shirreffs, S.M.; Maughan, R.J. Restoration of fluid balance after exercise-induced dehydration: Effects of alcohol consumption. *J. Appl. Physiol.* **1997**, *83*, 1152–1158. [CrossRef] [PubMed]
112. Flores-Salamanca, R.; Aragón-Vargas, L.F. Postexercise rehydration with beer impairs fluid retention, reaction time, and balance. *Appl. Physiol. Nutr. Metab.* **2014**, *39*, 1175–1181. [CrossRef]
113. Castro-Sepulveda, M.; Johannsen, N.; Astudillo, S.; Jorquera, C.; Álvarez, C.; Zbinden-Foncea, H.; Ramírez-Campillo, R. Effects of beer, non-alcoholic beer and water consumption before exercise on fluid and electrolyte homeostasis in athletes. *Nutrients* **2016**, *8*, 345. [CrossRef]
114. Wijnen, A.H.C.; Steennis, J.; Catoire, M.; Wardenaar, F.C.; Mensink, M. Post-exercise rehydration: Effect of consumption of beer with varying alcohol content on fluid balance after mild dehydration. *Front. Nutr.* **2016**, *3*, 45. [CrossRef]
115. Desbrow, B.; Murray, D.; Leveritt, M. Beer as a sports drink? Manipulating beer's ingredients to replace lost fluid. *Int. J. Sport Nutr. Exerc. Metab.* **2013**, *23*, 593–600. [CrossRef]
116. Desbrow, B.; Cecchin, D.; Jones, A.; Grant, G.; Irwin, C.; Leveritt, M. Manipulations to the alcohol and sodium content of beer for postexercise rehydration. *Int. J. Sport Nutr. Exerc. Metab.* **2015**, *25*, 262–270. [CrossRef]

Article

Phenolics Profile and Antioxidant Activity of Special Beers

Mirella Nardini * and Maria Stella Foddai

CREA, Research Centre for Food and Nutrition, via Ardeatina 546, 00178 Rome, Italy; mariastella.foddai@crea.gov.it
* Correspondence: mirella.nardini@crea.gov.it

Academic Editor: Fernando M. Nunes
Received: 6 May 2020; Accepted: 26 May 2020; Published: 26 May 2020

Abstract: The antioxidant activity and polyphenols content of beer associated with its low alcohol content are relevant factors for an evaluation of the nutritional quality of beer. To investigate the effect of adding foods on the nutritional quality of beer, seven special beers that were commercially available and produced adding natural foods (walnut, chestnut, cocoa, honey, green tea, coffee, and licorice) during the fermentation process were analyzed for their polyphenols and flavonoids contents, phenolics profile, and antioxidant activity. The results obtained showed that most of the special beers under study possessed antioxidant activity, as well as total polyphenols and flavonoids contents notably higher as compared with the five conventional beers analyzed. The highest polyphenols and flavonoids contents were exhibited in cocoa, walnut, chestnut, and licorice beers, followed by coffee, honey, and green tea beers. Antioxidant activity decreased in the order walnut, cocoa, chestnut, licorice, coffee, honey, and green tea. Most special beers were enriched in catechin, epicatechin, rutin, myricetin, quercetin, and resveratrol. The content of phenolic acids, especially ferulic, p-coumaric, syringic, and sinapic acids was generally higher in special beers as compared with conventional beers. Our findings showed that the addition of natural foods during the fermentation process remarkably increased antioxidant activity of beer and qualitatively and quantitatively improved its phenolics profile.

Keywords: beer; polyphenols; antioxidant activity; walnut; chestnut; green tea; coffee; cocoa; honey; licorice

---

1. Introduction

Oxidative stress is involved in the pathology of several human diseases, such as atherosclerosis, diabetes, neurodegenerative diseases, ageing, and cancer [1]. Dietary antioxidants can counteract the negative effects of oxidative stress. Polyphenols are the most abundant dietary antioxidants, due to their presence in all fruits and vegetables [1]. Polyphenol intake can be several hundreds of milligrams per day, up to 1 g/day, depending on dietary habits [2] and, in particular, in wine, coffee, beer, chocolate, and tea consumption; and it largely exceeds that of other antioxidants, such as vitamin E, vitamin C, and β-carotene [3]. Among polyphenols, phenolic acids account for about one-third of the total intake, while flavonoids account for the remaining two-thirds of the total intake [2]. Epidemiological studies have suggested associations between long-term consumption of polyphenols-rich foods and prevention of oxidative stress-related diseases such as cancer, cardiovascular diseases, diabetes, inflammation, and degenerative diseases [1,4–6].

Beer is one of the most popular alcoholic beverages consumed in large amounts all over the world, being a source of carbohydrates, amino acids, minerals, vitamins, and polyphenols. About 30% of beer polyphenols originate from hops, while the remaining 70% come from malt [7,8]. Moreover, hops provide compounds which become bitter acids (humulones) during the beer fermentation process [8].

The antioxidant activity and polyphenols content of beer associated with its low alcohol content are relevant factors in evaluating the nutritional quality of beer. Moderate beer drinking has been reported to increase plasma antioxidant and anticoagulant activities, to positively affect plasma lipid levels, and to exert protective effects on cardiovascular risk in humans [9–12].

In addition to the most familiar products, special beers produced with the addition of fruits, spices, or natural food during the fermentation process, have been becoming very popular throughout the world, responding to requests for new gustatory, olfactory, and visual stimuli from consumers. During re-fermentation and maturation of special beers, flavors and bioactive compounds, such as carotenoids and polyphenols, are extracted from fruits, spices, and natural food added to beer. Recently, the addition of fruits during the fermentation process has been reported to significantly increase the content of bioactive compounds and the antioxidant activity of beer [13]. Despite many studies describing the raw materials and the effects of technological processes, little is known about the healthy compounds and nutritional quality of commercially available beers [14–16].

In order to investigate the effect of several food additions on the nutritional quality of beer, we investigated total polyphenols and flavonoids contents, phenolics profile, and antioxidant activity of seven special beers produced with the addition of walnut, chestnut, cocoa, green tea, coffee, honey, or licorice during the fermentation process and compared our results with five conventional beers.

2. Results

2.1. Beers' Characterization

Conventional and special beers were examined in this study. The special beers were produced by the addition of the following different foods: walnut (*Juglans Regia* L., WALN), chestnut (*Castanea Sativa* L., CHES), green tea (*Camelia Sinensis* L., GTEA), coffee (*Coffea Arabica* and *Coffea Robusta* L., COFF), cocoa (*Theobroma Cacao* L., COCO), honey (*Wildflower honey*, HONE) and licorice (*Glycyrrhiza Glabra* L., LIQU), as shown in Table 1. The amount of the foods added varied in the different special beers from 2 to 62.5 g/L of beer.

Table 1. Ingredients of the special and conventional beers.

Beer Code	Food Added	Amount Added (g/L of Beer)	Ingredients
Special Beers:			
WALN	Walnut	35	Water, barley malt, oats, walnut, hops, yeast
CHES	Chestnut	40	Water, barley malt, dried chestnut, hops, yeast
GTEA	Green tea	9	Water, barley malt, wheat malt, hops, yeast, green tea
COFF	Coffee	35	Water, barley malt, oats, coffee (80% Arabica, 20% Robusta), hops, yeast
COCO	Cocoa beans	10	Water, barley malt, oats, carob, cocoa beans, hops, yeast
HONE	Honey	62	Water, barley malt, wildflower honey, hops, yeast
LIQU	Licorice	2	Water, barley malt, wheat malt, hops, licorice, sugar, yeast
Conventional Beers:			
ALE 1	-	-	Water, barley malt, corn, barley, hops, yeast
ALE 2	-	-	Water, barley malt, sugar, hops, yeast
ALE 3	-	-	Water, barley malt, caramelized barley malt, hops, yeast
LAGE 1	-	-	Water, barley malt, maize, hops, yeast
LAGE 2	-	-	Water, barley malt, barley, glucose syrup, hops, yeast

The characteristics of special and conventional beers are summarized in Table 2. All special beers were produced in Italy and were ale style beer (high fermentation beer), except one (CHES beer) which was a lager style beer (low fermentation beer). All conventional beers were produced in Italy, except one (ALE 1) which was produced in Belgium. Three conventional beers were ale style beers and two conventional beers were lager style beers. The alcoholic strength was in the range 4.5%–9.0% and 4.6%–6.6% for special and conventional beers, respectively (Table 2). Among the special beers, licorice (LIQU) and chestnut (CHES) beers exhibited the highest alcohol content (9.0% and 8.0%, respectively), while the alcohol content of the remaining special beers was quite close to that of conventional beers. The pH was in the range 4.04–4.64 and 4.29–4.87 for special and conventional beers, respectively (Table 2). International Bitterness Unit (IBU) values were in the range 7–30 for special beers, with the highest value reported for walnut beer (WALN), and in the range 15–35 for conventional beers, with the highest value reported for ALE 3 beer (Table 2). European Brewery Convention (EBC) values, referring to the color intensity of beer, were in the range 5–110 for special beers, with the highest value reported for coffee (COFF) and cocoa (COCO) beers, and in the range 4–20 for conventional beers, with the highest value reported for ALE 3 beer (Table 2).

Table 2. Characteristics, bitterness, pH, and color measurements of special and conventional beers.

Beer Code	Style	Country of Production	Alcohol Strength (% vol)	pH [a]	IBU	Color EBC
Special Beers:						
WALN	Ale	Italy	4.7	4.47	30	87
CHES	Lager	Italy	8.0	4.64	7	61
GTEA	Ale	Italy	4.5	4.54	12	5
COFF	Ale	Italy	4.5	4.04	15	110
COCO	Ale	Italy	7.0	4.41	10	110
HONE	Ale	Italy	6.8	4.34	8	18
LIQU	Ale	Italy	9.0	4.60	22	70
Conventional Beers:						
ALE 1	Ale	Belgium	6.6	4.39	28	15
ALE 2	Ale	Italy	5.2	4.61	25	8
ALE 3	Ale	Italy	5.2	4.29	35	20
LAGE 1	Lager	Italy	4.6	4.43	15	7
LAGE 2	Lager	Italy	4.8	4.87	20	4

Alcohol strength, IBU, and EBC values were provided by manufacturers. [a] Values are mean of three independent experiments. Standard error was < 0.02.

2.2. Total Polyphenols and Flavonoids Contents of Beers

Most special beers (six out of seven) showed total polyphenols content considerably and significantly ($p < 0.05$) higher (range 464–1026 mg/L of beer) as compared with that of the conventional beers (range 274–446 mg/L of beer) (Table 3). The highest polyphenols level was measured in cocoa (COCO) beer, followed by walnut (WALN), chestnut (CHES), licorice (LIQU), coffee (COFF), honey (HONE), and green tea (GTEA) beers. The polyphenols content of conventional beers was in the same order of that reported in our previous studies and in the literature [13,16–18].

Table 3. Antioxidant activity, total polyphenols and total flavonoids contents of special and conventional beers.

Beer Code	Total Polyphenols Gallic acid Eq. mg/L	Total Flavonoids Catechin Eq. mg/L	FRAP Fe_2SO_4 Eq. mM	ABTS Trolox Eq. mM
Special Beers:				
WALN	964.7 ± 9.6 [a]	90.1 ± 1.8 [a]	10.2 ± 0.02 [a]	5.2 ± 0.05 [a]
CHES	883.4 ± 10.9 [b]	71.7 ± 0.9 [b]	6.2 ± 0.08 [b]	3.4 ± 0.03 [b]
GTEA	464.4 ± 3.9 [f]	42.0 ± 0.3 [e]	3.6 ± 0.05 [d]	2.4 ± 0.03 [e]
COFF	582.7 ± 6.4 [d]	69.5 ± 1.0 [b]	5.0 ± 0.14 [e]	2.9 ± 0.03 [f]
COCO	1026.4 ± 3.0 [a]	96.4 ± 2.0 [c]	8.1 ± 0.10 [c]	3.9 ± 0.04 [c]
HONE	538.3 ± 8.3 [e]	48.7 ± 1.0 [f]	3.9 ± 0.01 [f]	2.5 ± 0.03 [d]
LIQU	819.7 ± 6.9 [c]	81.4 ± 1.3 [d]	6.1 ± 0.04 [b]	3.4 ± 0.01 [b]

Table 3. *Cont.*

Beer Code	Total Polyphenols Gallic acid Eq. mg/L	Total Flavonoids Catechin Eq. mg/L	FRAP Fe_2SO_4 Eq. mM	ABTS Trolox Eq. mM
Conventional Beers:				
Ale 1	446.1 ± 12.6 [f,i]	51.9 ± 1.1 [g]	3.7 ± 0.17 [d,f,h]	1.7 ± 0.03 [g,h]
Ale 2	382.7 ± 6.6 [l]	59.0 ± 0.9 [h]	3.4 ± 0.04 [h]	1.5 ± 0.02 [i]
Ale 3	424.4 ± 8.7 [g,f]	51.9 ± 1.3 [f,g]	3.9 ± 0.01 [f]	2.6 ± 0.02 [d]
LAGE 1	273.8 ± 4.1 [h]	26.6 ± 0.1 [l]	1.7 ± 0.02 [g]	1.8 ± 0.03 [g]
LAGE 2	320.6 ± 8.6 [m]	63.5 ± 0.8 [i]	2.8 ± 0.04 [i]	1.5 ± 0.06 [h,i]

FRAP, ferric reducing antioxidant power assay; ABTS, 2,2′-azino-bis (3-ethylbenzothiazoline-6-sulfonic acid) assay. Values are means ± SE (polyphenols content, n = 5; flavonoids content, n = 6; FRAP and ABTS, n = 3). Within each column, values with different superscript are significantly different ($p < 0.05$, one-way ANOVA, Fisher method).

Total flavonoids content of special beers was in the range 42–96 mg/L of beer. These values are somewhat higher as compared those measured in conventional beers (range 27–63 mg/L of beer) (Table 3). Among the special beers, the highest flavonoids content was measured in cocoa beer (COCO), followed by walnut (WALN), licorice (LIQU), chestnut (CHES) and coffee (COFF) beers, whereas honey (HONE) and green tea (GTEA) exhibited a total flavonoids content close to that of conventional beers.

A significant correlation was found between EBC values and total polyphenols (Figure 1a) or flavonoids (Figure 1b) contents ($p < 0.001$, R = 0.82878 and $p < 0.005$, R = 0.81706, respectively).

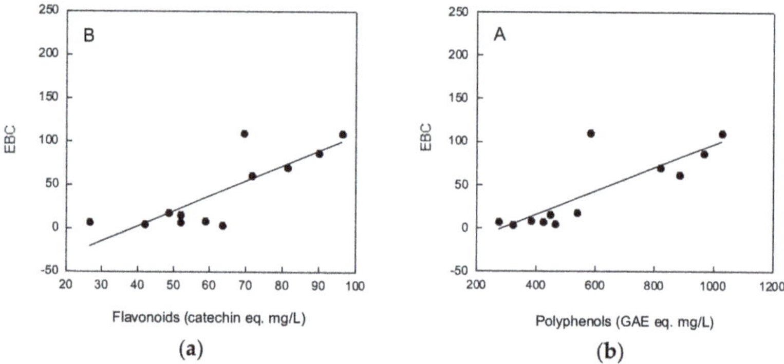

(a) (b)

Figure 1. Relationship between beer EBC values and polyphenols (**a**) or flavonoids (**b**) contents. Data were analyzed for correlation by Student's *t*-test.

2.3. Beers Antioxidant Activity

The antioxidant activity measured with ferric reducing antioxidant power (FRAP) assay was considerably higher in special beers (FRAP range 3.6–10.2 mM Fe_2SO_4/L of beer) as compared with that of conventional beers (range 1.7–3.9 mM Fe_2SO_4/L of beer) (Table 3). In the same way, the antioxidant activity evaluated by 2,2′-azino-bis(3-ethylbenzothiazoline-6-sulfonic) acid (ABTS) radical cation decolorization assay showed higher values in special beers (range 2.4–5.2 mM Trolox/L of beer) as compared with those of conventional beers (range 1.5–2.6 mM Trolox/L of beer). The highest antioxidant activity was measured in walnut (WALN) beer, followed by cocoa (COCO), chestnut (CHES), licorice (LIQU), and coffee (COFF) beers. Honey (HONE) and green tea (GTEA) beers showed antioxidant activity values close to those measured in conventional beers (Table 3). The antioxidant activity values measured in conventional beers were consistent with our previous results and with data from the literature [10,13,16,19].

As shown in Figure 2a, a strong correlation between total polyphenols content, measured by Folin–Ciocalteu assay, and antioxidant activity of beers, measured by both the FRAP and ABTS assays, was found ($r = 0.93815$, $p < 0.0001$ for FRAP assay and $r = 0.90592$, $p < 0.0001$ for ABTS assay). Furthermore, a strict correlation was observed between the total flavonoids content and the antioxidant activity of beers, measured by both the FRAP and ABTS methods ($r = 0.87913$, $p < 0.0002$ for FRAP assay and $r = 0.75286$, $p < 0.005$ for ABTS assay) (Figure 2b).

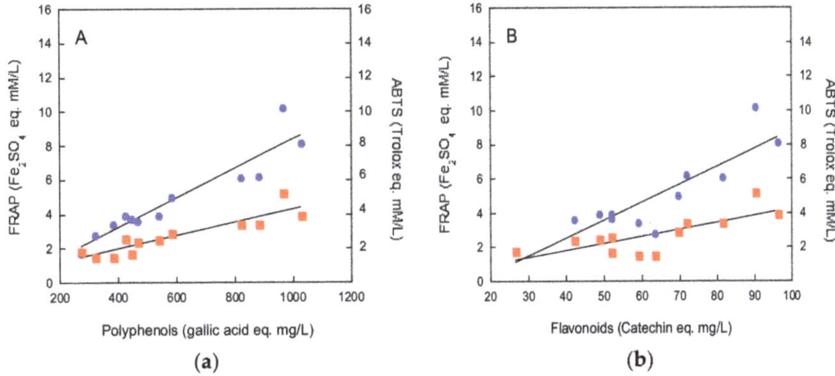

Figure 2. Relationship between beer antioxidant activity, measured by both FRAP (full circles) and ABTS (full squares) assays, and polyphenols (**a**) or flavonoids (**b**) contents. Data were analyzed for correlation by Student's *t*-test.

2.4. Phenolics Profile Analyses

Due to the role of polyphenols in determining beer quality, the hydroxycinnamic acid derivatives chlorogenic, vanillic, caffeic, *p*-coumaric, and ferulic acids; the hydroxybenzoic acid derivatives syringic and sinapic acids; the flavonoids catechin, epicatechin, rutin, myricetin, and quercetin; and the stilbene derivative resveratrol were measured by high-performance liquid chromatography (HPLC). As most phenolic acids are present in beer as esterified forms, we measured the level of both free and total (free plus conjugated forms) phenolic acids [16]. The content of single phenolic compounds, representative of the different classes of polyphenols, are shown in Tables 4 and 5 for conventional and special beers, respectively.

As a basis for comparison, first, conventional beers were analyzed. The total phenolic acids content of conventional beers, obtained by alkaline hydrolysis, varied in the range 21.78–38.89 mg/L of beer (Table 4). Total ferulic acid was by far the most abundant phenolic acid in conventional beers, regardless of the beer style, ranging from 10.27 to 21.66 mg/L of beer, followed by caffeic (range 1.61–5.99 mg/L of beer), sinapic (range 2.19–4.80 mg/L of beer), vanillic (range 2.30–4.65 mg/L of beer), and *p*-coumaric (range 0.77–2.77 mg/L of beer) acids, whereas syringic acid exhibited the lowest concentration (range 0–0.71 mg/L of beer). Lager style beers (LAGE 1 and LAGE 2) showed the lowest caffeic, syringic, and *p*-coumaric acids content as compared with ale style beers. The total amount of each phenolic acid, measured after alkaline hydrolysis, was higher with respect to the content of the respective free form, indicating that phenolic acids were present in beer mainly as conjugated forms. Free and total phenolic acids contents of conventional beers was in the same order of magnitude as that reported in our previous studies [13,16,20]. Free phenolic acids content measured in conventional beer was also in agreement with other data from the literature [21–25], whereas total phenolic acids content is usually not routinely measured. Noteworthily, chlorogenic acid; the flavonoids catechin, epicatechin, rutin, myricetin, and quercetin; and the stilbene derivative resveratrol were undetectable in all conventional beers in our experimental conditions, regardless of the beer style (Table 4).

Table 4. Phenolic acids, flavonoids, and resveratrol contents of conventional beers by high performance liquid chromatography with diode array detector (HPLC-DAD) (mg/L).

Beer Code	ALE 1	ALE 2	ALE 3	LAGE 1	LAGE 2
Phenolic Acids:					
Chlorogenic	nd	nd	nd	nd	nd
Vanillic					
Free	nd	nd	2.09 ± 0.08	nd	nd
Total	2.80 ± 0.05	3.58 ± 0.07	4.65 ± 0.06	4.46 ± 0.12	2.3 ± 0.07
Caffeic					
Free	nd	nd	1.24 ± 0.10	nd	nd
Total	3.00 ± 0.20	3.38 ± 0.01	5.99 ± 0.11	1.70 ± 0.08	1.61 ± 0.04
Syringic					
Free	nd	nd	0.25 ± 0.01	nd	nd
Total	0.71 ± 0.09	0.67 ± 0.04	0.51 ± 0.03	nd	0.32 ± 0.01
p-Coumaric					
Free	0.53 ± 0.03	1.12 ± 0.05	0.68 ± 0.02	1.06 ± 0.05	0.35 ± 0.04
Total	2.00 ± 0.10	2.77 ± 0.09	2.13 ± 0.04	1.56 ± 0.08	0.77 ± 0.01
Ferulic					
Free	0.90 ± 0.03	11.03 ± 0.54	2.91 ± 0.11	2.12 ± 0.06	1.81 ± 0.03
Total	10.27 ± 1.00	19.90 ± 0.21	21.66 ± 0.55	11.0 ± 0.07	13.71 ± 0.49
Sinapic					
Free	0.41 ± 0.01	0.44 ± 0.21	0.98 ± 0.10	0.36 ± 0.04	1.07 ± 0.04
Total	4.80 ± 0.05	2.19 ± 0.07	3.95 ± 0.11	3.53 ± 0.03	3.07 ± 0.06
Total Phenolic Acids [a]	23.58 ± 1.56	32.49 ± 0.49	38.89 ± 0.90	22.25 ± 0.38	21.78 ± 0.68
Flavonoids:					
Catechin	nd	nd	nd	nd	nd
Epicatechin	nd	nd	nd	nd	nd
Rutin	nd	nd	nd	nd	nd
Myricetin	nd	nd	nd	nd	nd
Quercetin	nd	nd	nd	nd	nd
Stilbenes:					
Resveratrol	nd	nd	nd	nd	nd

Values are means ± SE (n = 3). nd, not detectable. [a] Total phenolic acids content was calculated by the sum of single phenolic acids content obtained after alkaline hydrolysis.

The phenolic profile of special beers is shown in Table 5. The content of single phenolic acids differs considerably among the different special beers. Total phenolic acids content obtained after alkaline hydrolysis varied in the range 20.54–45.45 mg/L of beer, with chestnut (CHES) beer exhibiting the highest value, followed by cocoa (COCO), licorice (LIQU), coffee (COFF), honey (HONE), green tea (GTEA), and walnut (WALN) beers. Ferulic acid was by far the most abundant phenolic acid in all special beers, while syringic acid showed the lowest values, as found in conventional beers. In detail, total ferulic and vanillic acids varied in the ranges 8.22–27.55 and 2.03–5.09 mb/L of beer, respectively, with the highest value measured in chestnut (CHES) beer. Total caffeic acid content ranged from 1.48 to 9.20 mg/L of beer, with the highest value measured in coffee (COFF) beer and the lowest value in green tea (GTEA) beer. The total *p*-coumaric content ranged from 1.75 to 4.32 mg/L of beer, with the highest content in walnut (WALN) beer. The total sinapic acid content of special beers varied in the range 2.52–6.73 mg/L of beer, the highest values found in honey (HONE) beer, followed by licorice (LIQU), cocoa (COCO), chestnut (CHES), green tea (GTEA), walnut (WALN), and coffee (COFF) beers (Table 5). The total syringic acid content ranged between 0.67–1.42 mg/L of beer, with the highest value found in cocoa (COCO) beer, while it was undetectable in walnut (WALN) and coffee (COFF) beers, in our experimental conditions. The total amount of each phenolic acid measured after alkaline hydrolysis was higher with respect to the content of the respective free form, also indicating that, in the special beers, phenolic acids were mainly present as conjugated forms. Noteworthily, among the special beers, chlorogenic acid was detected only in coffee (COFF) beer.

Table 5. Phenolic acids, flavonoids, and resveratrol contents of special beers by HPLC-DAD (mg/L).

Beer Code	WALN	CHES	GTEA	COFF	COCO	HONE	LIQU
Phenolic Acids:							
Chlorogenic	tr	nd	nd	1.56 ± 0.10	nd	nd	nd
Vanillic							
Free	0.92 ± 0.12	1.57 ± 0.03	0.87 ± 0.04	0.78 ± 0.03	1.14 ± 0.09	0.80 ± 0.05	1.03 ± 0.10
Total	2.16 ± 0.26	5.09 ± 0.06	2.82 ± 0.15	2.03 ± 0.14	3.39 ± 0.17	3.09 ± 0.22	2.32 ± 0.11
Caffeic							
Free	0.52 ± 0.01	0.24 ± 0.01	tr	0.57 ± 0.02	0.50 ± 0.02	tr	0.56 ± 0.07
Total	3.16 ± 0.15	3.47 ± 0.03	1.48 ± 0.18	9.20 ± 0.21	3.69 ± 0.01	2.37 ± 0.17	3.71 ± 0.04
Syringic							
Free	tr	0.40 ± 0.03	0.62 ± 0.02	tr	0.54 ± 0.02	0.27 ± 0.01	0.40 ± 0.03
Total	tr	1.24 ± 0.05	0.96 ± 0.04	tr	1.42 ± 0.05	1.24 ± 0.10	0.67 ± 0.03
***p*-Coumaric**							
Free	0.68 ± 0.01	1.02 ± 0.06	0.11 ± 0.01	0.36 ± 0.02	1.38 ± 0.08	0.21 ± 0.01	1.06 ± 0.06
Total	4.32 ± 0.24	3.36 ± 0.07	2.24 ± 0.16	1.93 ± 0.08	3.26 ± 0.13	1.75 ± 0.03	2.95 ± 0.14
Ferulic							
Free	1.05 ± 0.02	1.81 ± 0.20	0.15 ± 0.01	0.63 ± 0.02	1.16 ± 0.03	0.43 ± 0.01	1.32 ± 0.08
Total	8.22 ± 0.17	27.55 ± 0.43	14.30 ± 0.40	20.50 ± 0.64	22.10 ± 0.73	19.20 ± 0.33	20.63 ± 0.87
Sinapic							
Free	0.45 ± 0.01	0.97 ± 0.12	0.49 ± 0.01	0.24 ± 0.02	0.44 ± 0.01	0.55 ± 0.01	1.03 ± 0.04
Total	2.68 ± 0.06	4.74 ± 0.04	4.48 ± 0.08	2.52 ± 0.02	4.89 ± 0.05	6.73 ± 0.03	6.66 ± 0.07
Totalphenolic Acids [a]	20.54 ± 0.88	45.45 ± 0.68	26.28 ± 1.01	36.18 ± 1.09	38.75 ± 1.14	34.38 ± 0.88	36.94 ± 1.26
Flavonoids:							
Catechin	tr	4.65 ± 0.13	2.98 ± 0.09	tr	4.58 ± 0.02	tr	tr
Epicatechin	1.80 ± 0.11	3.68 ± 0.12	3.09 ± 0.05	1.30 ± 0.07	1.83 ± 0.11	0.94 ± 0.05	tr
Rutin	nd	nd	0.68 ± 0.02	nd	nd	1.29 ± 0.02	0.92 ± 0.10
Myricetin	4.44 ± 0.27	tr	1.69 ± 0.05	0.39 ± 0.03	0.65 ± 0.02	2.67 ± 0.18	8.82 ± 0.07
Quercetin	6.55 ± 0.31	tr	1.17 ± 0.09	0.54 ± 0.02	1.52 ± 0.06	4.67 ± 0.23	2.63 ± 0.15
Stilbenes:							
Resveratrol	0.26 ± 0.20	0.35 ± 0.02	0.32 ± 0.02	0.23 ± 0.01	0.31 ± 0.01	0.24 ± 0.01	0.20 ± 0.01

Values are means ± SE ($n = 3$). Nd, not detectable and tr, traces amount.[a] Total phenolic acids content was calculated by the sum of single phenolic acids content obtained after alkaline hydrolysis.

Unlike conventional beers, the special beers under study exhibited detectable levels of the flavonoids catechin, epicatechin, rutin, myricetin, quercetin, as well as the stilbene resveratrol. The flavonoids epicatechin, myricetin, quercetin, as well as the stilbene resveratrol were present in almost every special beer under study, whereas the flavonoids catechin and rutin were detectable in three out of the seven special beers, in our experimental conditions (Table 5). The epicatechin content varied in the range 0.94–3.68 mg/L of beer, with the highest values measured in chestnut (CHES) and green tea (GTEA) beers, while only traces were found in licorice beer (LIQU). Myricetin and quercetin content varied in the range 0.39–8.82 and 0.54–6.55 mg/L of beer, respectively, with the highest myricetin value found in licorice (LIQU) beer and the highest quercetin level measured in walnut (WALN) beers, whereas only trace amount of both flavonoids were measured in chestnut (CHES) beer. Catechin was present in the range 2.98–4.65 mg/L of beer in chestnut (CHES), green tea (GTEA), and cocoa (COCO) beers, whereas only traces were detected in the remaining special beers. Rutin was detected in green tea (GTEA), honey (HONE), and licorice (LIQU) beers, ranging from 0.68 to 1.29 mg/L of beer (Table 5).

In regard to the stilbene derivative resveratrol, it was present in all special beers under study, although at a low level (range 0.20–0.35 mg/L of beer).

3. Discussion

The antioxidant activity and phenolics content of beer rely on the quantity and quality of starting material, as well as on the industrial brewing process. Beer exhibiting high phenolics content and high antioxidant activity display better quality, more stable flavor and aroma, foam stability, and longer shelf life as compared with beer with lower phenolics levels and weaker antioxidant properties [7,26–29].

In our study, total polyphenols and flavonoids contents of most special beers was remarkably higher as compared with conventional beers. In particular, the flavonoids catechin, rutin, myricetin, quercetin, as well as the stilbene, resveratrol, were undetectable under our experimental conditions, in all conventional beers analyzed.

Undoubtedly, beer color has an impact on beer taste and experience. The EBC values were remarkably higher in special beers as compared with conventional beers. The strong correlation found between EBC values and total polyphenols and flavonoids contents suggest a relevant contribution of plant food phenolics to special beer color, in addition to that of malt. A similar correlation between beer total polyphenols content and EBC values has been recently reported, which suggested that beer color is correlated to the total amount of phenolic compounds [30]. Instead, the IBU values gave similar bitterness values in both special and conventional beers.

Recently, the addition of fresh fruits during the fermentation process has been reported to increase antioxidant activity, total polyphenols and flavonoids contents, and to qualitatively and quantitatively improve the phenolics profile with respect to conventional beers [13]. In this study, the special beers produced with food addition during the fermentation step exhibited total polyphenols content and antioxidant activity even higher than those reported for fruit beers. Notably, the specific foods involved in the present study have been reported to contain high polyphenols levels and to possess strong antioxidant activity [31]. The strict correlation observed between antioxidant activity and total polyphenols and flavonoids contents suggest a central role of phenolics in the antioxidant properties of beers.

Our results showed that cocoa (COCO), walnut (WALN), chestnut (CHES), and licorice (LIQU) beers exhibited the higher polyphenols and flavonoids contents, followed by coffee (COFF), honey (HONE), and green tea (GTEA) beers. Antioxidant activity decreased in the order walnut (WALN), cocoa (COCO), chestnut (CHES), licorice (LIQU), coffee (COFF), honey (HONE), and green tea (GTEA) beers. The phenolic profile obtained by HPLC showed that most special beers are enriched in catechin, epicatechin, rutin, myricetin, quercetin, and resveratrol. Phenolic acids content, especially ferulic, *p*-coumaric, syringic, and sinapic acids, was generally higher in special beers as compared with the conventional beers.

Walnut beer (WALN) showed the highest antioxidant activity, measured by both FRAP and ABTS assays and high total flavonoids level. The HPLC analyses demonstrated the highest quercetin content among special beers, in addition to high levels of epicatechin and myricetin. In this regard, walnuts have been reported to contain many phytochemicals, including the highest known levels of phenolic antioxidants (phenolic acids, flavonoids, and tannins) with respect to other nut species [32–34].

Chestnut beer (CHES) exhibited the highest catechin, epicatechin, and resveratrol levels. Accordingly, chestnuts have been reported to contain high levels of catechin, and epicatechin, in addition to phenolic acids and tannins [35–37]. Moreover, chestnuts have been recognized as one of the richest foods with respect to polyphenols content, exhibiting very high antioxidant activity [31].

Among special beers, coffee (COFF) beer showed the lowest catechin, rutin, myricetin, quercetin, epicatechin, and resveratrol levels. However, coffee beer contained chlorogenic acid and the highest caffeic acid level among the special beers. Accordingly, both caffeic and chlorogenic acids have been reported to be present in high amounts in coffee [38,39] and are extracted from coffee during the fermentation process of coffee beer.

Cocoa and green tea are known to possess high polyphenols and flavonoids contents, especially catechin and epicatechin [31,40,41]. In agreement, high levels of catechin and epicatechin together with quercetin and myricetin were detected in both cocoa (COCO) and green tea (GTEA) beers, indicating, once again, that these compounds are extracted from cocoa and green tea during the fermentation process.

Despite the low amount of licorice added to beer (2 g/L of beer) during the fermentation process, licorice (LIQU) beer exhibited the highest myricetin content as compared with the other special beers together with high levels of quercetin, caffeic, *p*-coumaric, ferulic, and sinapic acids. Licorice has been reported to contain many bioactive compounds, particularly flavonoids, which are responsible for its yellow color [42,43]. Various biological activities have been associated with licorice extracts, particularly with its flavonoids and triterpenic saponins contents, such as antiviral, antimicrobial, antioxidant, anti-inflammatory, and anticancer effects [44,45].

Honey beer (HONE) showed the highest sinapic acid and rutin contents as compared with the other special beers, and high levels of mirycetin and quercetin. According to our results, the occurrence of caffeic acid, *p*-coumaric acid, ferulic acid, vanillic acid, sinapic acid, syringic acid, rutin, quercetin, myricetin, resveratrol, and epicatechin in honey has been reported by several studies [46,47]. Honey is one of the most renowned natural foods. Although its composition is extremely variable, depending on its botanical and geographical origins, the abundant presence of phenolic compounds, especially phenolic acids and flavonoids, and the antioxidant properties of honey have renewed interest toward this natural food.

In our previous study [16], we demonstrated that phenolic acids strongly contribute to the antioxidant activity of beer. Flavonoids have been reported to be free radical scavengers, metal chelators, and strong antioxidants [48,49]. Therefore, the enrichment in flavonoids observed in the special beers could account, at least in part, for the higher antioxidant activity measured in most of the special beers as compared with the conventional beers. The stilbene derivative resveratrol was also detected in the special beers, although at low levels, and it could contribute to the antioxidant activity of beers.

Antioxidant activity and polyphenols content of beer associated with its low alcohol content are relevant factors in determining the nutritional quality of beer. Total polyphenols content of conventional beer is quite low as compared with that of red wine. In fact, the total amount of polyphenols in red wine has been estimated to be in the range 2000–6000 mg/L of wine [50–53], whereas that of conventional beers has been reported to vary in the range 300–500 mg/L of beer for the most common beer styles [13,16]. Higher values (622 ± 77 and 875 ± 168 mg/L, respectively) have been reported only for abbey and bock beer styles [16]. However, the polyphenols content of conventional beers has been reported to be similar or even higher with respect to that reported for white wine (range 50–350 mg/L) [50–52]. Recently, a total polyphenols content of up to 770 mg/L of beer has been reported for fruit beers, produced through the addition of fresh fruits during the fermentation process [13]. The special beers examined in this study exhibited total polyphenols content in the range 464–1026 mg/L, even higher than that reported for fruit beers. These values are substantially higher as compared with those of the conventional beers, as well as compared with those of white wine. A similar trend could be observed for antioxidant activity. The FRAP values have been reported to range from 15 to 31 mM and from 2.2 to 5.5 mMFe$_2$SO$_4$ eq./L of red and white wine, respectively [53,54]. These values should be compared with those found in special beers (3.9–10.2 mM Fe$_2$SO$_4$ eq./L) and conventional beers (1.7–3.9 mM Fe$_2$SO$_4$ eq./L). Again, the antioxidant activity of special beers was comparable or even higher than that reported for white wines, although lower with respect to the antioxidant activity reported for red wines.

From our data, food addition during the fermentation step resulted in considerable improvement of the nutritional quality of beer, in terms of bioactive compounds content and antioxidant activity as compared with conventional beers. The increased amounts of polyphenols, particularly phenolic acids; flavonoids; and resveratrol in special beers have beneficial effects on beer drinkers.

Phenolic acids are small molecules with known antioxidant activity, acting as free radical acceptors and chain breakers. The antioxidant and biological effects, such as anti-inflammatory, cardioprotective, neuroprotective, antimicrobial, antiviral, anticancer effect of phenolic acids have been widely studied and reported in the literature, particularly for caffeic, ferulic, *p*-coumaric, and vanillic acids. Phenolic acids from beer have been described as being quickly absorbed and extensively metabolized in humans to the form of glucuronide and sulfate derivatives [55,56], which have been reported to retain antioxidant activity [57]. Flavonoids, the most abundant phenolic antioxidants in human diets, have been reported to be absorbed in humans, circulate in plasma, and are excreted in urine [2]. Flavonoids have been reported to display antioxidant activity, free radical scavenging capacity, metal chelation activity, coronary heart disease prevention, hepatoprotective, anti-inflammatory, and anticancer activities [49,58]. In regard to the stilbene resveratrol, bioavailability studies in humans have demonstrated its absorption and rapid metabolism to glucuronides and sulfates conjugates, the major plasma and urine metabolites [6]. Resveratrol has been reported to have several health-promoting effects in both animals and humans such as antioxidant, anti-inflammatory, antidiabetic, and antiproliferative properties [59,60].

A renewed interest has been focused on beer, due to its phenolic antioxidant component coupled with low ethanol content. Moderate beer drinking has been reported to increase plasma antioxidant and anticoagulant activities, to positively affect plasma lipid levels, and to exert protective effects on cardiovascular risk in humans [9–12,61]. Moreover, beer drinking seems to have no effect or even an inverse effect on total homocysteine concentration [62]. In conclusion, beer can contribute to the overall dietary intake of antioxidants and food addition to beer can significantly strengthen this contribution.

In addition to polyphenols, the barley, hops, and plant food contained other antioxidants, such as carotenoids, tocopherols, and ascorbic acid. All these compounds could contribute to some extent to the overall antioxidant activity of beers.

4. Materials and Methods

4.1. Materials

Caffeic acid, vanillic acid, sinapic acid, syringic acid, *p*-coumaric acid, ferulic acid, *o*-coumaric acid, chlorogenic acid (5-*O*-caffeoylquinic acid), catechin, epicatechin, resveratrol, myricetin, quercetin, trolox, gallic acid, ferric chloride, ferrous sulfate, sodium nitrite, aluminium chloride, potassium peroxodisulfate, 2,4,6-tripyridyl-S-triazine (TPTZ), 2,2′-azino-bis(3-ethylbenzothiazoline-6-sulfonic acid) diammonium salt (ABTS), and EDTA were from Sigma (St. Louis, MO, USA). Rutin was from Extrasynthese (Genay Cedex, France). Ascorbic acid and all organic solvents were obtained from Carlo Erba (Milano, Italy). Standard phenolics were dissolved in methanol (1 mg/mL), stored at −80 °C, and used within 1 week. Working standard solutions were obtained daily by dilution in sample buffer (1.25% glacial acetic acid, 7% methanol in twice-distilled water).

4.2. Beers

The conventional and special beers used in this study were purchased at local markets and beer shops. All special beers were produced by manufacturers with food addition during the first step of the fermentation process.

Special beers from the following different food typologies were explored: walnut (*Juglans regia* L. from Sorrento, Italy), chestnut (*Castanea Sativa* L. from Val Mongia, Italy), cocoa (*Theobroma Cacao* L.), honey (*Wildflower honey*), green tea (*Camelia Sinensis* L.), coffee (*Coffea Arabica* L., *Coffea Robusta* L.), and licorice (*Glycyrrhiza Glabra* L.). Table 1 showed the ingredients used for the beers' production and the amount of foods added during the first fermentation step.

Beer bottles were stored in the dark and analyzed immediately after opening. Aliquots were frozen at −80 °C for phenolics profile determination and analyzed within one week.

4.3. Beers' Analyses

The total polyphenols content was measured on 0.02 mL aliquots by the Folin–Ciocalteu method [63], using gallic acid as a reference compound. Briefly, beer samples were diluted with distilled water to give a final volume of 1 mL, then 0.1 mL of Folin–Ciocalteu's reagent was added. After 5 min, 0.2 mL sodium carbonate (35% *w/v*) was added. Final volume was adjusted to 2 mL with distilled water. After 1 h in the dark, absorbance at 765 nm was measured against an appropriate blank reagent. The results were expressed as milligrams of gallic acid equivalents per liter of beer.

The total flavonoids content was measured on 0.05 mL aliquots by a colorimetric method previously described [64], using catechin as the reference standard to obtain the calibration curve. Briefly, beer samples were diluted with distilled water to a final volume of 1.5 mL, and then 0.075 mL of 5% $NaNO_2$ solution was added. After 6 min, 0.15 mL of 10% $AlCl_3$ hexahydrate was added and allow to stand for another 5 min, before 0.5 mL 1 M NaOH was added. The volume was adjusted to 2.5 mL with distilled water, mixed, and absorbance at 510 nm was measured. The results are expressed as milligrams of catechin equivalents per liter of beer.

The total antioxidant activity of beers was evaluated by both the ferric reducing antioxidant power (FRAP) assay [65] and by the ABTS radical cation decolorization (ABTS) assay [66] on 0.01 mL of beer aliquots. FRAP assay is a colorimetric method that measures the reduction of a ferric-tripyridyltriazine complex to its ferrous colored form, in the presence of antioxidants. The reaction was monitored for 6 min after the addition of beer to the FRAP reagent and the 6 min absorbance readings used for calculation referring to the iron sulfate calibration curve (range 0–100 μM) and reported as mM Fe_2SO_4 equivalent/L of beer. The ABTS assay is based on free radical scavenging capacity. The ABTS radical cation was produced by reacting ABTS solution (7 mM) with potassium persulfate (2.45 mM final concentration) in distilled water at room temperature, in the dark, for 16 h before use. A working solution (ABTS reagent) was diluted to obtain absorbance values between 1.4 and 1.5 AU at 734 nm and prewarmed at 30 °C. The percentage inhibition of absorbance was calculated with reference to a Trolox calibration curve (0–15 μM and expressed as mM Trolox equivalent/L of beer. All solutions were prepared daily.

International Bitterness Unit (IBU) and European Brewery Convention (EBC) values were supplied by the manufacturer. The IBU values measure the bitterness of beer, due to the amount of iso-alpha-acids. The EBC values refer to the color intensity, roughly darkness of the beer.

4.4. Beer Treatment for Phenolics Profile Determination by High Performance Liquid Chromatography (HPLC)

Beer aliquots (1 mL) were added with *o*-coumaric acid (10 μg) as internal standard and NaCl (300 mg). Phenolics were extracted with diethylether and diethylacetate, as described by Pozo-Bayon et al. [67]. Pooled extracts were evaporated under vacuum at 30 °C by rotatory evaporator. For total phenolic acids determination, beer samples were added with *o*-coumaric acid (20 μg) as internal standard and hydrolyze by alkaline treatment in the presence of ascorbate and EDTA [20,39]. After hydrolysis, the samples were acidified at pH 3.0 with 4 N HCl, added with NaCl (300 mg) and extracted as above reported. The dried residues, obtained by the above reported procedures, were dissolved in 0.1 mL methanol, vortexed for 5 min, and then EDTA (0.5 M, 40 μL), ascorbic acid (5% *w/v*, 0.2 mL), and twice-distilled water up to 1 mL final volume, were added. Samples were vortexed for 5 min, filtered, and analyzed by HPLC after appropriate dilution. Quantification of phenolic compounds was calculated with reference to calibration curves obtained with pure standard phenolics (range 0.1–10 μg injected).

Recovery experiments were performed adding known amounts of pure phenolic compounds to beer samples, followed by the above reported extraction protocol. An almost complete recovery of all phenolics under study was measured (range 91.0%–105.8%). When samples were submitted to alkaline hydrolysis, prior to the extraction procedure for the total phenolic acids evaluation, the recovery of the phenolic acids under study was in the range 95.9%–104.6%.

4.5. HPLC Instrumentation

In our laboratory, phenolic compounds are routinely assayed in foods, beverages, human plasma, and cell extracts using high performance liquid chromatography (HPLC) [13,16,39,53,68]. The high performance liquid chromatograph is a PerkinElmer Series 200 Liquid Chromatography (PerkinElmer, Norwalk, CT, USA) with gradient pump, column thermoregulator, auto-sampling injector, and diode array detector (DAD) (PerkinElmer Norwalk, CT, USA). The operating conditions used were as follows: column temperature 30 °C, flow rate 1 mL/min, injection volume 50 μL, and detector at 280 nm. Chromatographic separation was obtained on a Supelcosil LC-18 column (5.0 μm particle size, 250 × 4.6 mm ID), equipped with a guard column (C_{18}, 5.0 μm particle size, 20 × 4.0 mm ID; both Supelco, Bellefonte, PA, USA).To separate phenolic compounds, a gradient elution was performed using the following two mobile phases: solution A, consisting of 1.25% glacial acetic acid in twice-distilled water and solution B, absolute methanol. The gradient used was as follows: 0–30 min, from 98% A,

2% B to 94% A, 6% B, linear gradient; 31–60 min, from 94% A, 6% B to 88% A, 12% B, linear gradient; 61–80 min, from 88% A, 12% B to 74% A, 26% B, linear gradient; 81–95 min, from 74% A, 26% B to 65% A, 35% B, linear gradient; 96–105 min, from 65% A, 35% B to 60% A, 40% B, linear gradient; and 106–120 min, 45% A, 55% B; 121–150 min, 98% A, 2% B.

4.6. Statistical Analysis

Data presented are means ± standard error. All measurements were made at least in triplicate. Statistical analysis was performed using a statistical package running on a PC (KaleidaGraph 4.0, Synergy Software, Reading, PA, USA). The Student's *t* test was used for regression analyses. The probability of $p < 0.05$ was considered to be statistically significant.

Author Contributions: Conceptualization, study design, and supervision M.N.; methodology M.N. and M.S.F.; software M.N. and M.S.F.; analysis of the data M.N.; writing—original draft preparation-editing M.N.; supervision M.N. All authors have red and agreed to the published version of the manuscript.

Funding: This research received no external funding.

Conflicts of Interest: The authors declare no conflict of interest.

References

1. Aruoma, O. Free radicals, oxidative stress and antioxidants in human health and diseases. *J. Am. Oil Chem. Soc.* **1998**, *75*, 199–212. [CrossRef] [PubMed]
2. Scalbert, A.; Williamson, G. Dietary intake and bioavailability of polyphenols. *J. Nutr.* **2000**, *130*, 2073S–2085S. [CrossRef] [PubMed]
3. Pulido, R.; Hernandez-Garcia, M.; Saura-Calixto, F. Contribution of beverages to the intake of lipophilic and hydrophilic antioxidants in the Spanish diet. *Eur. J. Clin. Nutr.* **2003**, *57*, 1275–1282. [CrossRef] [PubMed]
4. Rienks, J.; Barbaresko, J.; Nothlings, U. Association of polyphenol biomarkers with cardiovascular disease and mortality risk: A systematic review and meta-analysis of observational studies. *Nutrients* **2017**, *9*, e415. [CrossRef]
5. Grosso, G.; Micek, A.; Godos, J.; Pajak, A.; Sciacca, S.; Galvano, F.; Giovannucci, E.L. Dietary flavonoid and lignan intake and mortality in prospective cohort studies: Systematic review and dose-response meta-analysis. *Am. J. Epidemiol.* **2017**, *185*, 1304–1316. [CrossRef] [PubMed]
6. Del Rio, D.; Rodriguez-Mateos, A.; Spencer, J.P.; Tognolini, M.; Borges, G.; Crozier, A. Dietary (poly)phenolics in human health: Structures, bioavailability, and evidence of protective effects against chronic diseases. *Antioxid. Redox Signal.* **2013**, *18*, 1818–1892. [CrossRef]
7. Callemien, D.; Jerkovic, V.; Rozenberg, R.; Collin, S. Hop as an interesting source of resveratrol for brewers: Optimization of the extraction and quantitative study by liquid chromatography/atmosferic pressure chemical ionization tandem mass spectrometry. *J. Agric. Food Chem.* **2005**, *53*, 424–429. [CrossRef]
8. De Keukeleire, D.; de Cooman, L.; Rong, H.; Heyerick, A.; Kalita, J.; Milligan, S.R. Functional properties of hop polyphenols. *Basic Life Sci.* **1999**, *66*, 739–760.
9. Gronbaek, M.; Deis, A.; Sorensen, T.I.; Becker, U.; Schnohr, P.; Jensen, G. Mortality associated with moderate intakes of wine, beer and spirits. *Br. Med. J.* **1995**, *310*, 1165–1169. [CrossRef]
10. Gorinstein, S.; Caspi, A.; Libman, E.; Leontowicz, H.; Leontowicz, M.; Tahsma, Z.; Katrich, E.; Jastrzebski, Z.; Trakhtenberg, S. Bioactivity of beer and its influence on human metabolism. *Int. J. Food Sci. Nutr.* **2007**, *58*, 94–107. [CrossRef] [PubMed]
11. Kaplan, N.M.; Palmer, B.F. Nutritional and health benefits of beer. *Am. J. Med. Sci.* **2000**, *320*, 320–326. [CrossRef] [PubMed]
12. Costanzo, S.; Di Castelnuovo, A.; Donati, M.B.; Iacoviello, L.; de Gaetano, G. Wine, beer or spirit drinking in relation to fatal and non-fatal cardiovascular events: A meta-analysis. *Eur. J. Epidemiol.* **2011**, *26*, 833–850. [CrossRef] [PubMed]
13. Nardini, M.; Garaguso, I. Characterization of bioactive compounds and antioxidant activity of fruit beers. *Food Chem.* **2020**, *305*, 125437. [CrossRef] [PubMed]
14. Lugasi, A. Polyphenol content and antioxidant properties of beer. *Acta Alimentaria* **2003**, *32*, 181–192. [CrossRef]

15. Granato, D.; Branco, G.F.; FariaJde, A.; Cruz, A.G. Characterization of Brazilian lager and brown ale beers based on color, phenolic compounds, and antioxidant activity using chemometrics. *J. Sci. Food Agric.* **2011**, *91*, 563–571. [CrossRef]

16. Piazzon, A.; Forte, M.; Nardini, M. Characterization of phenolics content and antioxidant activity of different beer types. *J. Agric. Food Chem.* **2010**, *58*, 10677–10683. [CrossRef]

17. Vinson, J.A.; Mandarano, M.; Hirst, M.; Trevithick, J.R.; Bose, P. Phenol antioxidant quantity and quality in foods: Beers and the effect of two types of beer on an animal model of atherosclerosis. *J. Agric. Food Chem.* **2003**, *51*, 5528–5533. [CrossRef]

18. Gorjanovic, S.; Novakovic, M.; Potkonjak, N.; Leskosek-Cukalovic, I.; Suznjevic, D. Application of a novel antioxidative assay in beer analysis and brewing process monitoring. *J. Agric. Food Chem.* **2010**, *58*, 744–751. [CrossRef]

19. Zhao, H.; Li, H.; Sun, G.; Yang, B.; Zhao, M. Assessment of endogenous antioxidative compounds and antioxidant activities of lager beers. *J. Sci. Food Agric.* **2013**, *93*, 910–917. [CrossRef]

20. Nardini, M.; Ghiselli, A. Determination of free and bound phenolic acids in beer. *Food Chem.* **2004**, *84*, 137–143. [CrossRef]

21. Montanari, L.; Perretti, G.; Natella, F.; Guidi, A.; Fantozzi, P. Organic and phenolic acids in beer. *Lebensm. Wiss. Technol.* **1999**, *32*, 535–539. [CrossRef]

22. Floridi, S.; Montanari, L.; Marconi, O.; Fantozzi, P. Determination of free phenolics in wort and beer by coulometric array detection. *J. Agric. Food Chem.* **2003**, *51*, 1548–1554. [CrossRef]

23. Jandera, P.; Skerikova, V.; Rehova, L.; Hajek, T.; Baldrianova, L.; Skopova, G.; Kellner, V.; Horna, A. RP-HPLC analysis of phenolic compounds and flavonoids in beverages and plant extracts using a CoulArray detector. *J. Sep. Sci.* **2005**, *28*, 1005–1022. [CrossRef] [PubMed]

24. Vanbeneden, N.; Delvaux, F.; Delvaux, R. Determination of hydroxycinnamic acids and volatile phenols in wort and beer by isocratic high-performance liquid chromatography using electrochemical detection. *J. Chromatogr. A* **2006**, *1136*, 237–242. [CrossRef]

25. McMurrough, I.; Roche, G.P.; Cleary, K.G. Phenolic acids in beers and worts. *J. Inst. Brew.* **1984**, *90*, 181–187. [CrossRef]

26. Woffenden, H.M.; Ames, J.M.; Chandra, S. Relationships between antioxidant activity, color, and flavor compounds of crystal malt extract. *J. Agric. Food Chem.* **2001**, *49*, 5524–5530. [CrossRef]

27. Guido, L.F.; Curto, A.F.; Boivin, P.; Benismail, N.; Goncalves, C.R.; Barros, A.A. Correlation of malt quality parameters and beer flavor stability: Multivariate analysis. *J. Agric. Food Chem.* **2007**, *55*, 728–733. [CrossRef]

28. McMurrough, I.; Madigan, D.; Kelly, R.J. The role of flavonoid polyphenols in beer stability. *J. Am. Soc. Brew. Chem.* **1996**, *54*, 141–148.

29. Drost, B.W.; Van der Berg, R.; Freijee, F.J.M.; Van der Velde, E.G.; Hollemans, M. Flavor stability. *J. Am. Soc. Brew. Chem.* **1990**, *48*, 124–131. [CrossRef]

30. Bertuzzi, T.; Mulazzi, A.; Rastelli, S.; Donadini, G.; Rossi, F.; Spigno, G. Targeted healthy compounds in small and large-scale brewed beers. *Food Chem.* **2020**, *310*, 125935. [CrossRef]

31. Perez-Jimenez, J.; Neveu, V.; Vos, F.; Scalbert, A. Identification of the 100 richest dietary sources of polyphenols: An application of the Phenol-Explorer database. *Eur. J. Clin. Nutr.* **2010**, *64*, S112–S120. [CrossRef] [PubMed]

32. Hayes, D.; Angove, M.J.; Tucci, J.; Dennmis, C. Walnuts (*Juglans regia*) chemical composition and research in human health. *Crit. Rev. Food Sci. Nutr.* **2016**, *56*, 1231–1241. [CrossRef] [PubMed]

33. Alasalvar, C.; Shahidi, F. Tree nuts: Composition, phytochemicals, and health effects: An overview. In *Tree Nuts: Composition, Phytochemicals, and Health Effects*; Alasalvar, C., Shahidi, F., Eds.; CRC Press: Boca Raton, FL, USA, 2009; pp. 1–10.

34. Harnly, J.M.; Doherty, R.F.; Beecher, G.R.; Holden, J.M.; Haytowitz, D.B.; Bhagwat, S.; Gebhardt, S. Flavonoid content of US fruits, vegetables and nuts. *J. Agric. Food Chem.* **2006**, *54*, 9966–9977. [CrossRef] [PubMed]

35. De Vasconcelos, M.C.B.M.; Bennett, R.N.; Rosa, E.A.S.; Ferreira-Cardoso, J.V. Industrial processing effects on chesnutnfruits (*Castanea Sativa Mill.*). 2. Crude protein, free amino acids and phenolic phytochemicals. *Int. J. Food Sci. Technol.* **2009**, *44*, 2613–2619.

36. De Vasconcelos, M.C.B.M.; Bennett, R.N.; Rosa, E.A.S.; Ferreira-Cardoso, J.V. Composition of European chestnut (*Castanea Sativa Mill.*) and association with health effects: Fresh and processed products. *J. Sci. Food Agric.* **2010**, *90*, 1578–1589. [CrossRef]

37. De Pascual-Teresa, S.; Santos-Buelga, C.; Rivas-Gonzalo, J.C. Quantitative analysis of flavan-3-ols in Spanish foodstuffs and beverages. *J. Agric. Food Chem.* **2000**, *48*, 5331–5337. [CrossRef]
38. Moreira, A.S.P.; Nunes, F.M.; Simoes, C.; Maciel, E.; Domingues, P.; Domingues, M.R.M.; Coimbra, M.A. Data on coffee composition and mass spectrometry analysis of mixtures of coffee related carbohydrates, phenolic compounds and peptides. *Data Brief* **2017**, *13*, 145–161. [CrossRef]
39. Nardini, M.; Cirillo, E.; Natella, F.; Scaccini, C. Absorption of phenolic acids in humans after coffee consumption. *J. Agric. Food Chem.* **2002**, *50*, 5735–5741. [CrossRef]
40. Rodriguez-Carrasco, Y.; Gaspari, A.; Graziani, G.; Santini, S.; Ritieni, A. Fast analysis of polyphenols and alkaloids in cocoa-based products by ultra-high performance liquid chromatography and Orbitrap high resolution mass spectrometry (UHPLC-Q-Orbitrap-MS/MS). *Food Res. Int.* **2018**, *111*, 229–236. [CrossRef]
41. Zhao, C.N.; Tang, G.Y.; Cao, S.Y.; Xu, X.Y.; Gan, R.; Liu, Q.; Mao, Q.Q.; Shang, A.; Li, H.B. Phenolic profiles and antioxidant activities of 30 tea infusion from green, black, oolong, white, yellow and dark teas. *Antioxidants* **2019**, *8*, 215. [CrossRef]
42. Pastorino, G.; Cornara, L.; Soares, S.; Rodrigues, F.; Oliveira, M.B.P.P. Liquorice (*Glycyrrhiza glabra*): A phytochemical and pharmacological review. *Phytother. Res.* **2018**, *32*, 2323–2339. [CrossRef]
43. Rizzato, G.; Scalabrin, E.; Radaelli, M.; Capodaglio, R.; Piccolo, O. A new exploration of licorice metabolome. *Food Chem.* **2017**, *221*, 959–968. [CrossRef] [PubMed]
44. Asl, M.N.; Hosseinzadeh, H. Review of pharmacological effects of Glycyrrhiza sp. and its bioactive compounds. *Phytother. Res.* **2008**, *22*, 709–724. [CrossRef] [PubMed]
45. Fiore, C.; Eisenhut, M.; Ragazzi, E.; Zanchin, G.; Armanini, D. A history of the therapeutic use of liquorice in Europe. *J. Ethnopharmacol.* **2005**, *99*, 317–324. [CrossRef] [PubMed]
46. Ciulu, M.; Spano, N.; Pilo, M.I.; Sanna, G. Recent advances in the analysis of phenolic compounds in unifloral honeys. *Molecules* **2016**, *21*, 45. [CrossRef]
47. Cianciosi, D.; Forbes-Hernandez, T.J.; Afrin, S.; Gasparrini, M.; Reboredo-Rodriguez, P.; Manna, P.P.; Zhang, J.; Bravo Lamas, l.; Martinez Florez, S.; Toyos, P.A.; et al. Phenolic compounds in honey and their associated health benefits: A review. *Molecules* **2018**, *23*, 2322. [CrossRef]
48. Cook, N.C.; Samman, S. Review: Flavonoids-chemistry, metabolism, cardioprotective effects and dietary sources. *J. Nutr. Biochem.* **1996**, *7*, 66–76. [CrossRef]
49. Kumar, S.; Pandey, A.K. Chemistry and biological activities of flavonoids: An overview. *Sci. World J.* **2013**, *2013*, 162750. [CrossRef]
50. Quideau, S.; Deffieux, D.; Douat-Casassus, C.; Pouysegu, L. Plant polyphenols: Chemical properties, biological activities and synthesis. *Angew. Chem. Int. Ed. Engl.* **2011**, *50*, 586–621. [CrossRef] [PubMed]
51. Chamkha, M.; Cathala, B.; Cheynier, V.; Douillard, R. Phenolic compositionof champagnes from Chardonnay and Pinot Noir vintages. *J. Agric. Food Chem.* **2003**, *51*, 3179–3184. [CrossRef]
52. Stockham, K.; Sheard, A.; Paimin, R.; Buddhadasa, S.; Duong, S.; Orbell, J.D.; Murdoch, T. Comparative studies on the antioxidant properties and polyphenolic content of wine from different growing regions and vintages, a pilot study to investigate chemical markers for climate changes. *Food Chem.* **2013**, *140*, 500–506. [CrossRef] [PubMed]
53. Garaguso, I.; Nardini, M. Polyphenols content, phenolics profile and antioxidant activity of organic red wines produced without sulfur dioxide/sulfites addition in comparison to conventional red wines. *Food Chem.* **2015**, *179*, 336–342. [CrossRef] [PubMed]
54. Nardini, M.; Garaguso, I. Effect of sulfites on antioxidant activity, total polyphenols and flavonoids measurements in white wine. *Foods* **2018**, *7*, 35. [CrossRef] [PubMed]
55. Bourne, L.; Paganga, G.; Baxter, D.; Hughes, P.; Rice-Evans, C. Absorption of ferulic acid from low-alcohol beer. *Free Radic. Res.* **2000**, *32*, 273–280. [CrossRef] [PubMed]
56. Nardini, M.; Natella, F.; Scaccini, C.; Ghiselli, A. Phenolic acids from beer are absorbed and extensively metabolized in humans. *J. Nutr. Biochem.* **2006**, *17*, 14–22. [CrossRef] [PubMed]
57. Piazzon, A.; Vrhovsek, U.; Masuero, D.; Mattivi, F.; Mandoj, F.; Nardini, M. Antioxidant activity of phenolic acids and their metabolites: Synthesis and antioxidant properties of the sulfate derivatives of ferulic and caffeic acids and of the acyl glucuronide of ferulic acid. *J. Agric. Food Chem.* **2012**, *60*, 12312–12323. [CrossRef]
58. Williamson, G.; Manach, C. Bioavailability and bioefficacy of polyphenols in humans. II. Review of 93 intervention studies. *Am. J. Clin. Nutr.* **2005**, *81*, 243S–255S. [CrossRef]

59. Tome-Carneiro, J.; Larrosa, M.; Gonzales-Sarrias, A.; Tomas-Barberan, F.A.; Garcia-Conesa, M.T.; Espin, J.C. Resveratrol and clinical trials: The crossroad from in vitro studies to human evidence. *Curr. Pharm. Design* **2013**, *19*, 6064–6093. [CrossRef]

60. Smoliga, J.M.; Baur, J.A.; Hausenblas, H.A. Resveratrol and health—A comprehensive review of human clinical trials. *Mol. Nutr. Food Res.* **2011**, *55*, 1129–1141. [CrossRef]

61. Arranz, S.; Chiva-Blanch, G.; Valderas-Martinez, P.; Medina-Remon, A.; Lamuela-Raventos, R.M.; Estruch, R. Wine, beer, alcohol and polyphenols on cardiovascular disease and cancer. *Nutrients* **2012**, *4*, 759–781. [CrossRef] [PubMed]

62. Van der Gaag, M.S.; Ubbink, J.B.; Sillanaukee, P.; Nikkari, S.; Hendriks, H.F.J. Effect of consumption of red wine, spirits and beer on serum homocysteine. *Lancet* **2000**, *335*, 1522. [CrossRef]

63. Singleton, V.L.; Rossi, J.A. Colorimetry of total phenolics with phosphomolybdic-phosphotungstic acid reagents. *Am. J. Enol. Vitic.* **1965**, *16*, 144–158.

64. Dewanto, V.; Wu, X.; Adom, K.K.; Liu, R.H. Thermal processing enhances the nutritional value of tomatoes by increasing total antioxidant activity. *J. Agric. Food Chem.* **2002**, *50*, 3010–3014. [CrossRef]

65. Benzie, I.F.F.; Strain, J.J. The ferric reducing ability of 1 plasma (FRAP) as a measure of "Antioxidant power": The FRAP assay. *Anal. Biochem.* **1996**, *239*, 70–76. [CrossRef] [PubMed]

66. Re, R.; Pellegrini, N.; Proteggente, A.; Pannala, A.; Yang, M.; Rice-Evans, C. Antioxidant activity applying an improved ABTS radical cation decolorization assay. *Free Radic. Biol. Med.* **1999**, *26*, 1231–1237. [CrossRef]

67. Pozo-Bayon, M.A.; Hernandez, M.T.; Martin-Alvarez, P.J.; Polo, M.C. Study of low molecular weight phenolic compounds during the aging of sparkling wines manufactured with red and white grape varieties. *J. Agric. Food Chem.* **2003**, *51*, 2089–2095. [CrossRef]

68. Nardini, M.; Forte, M.; Vrhovsek, U.; Mattivi, F.; Viola, R.; Scaccini, C. White wine phenolics are absorbed and extensively metabolized in humans. *J. Agric. Food Chem.* **2009**, *57*, 2711–2718. [CrossRef]

Sample Availability: Samples of the compounds are not available from the authors.

MDPI

St. Alban-Anlage 66

4052 Basel

Switzerland

Tel. +41 61 683 77 34

Fax +41 61 302 89 18

www.mdpi.com

Molecules Editorial Office

E-mail: molecules@mdpi.com

www.mdpi.com/journal/molecules

www.ingramcontent.com/pod-product-compliance
Lightning Source LLC
LaVergne TN
LVHW071030170726
843515LV00006B/1252